DATE DUE

Second Edition

Basic Concepts *of* Environmental Chemistry

Des W. Connell

CRC Taylor & Francis
Taylor & Francis Group

Boca Raton London New York Singapore

A CRC title, part of the Taylor & Francis imprint, a member of the Taylor & Francis Group, the academic division of T&F Informa plc.

Published in 2005 by
CRC Press
Taylor & Francis Group
6000 Broken Sound Parkway NW, Suite 300
Boca Raton, FL 33487-2742

Printed in the United States of America on acid-free paper
10 9 8 7 6 5 4 3 2 1

International Standard Book Number-10: 1-56670-676-9 (Hardcover)
International Standard Book Number-13: 978-1-5667-0676-6 (Hardcover)
Library of Congress Card Number 2005041277

Library of Congress Cataloging-in-Publication Data

Connell, D. W.
Basic concepts of environmental chemistry / by Des W. Connell.--2nd ed.
p. cm.
Rev. ed. of: Basic concepts of environmental chemistry / Des W. Connell...[et al.]. 1997.
Includes bibliographical references and index.
ISBN 1-56670-676-9 (alk. paper)
1. Environmental chemistry. I. Basic concepts of environmental chemistry. II. Title.

TD193.C67 2005
628.5--dc22 2005041277

T&F informa

Taylor & Francis Group
is the Academic Division of T&F Informa plc.

Visit the Taylor & Francis Web site at
http://www.taylorandfrancis.com

and the CRC Press Web site at
http://www.crcpress.com

Preface

In recent years, there has been an extraordinary development in our knowledge of the behavior and effects of chemicals in the environment. In the 1960s, our knowledge of environmental chemistry was confined principally to sets of data on the concentrations of chemicals in the environment. Now a body of specific knowledge has been developed that provides a theoretical basis for understanding distribution, transformation, toxicity, and other environmental properties of chemicals. Out of this has emerged a new branch of the discipline of chemistry that can be defined as follows:

> Environmental chemistry is the study of sources, reactions, transport and fate of chemical entities in the air, water and soil environments as well as their effects on human health and the natural environment.

Environmental chemistry is probably the most interdisciplinary of all the branches of chemistry since it includes aspects of all the traditional chemistry branches, as well as biochemistry, toxicology, limnology, ecology, and so on. It has a practical focus on the environmental management of chemicals since that is a major area of concern. It also includes behavior of natural chemical entities in natural systems.

Environmental chemistry now has sufficient intellectual depth to provide the scope for undergraduate and graduate courses in chemistry. This book utilizes the fundamental properties of bonds and molecules as a framework for understanding the behavior and effects of environmental chemicals. The properties of common contaminants are illustrated using environmental behavior, rather than laboratory behavior, as the expression of the underlying bonding and molecular characteristics. It goes on to link these characteristics to biological effects in the environment, such as toxicity. Also, these fundamental aspects are utilized in considering the great global environmental chemistry processes, including respiration, photosynthesis, and so on. Recent developments in the management of hazardous chemicals, such as ecotoxicology and risk assessment, are treated as an aspect of environmental chemistry.

This book assumes only a basic prior knowledge of chemistry and is designed as a textbook for courses in environmental chemistry. Its objective is to provide knowledge of environmental chemistry based on a series of theoretical principles, rather than as a set of disjointed facts. It aims to give students not just knowledge of environmental chemistry, but an understanding of how and why processes in the environment occur. It can also be used as a basis for understanding chemistry, in general, since the principles apply generally as well as to the environment.

We have developed this textbook based on our experience of teaching graduate and undergraduate courses on environmental chemistry over the last 25 years. To some extent, this represents a culmination of this experience and a distillation of the knowledge we have acquired in that time.

Many people have assisted us in preparing the first and second editions of this book. Greg Miller was involved in discussions of the content and scope, Mrs. Rahesh Garib typed the difficult manuscript, and Chris Clayton and Aubrey Chandica prepared the diagrams. To all of these people, we are extremely grateful.

Des W. Connell

The Author

Des W. Connell is Professor Emeritus of Environmental Chemistry in the Faculty of Environmental Sciences at Griffith University in Brisbane, Australia. His broad interests are in environmental chemistry, ecotoxicology, and water pollution, and he has published several books on these topics. His research interests are in bioconcentration, quantitative structure-activity relationships (QSARs), and the mechanisms of nonspecific toxicity. Professor Connell was the President of the Australian Society for Ecotoxicology and the Inaugural Chairperson of the Environmental Chemistry Division of the Royal Australian Chemical Institute and was awarded the Institute's Environment Medal in 1993. Professor Connell received his B.Sc., M.Sc., and Ph.D. in organic chemistry from the University of Queensland and a D.Sc. in environmental chemistry from Griffith University.

The Co-authors

Darryl W. Hawker is a Senior Lecturer in environmental chemistry in the Faculty of Environmental Sciences at Griffith University in Brisbane, Australia. His research interests are in the physicochemical properties of chemicals, particularly organometallic compounds, and how these properties influence environmental behavior. Dr. Hawker received his B.Sc. (Honors) and Ph.D. degrees in organic chemistry from the University of Queensland.

Michael St. J. Warne is a Research Theme Leader in the Commonwealth Scientific and Industrial Research Organization (CSIRO) Land and Water and was a Senior Research Ecotoxicologist in the New South Wales Environment Protection Authority (NSW EPA). He has been president of the Australasian Society for Ecotoxicology. His research interests include the toxicity of mixtures and phosphate-free detergents, quantitative structure-activity relationships (QSARs), and earthworm ecotoxicology. He is involved in developing the national ecological risk assessment framework for contaminated sites and is developing an Australian ecotoxicology database. Before joining the NSW EPA, he was a Lecturer in environmental chemistry/toxicology in the Faculty of Environmental Sciences, Griffith University. Dr. Warne has a B.Sc. (Honors) in biology from Newcastle University and a Ph.D. in ecotoxicology from Griffith University.

Peter P. Vowles, retired, was a Lecturer in environmental chemistry in the Faculty of Environmental Sciences at Griffith University in Brisbane, Australia. His research interests are in physicochemical factors influencing the fate of persistent organic chemicals and the identification and source apportionment of urban aerosols. He is also interested in approaches and techniques for teaching chemistry, particularly environmental chemistry. Dr. Vowles has a B.Sc. (Honors) and Ph.D. in chemistry from the University of Melbourne and a Diploma in Education from the Melbourne State College.

Table of Contents

PART I Principles of Environmental Chemistry

PART II Basic Properties of Chemicals in the Environment

PART III Contaminants in the Environment

PART IV Processes in the Natural Environment

PART V Management of Hazardous Chemicals

Part I

Principles of Environmental Chemistry

1 The Role and Importance of Environmental Chemistry

1.1 ALCHEMY

Modern chemistry has its early beginnings in ancient Greece and Egypt. The Greeks were great philosophers and thinkers and debated the ultimate nature of the material world that surrounded them. On the other hand, the Egyptians were skilled in the practical arts of applied chemistry. They had an intimate knowledge of such matters as embalming of the dead, dyeing of clothing, and the isolation of some metals. In fact, the word *alchemy* is derived from the Arab word *al-kimiya*. This phrase was used to describe the art of transformation of materials and the practical use of chemicals in society. However, the lack of understanding of chemical processes led to alchemy being more of an art than a science. The art of alchemy was closely related to religion and involved the use of a set of hieroglyphics to represent the different metals. For example, gold was depicted as the sun, silver as the moon, copper as Venus, and so on. The occurrence of chemical changes was interpreted in a mysterious fashion and took on a mythological significance. The major area of concern of alchemy was not the development of a science, but the transmutation of metals, particularly the transmutation of base metals such as lead into gold.

The decline of alchemy was signaled by the works of such scientists as Agricola, in fact a German named Georg Bauer, who lived between 1494 and 1555. Agricola published a book titled *De Re Metallica* [*Of Metallurgy*], which took a practical approach and gathered together all of the knowledge available on metallurgy at the time. Also, the German scientist Andreas Libau (1540–1616), better known as Libavius, published *Alchemia* in 1597. Despite the name of this text, the book is written clearly and without resort to mysticism. The famous Swiss physician Paracelsus (1493–1541), although an alchemist, believed that the processes that occur in the human body are basically of a chemical nature and that chemical medicines could provide remedies to illness. So this marks the start of the development of chemotherapy, and doctors everywhere now use a vast array of chemicals to cure all kinds of illnesses.

Perhaps the end to alchemy occurred in 1661, when Robert Boyle, an Irishman, published his book *The Sceptical Chymist*. Boyle attacked the ideas and approach of the alchemists and advocated a rational scientific approach. To mark the change from alchemy, Boyle also advocated the use of the term *chemistry* to describe the science of materials. The well-known Boyle's law was developed and amply illustrates the scientific approach to chemistry that he promoted.

It could be said that **environmental chemistry** had its beginning with one of the great pioneers of chemistry, Antoine Lavoisier, who was born in Paris in 1743. Lavoisier's experiments on the atmosphere mark a great advance in chemistry and also a great advance in understanding the chemistry of the atmospheric environment. Not only did he discover fundamental information concerning the chemistry of air, but he also examined the use of air by animals and, by so doing, was investigating one of the major aspects of the chemistry of the environment. The use of oxygen by animals, and their consequent release of carbon dioxide, is a fundamental aspect of environmental chemistry. Lavoisier's *Elementary Book of Chemistry*, published in 1789, effectively marks the start of the systematic development of the then new science of chemistry.

1.2 THE CHEMICAL AGE

In the first half of the 19th century, early chemists were endeavoring to describe the nature of molecules. They were taking the first steps in assembling the structures of organic chemicals, which we now take for granted as common scientific knowledge. As they were carrying out this early work, they were concerned with the reactions between organic chemicals and the nature of the products generated, thus embarking on the path to modern synthetic organic chemistry.

One of these early chemists, William Perkin (1838–1907), became aware that some of these synthetic organic chemical processes could possibly be turned to commercial advantage. Perkin left his academic studies at the university and used money obtained from his family to start a factory to manufacture synthetic dyes. One dye he produced, aniline purple, was extremely popular in the textile industry and in heavy demand with textile manufacturers in Europe. In fact, Perkin had founded the first chemical industry to be based on studies of the nature of organic compounds conducted in a research environment. Not only had he founded the synthetic chemical industry, but he was also able to retire a very wealthy man at an early age. This marks the start of the synthetic chemical industry, which, from this small beginning, was to expand to a size and extent where its products now dominate modern society. This development had important implications for the environment, since it resulted in the preparation and discharge of relatively large quantities of synthetic organic substances, previously unknown to occur in nature and with unknown environmental effects.

Of course, chemical processes such as the smelting of ores and the manufacture of soap had been carried out on a large commercial scale prior to this time. But these processes were developed as a result of trial and error and were more in the nature of trade practices rather than the products of scientific chemical research.

Another chemist whose activities were to have important environmental implications was on the scene about this time: German scientist Fritz Haber (1868–1934). Haber demonstrated in the laboratory how nitrogen and hydrogen could be combined to yield ammonia. He further developed this on a commercial scale so that commercial plants operating in Germany in the early part of this century were fixing nitrogen gas from the atmosphere and producing ammonia. This ammonia could be used as a fertilizer for the production of food crops, thereby eliminating the need to rely on

TABLE 1.1
Use of Synthetic Chemicals in Human Society

Substance	Examples of Chemical Classification	Uses
Plastics	Polystyrene, polyvinylchloride polypropylene, nylon	Textiles, car tires, household goods, furniture, etc.
Pesticides	DDT, 2,4-dichlorophenoxyacetic acid (2,4-D), malathion, glyphosate	Control of weeds, insects, and other pests
Drugs	Aspirin, penicillin, Valium, sulfanilamide, barbiturates	Medicinal uses
Petroleum products	Hydrocarbons	Motor fuel, lubricant
Crop fertilizer	Ammonia, ammonium sulfate, ammonium nitrate, ammonium phosphate	Stimulate food crop production

the natural occurrence of nitrogen compounds in soil to stimulate plant growth. This process has continued to be used to the present day, thereby allowing the production of large quantities of plant fertilizer used to produce food crops and allowing rapid expansion of the world's population. Without this process, the production of food would be limited by the availability of natural nitrogen compounds in soil for food plant growth. Thus, this chemical process has probably had the greatest impact on the environment through the rapid expansion of the human population with the subsequent environmental changes that have resulted.

It is traditional for human history to be divided into a series of ages, each building on and improving the technology of the previous. Thus, we have a sequence of ages going from the Stone Age to the Bronze Age to the Industrial Age and the Atomic Age. However, during the last approximately 100 years, the age in which we are living could very well be described as the **Chemical Age**. Table 1.1 indicates the wide range of synthetic chemicals now used in human society. The use of these substances permeates almost every aspect of our society, from life in the home to transport, food production, availability of medicines, and so on. The production and use of these many substances have improved human conditions and caused an enormous increase in the human population. These factors have many implications for the chemical processes that occur in the environment.

1.3 ENVIRONMENTAL CHEMISTRY

Environmental chemistry is not new, but the use of this term to describe a body of chemical knowledge has become accepted by scientists in recent years. This principally reflects the importance attached to the behavior and possible adverse effects of chemicals discharged to the environment.

There are many scientific events that could now be identified as the starting points of environmental chemistry. Some of these are in the area of fundamental chemistry, while others lie in applied chemistry. Also, the chemical aspects of other disciplines have played a major role. One of the starting points has been mentioned previously: the work of Lavoisier. His analysis of air and determination of the nature of combustion are fundamental to the development of modern chemistry and atmospheric chemistry. Turning to the applications of chemistry, perhaps pollution of the Thames River during the 1800s can be identified as another starting point of environmental chemistry. The bad odors and diseases associated with the Thames were investigated by various British Royal Commissions, which weighed evidence on the chemistry of sewage treatment and water pollution. This culminated with the Royal Commission on Prevention of River Pollution, which reported in 1885 and recommended the use of the biochemical oxygen demand (BOD) test. This test is still extensively used today to evaluate the effects of sewage and other wastewaters on waterways.

The chemistry of the oceans, lakes, and freshwater areas also has a comparatively long history. In 1872, HMS *Challenger* commenced its historic voyages with many of the most noted men of science at that time aboard. They conducted many investigations of seawater composition and chemical processes in the oceans. Similarly in limnology, the key to understanding the ecology of freshwater areas lies in chemical transformations of carbon, oxygen, nitrogen, and phosphorus. This was recognized early in the development of this discipline, which has always had a strong emphasis on chemistry.

Since these early beginnings, environmental chemistry has expanded rapidly. Natural chemical processes in all sectors of the environment, particularly soil, water, and the atmosphere, have been subject to investigation, as well as the environmental behavior of contaminating chemicals. Not surprisingly, the management of chemicals discharged to the environment has become a major focus for environmental chemistry. Governmental agencies and industries employ large numbers of environmental chemists on a worldwide basis to monitor and manage the discharge of chemicals and their adverse effects.

Over recent years, a body of knowledge associated specifically with the behavior and effects of chemicals in the environment has been developing. A theoretical basis for understanding the distribution, transformation, toxicity, and other biological properties of chemicals in the environment is now becoming established. This means that environmental chemistry can be appropriately seen as a subbranch or subdiscipline of chemistry. This is reflected in the availability of excellent textbooks on the topic as illustrated by those by Manahan (2001) and Schwartzenbach et al. (2002). The recent development of environmental chemistry was initiated during the 1960s. Rachel Carson's book *Silent Spring*, published in 1962, can be identified as a significant event stimulating worldwide interest and concern regarding chemical residues in the environment. This event was made possible by the development of analytical techniques capable of detecting chemicals at very low concentrations. In 1952, Richard Synge and Archer Martin were awarded the Nobel Prize for inventing the chromatography technique, which has a principal application to the analysis of organic compounds. Later, the technique was extended and improved such that trace amounts of xenobiotic organic chemicals could be quantified in environmental samples. The flame ionization detector extended the sensitivity of the method so that levels of a few parts per million of organic compounds could be detected in samples. In addition, with the chlorinated hydrocarbons, the development of the electron capture detector further extended the sensitivity of the technique. In 1967, only 15 years after the Nobel Prize for chromatography was awarded, a considerable body of information was available on the concentrations of xenobiotic organic compounds in biota.

Environmental chemistry was dominated at this time by the collection of data on residues of synthetic compounds in biota, but there was little understanding of the mechanisms, how the residues accumulated, or their biological effects. Determination of residue levels remains an important aspect of chemical behavior in the environment, but now there is considerable interest in placing residues in a broader context of environmental effects. There have been major advances in understanding the distribution, transport, and transformation of contaminants, as well as exposure and uptake by biota. But human health effects as well as the responses of natural ecosystems require further expansion of knowledge (Boethling and Mackay, 2000; Lyman et al., 1990).

Over the last 15 years, some theoretical concepts have emerged that provide a sound conceptual basis for important aspects of environmental chemistry. Drawing on ideas already established in other fields, the introduction of partition and fugacity theory to explain environmental distribution of chemicals has occurred. In addition, the use of properties measured in the laboratory to assess behavior and effects in the environment has allowed the development of a clear understanding of many environmental processes. Theoretical methods to predict environmental properties of chemicals have been placed on a sound footing and provide a basis for expansion of knowledge in the future.

The global problem of the **greenhouse effect** now occupies the center stage of environmental management. The environmental chemistry of the processes leading to this problem provides the starting point in its resolution. A range of different scientists have pointed out the importance of developing a mechanistic chemical model of the cycles of carbon and other elements and their interaction with the land and sea in devising strategies for management of this global problem.

1.4 THE SCOPE OF ENVIRONMENTAL CHEMISTRY

It is difficult to precisely define environmental chemistry since the topic has not yet reached a stage where there is universal accord in the chemical community on its scope. However, this is developing and the following definition provides a reasonably acceptable statement at this stage:

> **Environmental chemistry** is the study of sources, reactions, transport, and fate of chemical entities in the air, water, and soil environments, as well as their effects on human health and the natural environment.

Environmental chemistry is probably the most interdisciplinary of the many branches of chemistry. It contains aspects of related branches of chemistry, such as organic chemistry, analytical chemistry, physical chemistry, and inorganic chemistry, as well as more diverse areas, such as biology, toxicology, biochemistry, public health, and epidemiology. Figure 1.1 represents a diagrammatic illustration of some of the major aspects of these relationships. To place the area in some perspective, a set of topics that fall wholly or partially within environmental chemistry are listed in Table 1.2. Many of these topics are concerned with chemical pollutants in the environment,

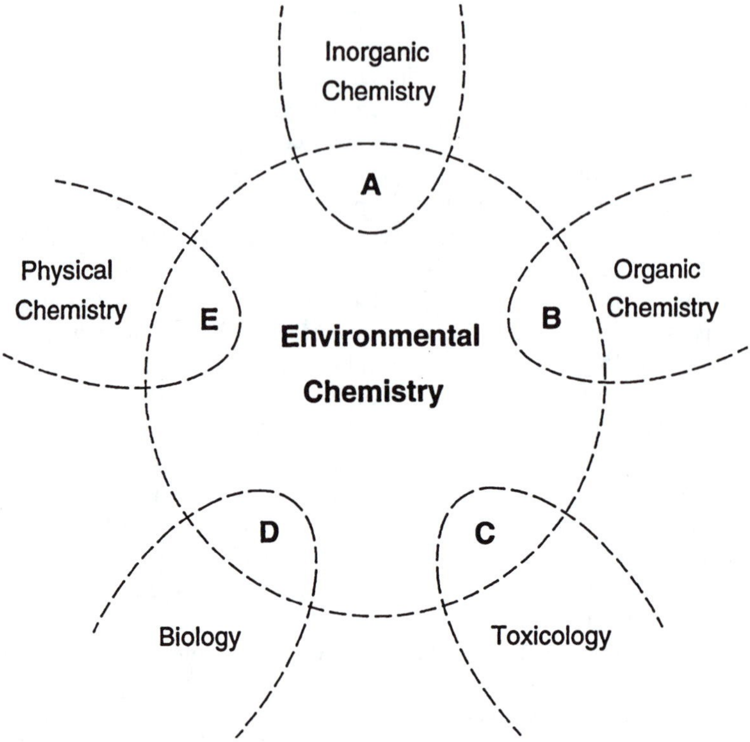

FIGURE 1.1 Diagrammatic illustration of some of the relationships between environmental chemistry and other areas of science.

but environmental chemistry is not only concerned with pollution, but also with the behavior of natural chemicals in natural systems. This is exemplified by topics such as oceanic and limnological chemistry, which are primarily concerned with natural systems.

Investigations within the scope of environmental chemistry provide a possible explanation for our very existence. The primitive Earth's atmosphere contained simple gases that, on equilibration, followed by subsequent complex reaction sequences, have led to the formation of proteins, carbohydrates, fats, and other substances that are the basic molecules needed for life to develop. Later processes in this sequence may have led to the formation of cells and then on to more complex life-forms. The mechanisms of these processes draw heavily on our present understanding of the fundamental properties of molecules and reaction mechanisms.

Environmental chemistry is basically concerned with developing an understanding of the chemistry of the world in which we live. Such investigations and knowledge have an intrinsic value of their own. Our world, with all of its many complex chemical processes, is a worthy topic for a well-rounded education and research. The chemical processes in soil, water, and the atmosphere are central to our existence, and through them we can better understand ourselves and our role in the environment. Thus, it could be argued that environmental chemistry should be a part

TABLE 1.2
Some Topics Considered to Be Wholly or Partially within Environmental Chemistry

Aspect of the Environment	Area within or Relevant to Environmental Chemistry
Evolution	Chemical evolution
Chemical processes in sectors of the abiotic natural environment	Oceanic, atmospheric, soil, and limnological chemistry, global chemical systems
Chemical influences in natural ecosystems	Chemical ecology, pheromones, allelochemistry
Behavior of hazardous chemicals in the environment	Mathematical modeling of environmental distribution, degradation processes, waste disposal
Effects of toxic chemicals on individuals, populations, and ecosystems	Environmental toxicology, ecotoxicology, quantitative structure-activity relationships (QSARs), environmental analysis
Effects of chemicals on human populations	Environmental health, safety, occupational health, epidemiology

of the chemistry curriculum wherever chemistry is taught. At a tertiary level, graduates in chemistry are in fact now expected to be familiar with many aspects of environmental chemistry. They are being asked by management in industry and within government to advise on the environmental and health effects of chemicals, management and safety of chemicals, and so on.

1.5 CHEMISTRY IN ENVIRONMENTAL MANAGEMENT

There is little doubt that the principal applications of environmental chemistry are in the development of an understanding of the behavior and effects of discharged chemicals on human health and the natural environment. When chemical use results in environmental contamination, it is necessary to set standards for acceptable concentrations in water, air, soil, and biota. Monitoring of these concentrations, and the resultant effects, must then be undertaken to ensure that the discharge standards are realistic and provide protection from adverse effects. Also, considerable attention is now being focused on regulation of the use of new chemicals and prevention of chemicals that may have adverse effects from entering the marketplace and environment. Most of these actions are undertaken by government and industry, for example, the disposal of chemical wastes generated in highly concentrated form, as well as occurring as trace contaminants in discharges. These chemical management issues involve political, social, and economic problems as well as technical problems, and many new approaches to management will be needed. On the other hand, new industrial processes that generate less chemical waste will be required. Some of the activities undertaken in government agencies and industry, within the scope of environmental chemistry, are outlined in Table 1.3. In addition, there is a considerable

TABLE 1.3
Some Activities of Government and Industry Related to Environmental Chemistry

Objective	Some Actions Taken
Management of industrial emissions	Setting emission standards, monitoring ambient concentrations, disposal of waste, modeling distribution of chemicals
Protection of workers' health	Biochemical and physiological testing, epidemiology, monitoring ambient concentrations, evaluation of adverse effects
Protection of the natural environment	Monitoring contaminants in water, air, soil, and biota, evaluation of adverse effects
Testing and evaluation of new chemicals	Modeling of potential distributions, testing toxicity and other effects

volume of research in environmental chemistry being undertaken. A number of well-established research institutions specializing in environmental chemistry are operating in countries throughout the world.

Environmental chemistry activities in government, industry, and education are growing, to some extent in concert with the concerns expressed regarding chemicals in the environment by the community. At present, despite the enormous benefits that accrue from the many uses of chemicals, the community often sees chemicals as having a negative impact. A negative impact has occurred in many situations, but many of these have been eliminated and increased knowledge has provided a basis for enlightened future management and control. The further development of environmental chemistry will provide access to chemical products that will enhance the lives of people without resulting in detrimental effects. In this way, environmental chemistry will have a significant effect on the future of chemistry, chemists, and the chemical industry, as well as the community in general.

1.6 BASIC CONCEPT OF THIS BOOK

In the early 1970s, a new approach to explaining the environmental effects of chemicals was initiated. Hamelink and coworkers (1971) proposed that the bioaccumulation of organic compounds by biota was governed by the properties of the chemical rather than ecological factors, as was previously thought. Since that time, this concept has been extended considerably, and now it is clear that the distribution of chemicals in the environment relates to the properties of the chemical as well as characteristics of the environment (see Mackay et al., 1992).

The relationship of biological effects, particularly toxicity, to properties of a chemical, such as the olive oil–water partition coefficient, was well established by Ernest Overton in the early 1900s and extended by Covin Hansch during a later period. These concepts were then applied, with considerable success, to explain

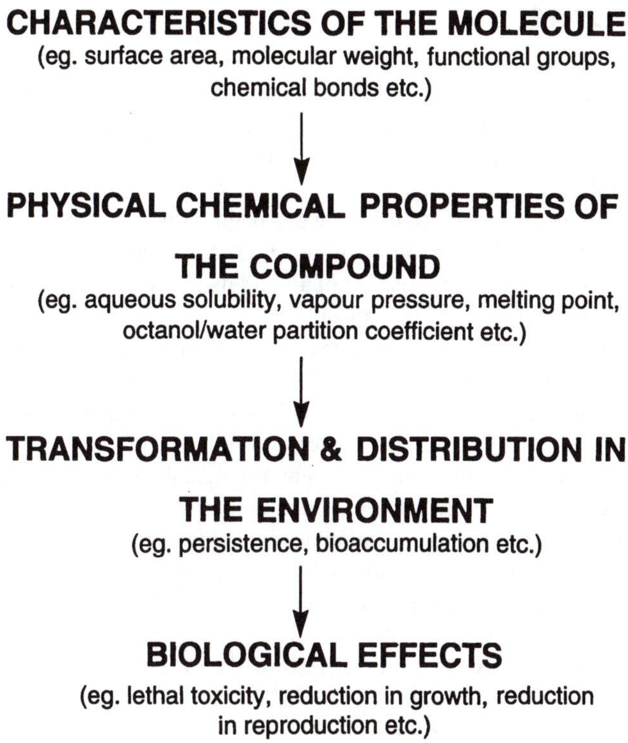

FIGURE 1.2 The environmental chemistry of a chemical can be seen as an interrelated set of characteristics, as shown.

toxicity and other effects of chemicals in the environment. In addition, the influence of characteristics of the molecule itself, such as molecular surface area, was found to be of considerable significance.

A concept of an interdependent set of properties of a chemical has now developed, as shown in Figure 1.2. This has been used as the basic underlying concept of this book. The basic characteristics of the molecule itself are the starting point for understanding the environmental chemistry of chemicals. These characteristics govern the physical-chemical properties of the compound, such as its aqueous solubility and vapor pressure, which in turn control transformation and distribution in the environment. Biological effects relate to the chemical in its transformed and dispersed state. This set of interrelationships also provides a basis for predicting the environmental properties of a chemical. The distribution, transformation, and some biological properties can be predicted in many situations utilizing physical-chemical properties.

This concept provides a framework into which aspects of the environmental characteristics of a chemical can be logically placed. It allows these many diverse environmental characteristics to be rationalized into a set of relationships, rather than be seen as a disorganized collection of facts.

1.7 KEY POINTS

1. The alchemists were the earliest practitioners of applied chemistry, but alchemy was more of an art than a science. Many alchemists were concerned with mysticism and the transmutation of base metals such as lead into gold.
2. Environmental chemistry had an early beginning with Antoine Lavoisier, a French scientist born in Paris in 1743. Lavoisier combined classical experiments on the composition of air and its use by animals to investigate the chemical nature of the atmosphere.
3. The commercial application of chemical knowledge derived from laboratory experiments was commenced by William Perkin in the late 1800s. Perkin founded the synthetic dye industry, which was the forerunner of the large and diverse synthetic chemical industry of today.
4. Perhaps one of the most important environmental applications of chemical knowledge has been the development, by Fritz Haber in the early 1900s, of a commercial process to combine atmospheric nitrogen with hydrogen to produce ammonia. This allowed the manufacture of artificial fertilizers to enhance food crop production and the expansion of the human population.
5. The term *environmental chemistry* has become common only in recent times, although environmental aspects of chemistry have been investigated since the beginning of chemistry itself.
6. Environmental chemistry is the study of the sources, reactions, transport, and fate of chemical entities in air, water, and soil environments, as well as their effects on human health and the natural environment.
7. Over the last 15 years, theoretical concepts of the distribution, behavior, and effects of chemicals in the environment have been developed that provide a fundamental understanding of these processes.
8. Environmental chemistry plays a major role in government and industry in relation to many areas, including management of industrial emissions, protection of workers' health, protection of the natural environment, and testing and evaluation of new chemicals.
9. The environmental properties of a chemical can be seen as an interrelated set of properties based on the characteristics of the molecule. These characteristics govern the physical-chemical properties of the compound, which in turn control transformation, distribution, and biological effects in the environment.

REFERENCES

Boethling, R.S. and Mackay, D., *Handbook of Property Estimation Methods: Environmental and Health Sciences*, Lewis Publishers, Boca Raton, FL, 2000.

Hamelink, J.L., Waybrant, R.C., and Ball, C., A proposal: exchange equilibria control the degree chlorinated hydrocarbons are biologically magnified in lentic environments, *Transactions of the American Fisheries Society*, 100, 207, 1971.

Lyman, W.J., Reehl, W.F., and Rosenblatt, D.H., *Handbook of Chemical Property Estimation Methods: Environmental Behavior of Organic Compounds*, American Chemical Society, Washington, DC, 1990.

Mackay, D., Shiu, W.Y., and Ma, K.C., *Illustrated Handbook of Physical-Chemical Properties and Environmental Fate for Organic Chemicals: Volume 1*, Lewis Publishers, Boca Raton, FL, 1992.

Manahan, S.E., *Environmental Chemistry*, 7th ed., CRC Press, Boca Raton, FL, 2001.

McGowen, T., *Chemistry: The Birth of a Science*, Franklin Watts, New York, 1989.

Schwartzenbach, R.P., Gschwend, P.M., and Imboden, D.M., *Environmental Organic Chemistry*, 2nd ed., John Wiley & Sons, New York, 2002.

QUESTIONS

1. The terms *Stone Age*, *Bronze Age*, *Industrial Age*, and *Atomic Age* are used to describe phases of knowledge and technology that human civilizations have passed through. Why is the term *Chemical Age* appropriate for today's society?
2. The applications of chemistry in industry have had a major effect on human society. Which chemical applications have had the greatest impact on the human and natural environment?
3. Environmental chemistry has developed from a diverse field of investigations conducted in relation to the environment. Which of these areas of investigation have been the most important in influencing the scope of environmental chemistry as it is perceived today?

ANSWERS

1. The chemical industry is one of the largest industries in our society.
   - Chemical products dominate the home, workplace, agriculture, transport, medicine, and other areas of human activity.
   - Waste chemical control is now a major aspect of environmental management for industry and government.
2. The artificial fixation of nitrogen to produce ammonia that is used as an agricultural fertilizer has allowed the expansion of the human population.
   - Commercial synthetic chemical operations produce products that are used in almost all human activities.
3. Development of analytical techniques for trace organic compounds and other substances has revealed widespread trace contamination.
   - Attribution of adverse biological effects to contamination by trace chemicals.

Part II

Basic Properties of Chemicals in the Environment

2 Bonds and Molecules: Their Influence on Physical-Chemical Properties in the Environment

2.1 INTRODUCTION

The behavior of chemicals in the environment is governed by their physical-chemical properties as well as transformation and degradation processes, which are discussed in Chapter 3. The physical-chemical properties of compounds include such characteristics as boiling point (bp), melting point (mp), solubility in water, and similar properties. These properties are, in fact, measurements made in the laboratory of environmentally relevant characteristics. For example, compounds with low boiling points evaporate rapidly into the atmosphere, whereas compounds that are highly soluble in water disperse readily in streams and rivers. Thus, an understanding of these properties would be expected to give a clearer perception of how compounds will behave in the environment. This means that the measurement of the physical-chemical properties of a compound could be used to provide an evaluation of its environmental distribution in air, water, sediments, soil, and animals.

2.2 STATES OF MATTER IN THE ENVIRONMENT

A cursory examination of the nature of our environment reveals that matter exists in basically three states: solid, liquid, and gas. Solids are present as soil, rocks, and so on, whereas liquids are represented by the great water bodies of the oceans, lakes, and rivers. The gaseous component of the environment is the atmosphere. These states exercise a basic influence on the nature and distribution of all substances, both natural and man-made, in the environment in basic chemical terms. **Solids** have a definite volume and shape and consist of large numbers of particles that could be atoms, ions, or molecules (Figure 2.1). In fundamental terms, solids are matter where attractive forces between the particles present are more powerful than disruptive forces, and these hold the particles in a fixed relationship to one another. This also means that the particles are closely packed together, with the consequent effect of making solids relatively dense. Some solids can be converted into liquids by heating.

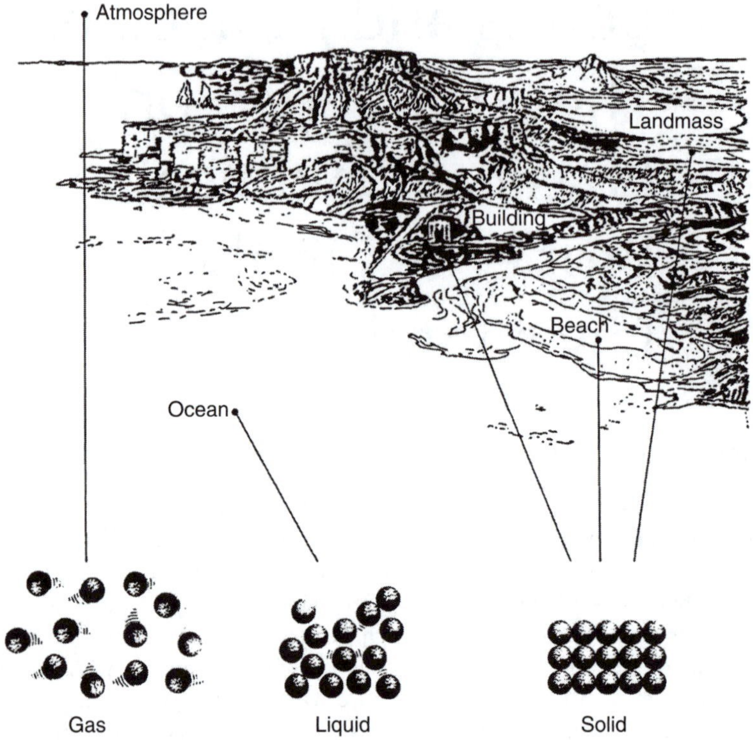

FIGURE 2.1 Illustration of the different states of matter in the natural environment.

For example, ice melts to form liquid water. Heating actually increases the vibrational motions of the particles within the solids until the particles can no longer be held in a fixed relationship to one another. At this point, the substance is converted into a liquid, and the temperature at which this occurs is referred to as the melting point (mp). For example, the melting point of ice at atmospheric pressure is 0°C.

Liquids have a definite volume but no definite shape. With liquids, the particles are close together, almost as close as solids, but are free to move relative to one another (Figure 2.1). This means that the cohesive forces holding the particles together are balanced by the dispersive forces. So the waves of the oceans and the flow of water in rivers represent the basic properties of liquids. An important characteristic of many liquids is their boiling point (bp), which is the temperature at which the liquid vaporizes to give a gas. For example, water has a boiling point of 100°C at atmospheric pressure. However, it should be kept in mind that water does not have to boil in order to vaporize. When the temperature is increased and the rate of vibration of the particles in a liquid increases, some particles bounce completely out of the liquid before the boiling point is reached. In this manner, liquids (and to a lesser extent solids) exhibit a pressure above the surface, referred to as the vapor pressure (vp). Water evaporates from the surface of the oceans, lakes, and rivers without reaching its boiling point. As the temperature increases, the rate of evaporation increases. Somewhat similarly, gases from the atmosphere may dissolve to some extent in water bodies, although the water bodies are at much higher temperatures than the boiling point of the gas. Thus, oxygen and nitrogen gases are present in low concentrations in seawater.

Gases and vapors are produced when the thermal energy input into the liquid reaches a level whereby the movements of the particles (kinetic energy) become greater than the cohesive energy. The particles move apart at this temperature, which is the boiling point, and move at relatively high velocity, colliding with one another and other substances present in the atmosphere. In this way, the molecules in the atmosphere are in constant motion and collide with one another and with the walls of any container, leading to the phenomenon known as pressure. At sea level, the pressure is about 1 atmosphere (atm), and this pressure falls with distance away from the Earth's surface. As both the density of the atmosphere and the pressure fall, the distance between the molecules in the atmosphere increases and the mean free path between collisions increases. So, at sea level, a molecule has a mean free path of about 10^{-6} cm. At a height from the Earth's surface of 100 km, the pressure has fallen to 3×10^{-7} atm, with a molecule having a mean free path of about 10.0 cm. At a height of 500 km, the mean free path has increased markedly, now exceeding 2×10^{6} cm (20 km), while the pressure is $<10^{-6}$ atm.

The properties of some substances in the environment are shown in Table 2.1. Oxygen and nitrogen have relatively low bp and mp values and are the major gaseous components in the atmosphere. Water has higher bp and mp values and exists as a liquid, with the molecules in close association but free to move in relation to one another. Quartz and common salt have very high bp and mp values, reflecting the close and fixed relationship the atoms present have to one another.

2.3 NATURE OF BONDS

To understand why matter exists as solids, liquids, or gases and how they physically evaporate, dissolve, and generally distribute in the environment, we must start by considering the molecular nature of chemical compounds and the way atoms are bound to one another. First, we will look at the nature of chemical bonds, and this

TABLE 2.1
Properties of Some Substances in the Environment

Substances	Occurrence in the Environment	Normal Physical State	Boiling Point (°C) (Atmospheric Pressure)	Melting Point (°C)
Oxygen	21% of atmosphere	Gas	–183	–218
Nitrogen	78% of atmosphere	Gas	–196	–210
Water	Oceans, lakes, rivers	Liquid	100	0
Common salt	3.5% of seawater	Solid	1413	801
Quartz	Rocks, sand, geological strata	Solid	2230	1610

will give an insight into physical-chemical properties that will provide an understanding of environmental properties.

The most important type of bonding is the covalent bond. Usually with this type of bonding, two atoms react together, with each contributing one electron to form a bond. Thus, two atoms of hydrogen can react to form a hydrogen molecule with one covalent bond. Thus,

$$H\cdot + \cdot H \rightarrow H\text{–}H$$

Each covalent bond consists of two electrons moving rapidly between the hydrogen atoms in a defined space, as illustrated in Figure 2.2. Each electron holds a full negative charge, and if the electrons spend equivalent times near the two atoms in the bond, this results in no difference in charge between the two ends of the bond. This can be interpreted as the electron density around the two atoms being symmetrical, leading to a nonpolar bond. However, a different situation applies with the hydrogen chloride bond. In this bond, the chlorine tends to attract electrons as shown in Figure 2.3. This attraction is not sufficient to cause the electron to remain permanently with the chlorine atom, but causes the electron to spend more of its time when it is moving between the two atoms toward the chlorine end of the bond. This results in a small partial negative charge (denoted as $\delta-$) occurring on the chlorine atom and leads to a small partial positive charge (denoted as $\delta+$) occurring at the hydrogen atom. These factors are illustrated with the hydrogen chloride bond in Figure 2.3. The outcome of this effect is that the bond becomes **polar** and has a **dipole moment** or **polarity**. Quantitatively, the polarity of bonds can be characterized by two factors: (1) the charge on each atom and (2) the distance between the two atoms.

The combination of these two factors allows the development of a quantitative characteristic described as the **dipole moment**, which is defined as follows:

Dipole moment (DM) = in Debye units (D)
Electrostatic charge × Distance between atoms

If both atoms, or ends of the bond, are the same, as with the hydrogen (H_2) and chlorine (Cl_2) molecules, then there can be no differences in the overall electron density in the bond and the dipole moment for such symmetrical bonds must be zero. However, most unsymmetrical bonds have a dipole moment. For example, the hydrogen chloride bond has a dipole moment of 1.03 Debyes (D); there are examples of the dipole moments of bonds contained in Table 2.2.

It can be expected that a bond will be polar if the two atoms involved differ in their ability to attract electrons. This property is described as the **electronegativity** of an atom and can be estimated quantitatively. The order of magnitude of the electronegativity of some common elements is shown below:

$$F > O > Cl, N > Br > C, H$$

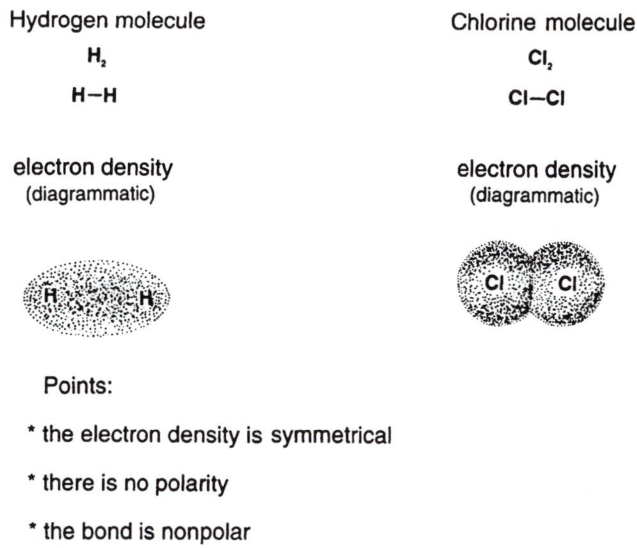

FIGURE 2.2 Illustration of the characteristics of nonpolar covalent bonds in the hydrogen and chlorine molecules.

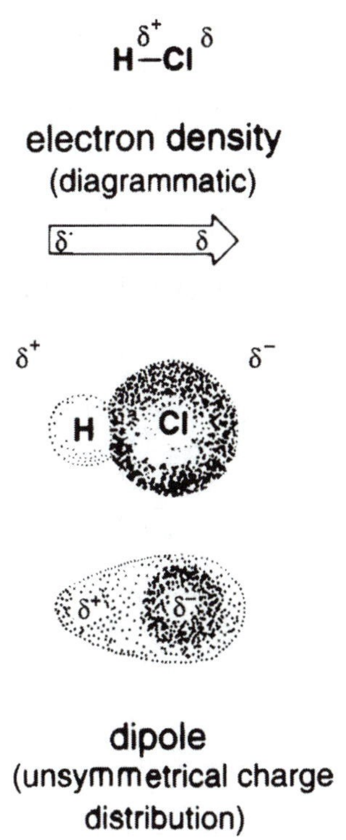

FIGURE 2.3 Illustration of the characteristics of polar covalent bonds using the hydrogen chloride molecule H-Cl as an example.

TABLE 2.2
Dipole Moments of Some Chemical Bonds

Bond	Dipole Moment (Debye, D)	Bond Description
H–F	1.91	Highly polar
H–Cl	1.08	Polar
H–Br	0.80	Polar
H–O	1.5	Polar
H–N	1.3	Polar
C–H	0.4	Weakly polar
C–Cl	0.5	Weakly polar
C–Br	0.4	Weakly polar
C—O	0.7	Weakly polar
C–N	0.2	Nonpolar

Quantitative measures of the electronegativities of various elements are shown in Table 2.3. There is an approximate correlation between the differences in the electronegativity of two atoms in a bond and its dipole moment. For example, the difference in electronegativity of H and F is 1.9, and the dipole moment is 1.91 D; the H and O difference is 1.4, and DM is 1.5; and the C and O difference is 1.0, and DM is 0.7. The partial negative charge resides at the F and O atoms in these bonds.

TABLE 2.3
Electronegativities of Some Elements

Element	Electronegativity	Element	Electronegativity
Li	1.0	C	2.5
Na	0.9	Pb	1.9
K	0.8	N	3.0
Cr	1.6	P	2.1
Fe	1.8	O	3.5
Cu	1.9	S	2.5
Ag	1.9	F	4.0
Au	2.4	Cl	3.0
Zn	1.6	Br	2.8
Cd	1.7	I	2.5
H	2.1		

2.4 POLARITY OF MOLECULES

The C–H bond is one of the most common in organic compounds. By looking at Table 2.3, it can be seen that the electronegativity difference between these two atoms is 0.4. On the other hand, if we look at the O and H atoms, we find that the electronegativity difference is 1.4. This leads to a description of the C–H bond as being **weakly polar** and the O–H bond as being **polar**. By this means we can provide descriptive terms for the polarity of the various bonds, as shown in Table 2.2.

If we join sets of bonds and atoms together to form covalent compounds, this results in the overall polarity of the compound being due to a combination of the dipole moments of the bonds present. The overall dipole moment for the molecule is due to the spatial resolution or vector sum of the various dipole moments of the bonds present in terms of their size and direction.

Figure 2.4 contains a diagrammatic representation of the methane molecule. Since the hydrogen atoms in this molecule are directed symmetrically toward the corners of a tetrahedron, the spatial resolution of the dipole moments of the C–H bonds would result in a dipole moment for methane of zero. So, even though the C–H bond is a weakly polar bond, the overall dipole moment is zero. However, if we replace one of the hydrogen atoms in methane with a chlorine atom to give monochloromethane (Figure 2.4), this gives an unsymmetrical molecule that has a dipole moment of 1.87 D. Similarly, the symmetrical compound tetrachloromethane, which is also known as carbon tetrachloride, has a dipole moment of zero since it is a symmetrical molecule, and trichloromethane, also known as chloroform, has a dipole moment of 1.02 D.

In this discussion we have looked at atoms in terms of their electronegativity, but common groupings of atoms and functional groups have consistent properties in terms of their effects on polarity of molecules. The polarity classifications of a variety of common groupings are shown in Table 2.4. It is interesting to note that the alkyl groups containing many C–H bonds are listed as nonpolar, whereas the C–H bond is listed as weakly polar. This is due to the spatial orientation of the C–H bonds present, which tends to reduce polarity. Thus, the n-alkanes such as hexane are nonpolar.

A variety of important chemicals widely distributed in the environment are weakly polar to nonpolar. For example, DDT with the structure shown in Figure 2.4 is weakly polar to nonpolar since the weakly polar C–Cl bonds are distributed in such a manner as to reduce the overall polarity of the molecule to a very low level. The polychlorinated biphenyls (PCBs) and most of the chlorohydrocarbon pesticides are also weakly polar to nonpolar. On the other hand, the herbicide 2,4-dichlorophenoxyacetic acid (2,4-D) (Figure 2.4) is a polar molecule because the carboxylic group (–COOH) is polar (Table 2.4).

One of the most important compounds in the environment is water since it is the major component of oceans, lakes, and rivers, and thus has a major impact on the distribution of chemicals. Water has a molecular structure that consists of two hydrogen atoms attached to an oxygen atom, with both O–H bonds residing in a plane at an angle of 104° to each other. Spatial analysis of the direction and strength of these bonds gives a resultant dipole moment of 1.80 D, which is in agreement

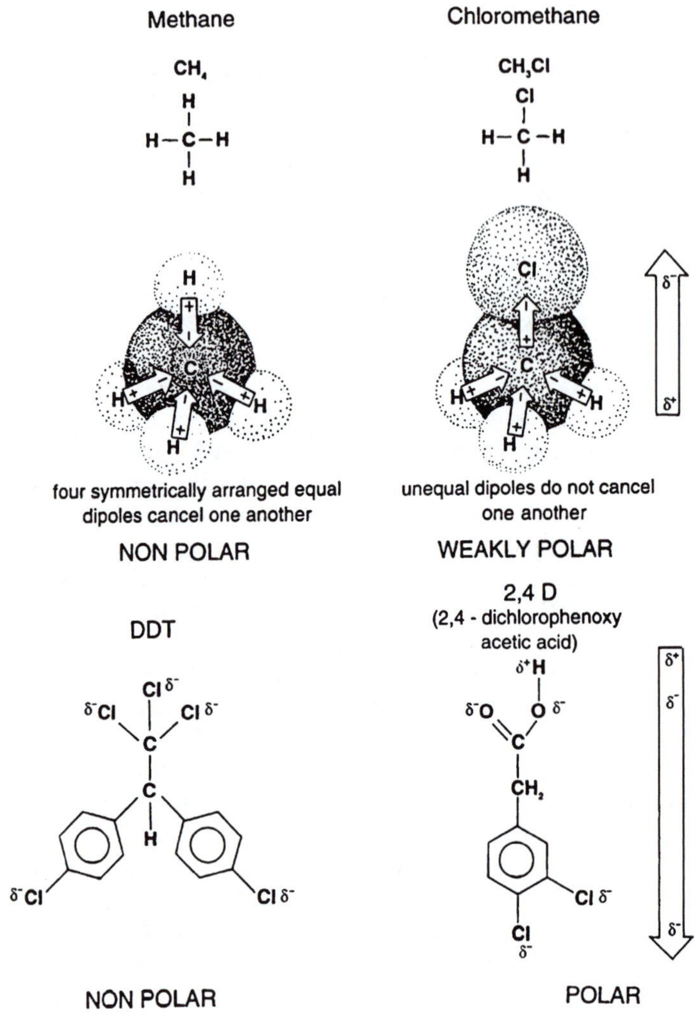

FIGURE 2.4 Chemical structures of some compounds with different polarities.

with the experimental value of 1.87 D (Figure 2.5). As a result, water is a polar molecule with physical properties, and its ability to dissolve other substances is strongly influenced by its polarity.

2.5 IONIC COMPOUNDS

With some elements, the difference in electronegativities can be so great that one atom completely loses an electron to the other atom in the bond. For example, in the hydrogen chloride molecule, there can be a transfer of an electron from the hydrogen to the chlorine, resulting in the formation of a chloride ion with a full

TABLE 2.4
Polarity of Common Groupings in Organic Molecules

Highly Polar	Polar	Weakly Polar	Nonpolar
$-COO^-$	$-O-H$	$>C-Cl$	$-CH_3$
$-O^-$	$-COOH$	$>C-Br$	$-CH_2CH_3$
$-NH^{3+}$	$-NH_2$	$>C-H$	
	$>C=O$		

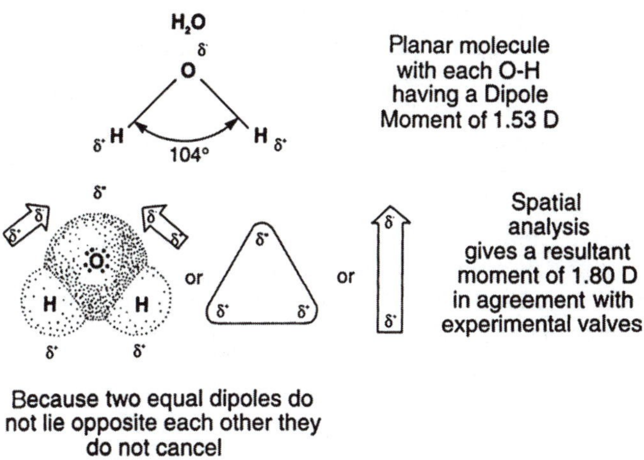

FIGURE 2.5 Representations of the water molecule and its polar characteristics.

negative charge and, consequently, a hydrogen ion with a full positive charge. This is illustrated below:

$$\overset{\delta+}{H}-\overset{\delta-}{Cl} \rightarrow H-Cl \rightarrow H^+ + Cl^-$$

In this transfer, the arrow indicates that both electrons that form the covalent bond have moved to the chlorine atom. In this situation, the chlorine atom in the HCl molecule, which held a small negative charge (δ–), now becomes a fully negatively charged free atom, described as an ion. The covalent bond has been broken and no

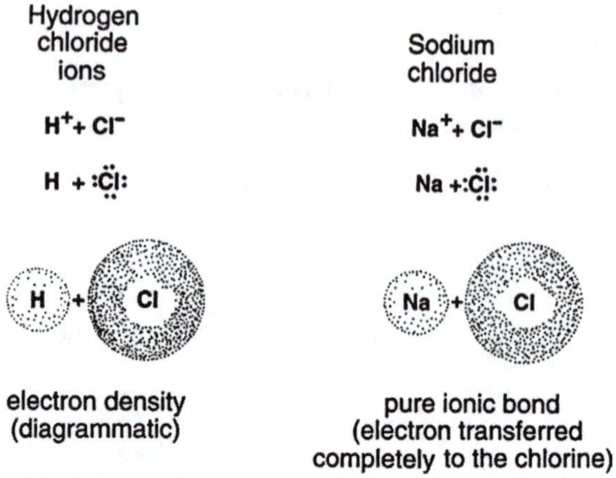

FIGURE 2.6 Formation of ions with hydrogen chloride and sodium chloride.

longer exists (Figure 2.6). With the formation of ions, the hydrogen ion and chloride ion can move apart depending on conditions. In fact, in hydrogen chloride there is a mixture of polar hydrogen chloride molecules together with hydrogen and chloride ions. Measurements indicate that in pure liquid hydrogen chloride, 17% of the compound is in the ionic form. With metal salts of some substances, there can be 100% of the substance in the ionic form. Examples of these substances are sodium chloride, potassium bromide, and so on. Similar processes occur with these substances as with hydrogen chloride, as illustrated in Figure 2.6, but to a greater extent.

Compounds that form ions in aqueous solutions will conduct electricity and are often referred to as **electrolytes**. Acids are compounds that release hydrogen ions into solution, and bases are compounds that release hydroxide ions (OH^-) into solution. For example,

$$\text{Na OH} \rightarrow Na^+ + OH^-$$

Sodium hydroxide (caustic soda)

$$NH_3 + H_2O \rightarrow NH_4^+ + OH^-$$

Ammonia

Salts are ionic compounds formed from the cation of a base and the anion of an acid that yield neither a hydrogen ion nor hydroxide ion on dissociation. Some typical salts are

$$\text{Cu } SO_4 \rightarrow Cu^{2+} + SO_4^{2-}$$

$$\text{Na Cl} \rightarrow Na^+ + Cl^-$$

Generally, complete dissociation of salts occurs in water. However, acids may only partially dissociate. For example, strong acids, such as the mineral acids $H_2\ SO_4$ (sulfuric) and NO_3 (nitric), and the organic acids $H_3C{\cdot}COOH$ (acetic) and $H{\cdot}COOH$ (formic), dissociate only partly. Thus, with acetic acid, an equilibrium is formed that favors the undissociated form

$$H_3\ C{\cdot}COOH \leftrightarrow H_3\ C{\cdot}COO^- + H^+$$

Reactant Products

A general equilibrium situation is established where K = [products]/[reactants], and in the case of acetic acid,

$$K = \frac{[H^+][H_3C \cdot COO^-]}{[H_3C \cdot COOH]}$$

where K is the equilibrium constant and $[H^+][H_3C{\cdot}COO^-]$ and $[H_3C{\cdot}COOH]$ are the molar concentrations of the various substances. The value of K can be calculated from actual measurements as

$$K = \frac{(0.00135)(0.00135)}{(0.0986)}$$

K is constant for dilute solutions of acetic acid at 1.85×10^{-5}. This means that the proportion of the compound that dissociates into ions is constant in dilute solution irrespective of the concentration. There is a similar situation with bases. For example, with ammonia an equilibrium reaction occurs in water; thus,

$$NH_3 + H_2O \rightarrow NH_4^+ + OH^-$$

Reactant Products

$$K' = \frac{[NH_4^+][OH^-]}{[NH_3][H_2O]}$$

where K′ is the equilibrium constant and $[NH_4^+]$, $[OH^-]$, $[NH_3]$, and $[H_2O]$ are the concentrations of the various substances. Since the concentration of water is very high and constant, it is included in K:

$$\therefore K'[H_2O] = K = \frac{[NH_4^+][OH^-]}{[NH_3]}$$

From actual measurement, $K = 1.8 \times 10^{-5}$.

With pure water a small proportion of the molecules dissociate and equal concentrations of H^+ and OH^- are formed:

$$H_2O \leftrightarrow H^+ + OH^-$$

$$K = \frac{[H^+][OH^-]}{[H_2O]}$$

where K is the equilibrium constant and $[H^+]$, $[OH^-]$, and $[H_2O]$ are the molar concentrations of the substances indicated.

Since the undissociated form $[H_2O]$ is present in close to 100% concentration and effectively constant, it is included in K. So,

$$K\,[H_2O] = K_w = [H^+][OH^-] \text{ and } [H^+] = [OH^-] = \sqrt{K_w}$$

and from actual measurement,

$$K_w = 1 \times 10^{-14}\ M$$

thus,

$$[H^+] = [OH^-] = 1 \times 10^{-7}\ M$$

The hydrogen ion concentration present in water is an important charcteristic influencing behavior of ions. For convenience, it is placed on a scale that is easy to use; thus,

$$pH = -\log\,[H^+]$$

This means that pure neutral water has a pH of 7. Since acids release H^+ into the solution, the molar concentrations will be greater than 10^{-7}, i.e., 10^{-3}, 10^{-4}, and so on. This means acid solutions will have pH < 7. Alkaline solutions have lower concentrations of H^+ than neutral solutions, i.e., 10^{-8}, 10^{-10}, and so on, and thus have pH > 7.

2.6 INTERMOLECULAR FORCES

In the previous section we considered charges that are generated within molecules. It can be appreciated that these charges within molecules will interact with adjacent

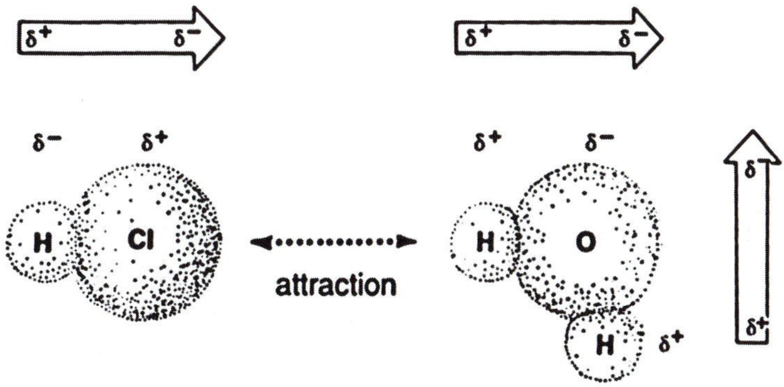

FIGURE 2.7 Attraction between the negative end of a hydrogen chloride molecule and the positive end of a water molecule.

molecules with the usual rules for the interaction of charges applying. This means that like charges result in repulsion and unlike charges result in attraction. Thus, positive-to-positive and negative-to-negative forces result in the generation of a disruptive or repulsive force, and positive-to-negative charges result in an attraction or a cohesive force. Intermolecular forces between covalent compounds are usually described as **van der Waals forces**. Van der Waal was a Dutch scientist who developed an understanding of intermolecular forces. These intermolecular forces have a strong influence on the physical-chemical properties of compounds and how substances behave in the environment.

Molecules that are polar can be attracted to one another so that particular orientations of the positive and negative charges are formed. Examples of these interactions, referred to as **dipole bonding**, are shown in Figure 2.7 and Figure 2.8. Here the δ– charge on the chlorine in hydrogen chloride is oriented toward the δ+ on the hydrogen atom in water (Figure 2.7). In water, alcohols, and acids, a network of orientations of the δ+ hydrogen toward the δ– of the oxygen in another molecule of the same compound is set up. Of course, in mixtures, these interactions occur between molecules of different compounds. The **hydrogen bond** is a special form of dipole bonding that is of particular significance because it is relatively common and strong. Hydrogen bonding with water, alcohol, and acid molecules is shown diagrammatically in Figure 2.8. Somewhat similarly, polar molecules with dipole moments may interact in various ways with ions.

Another important group of intermolecular interactions are described as **London forces** or **dispersion forces**. These operate over relatively short distances and are responsible for the aggregation of molecules that possess neither free charges nor dipole moments. London forces increase rapidly with molecular weight and are particularly strong with aromatic compounds, but with relatively low molecular weight compounds, these are often relatively weak when compared with dipole interactions.

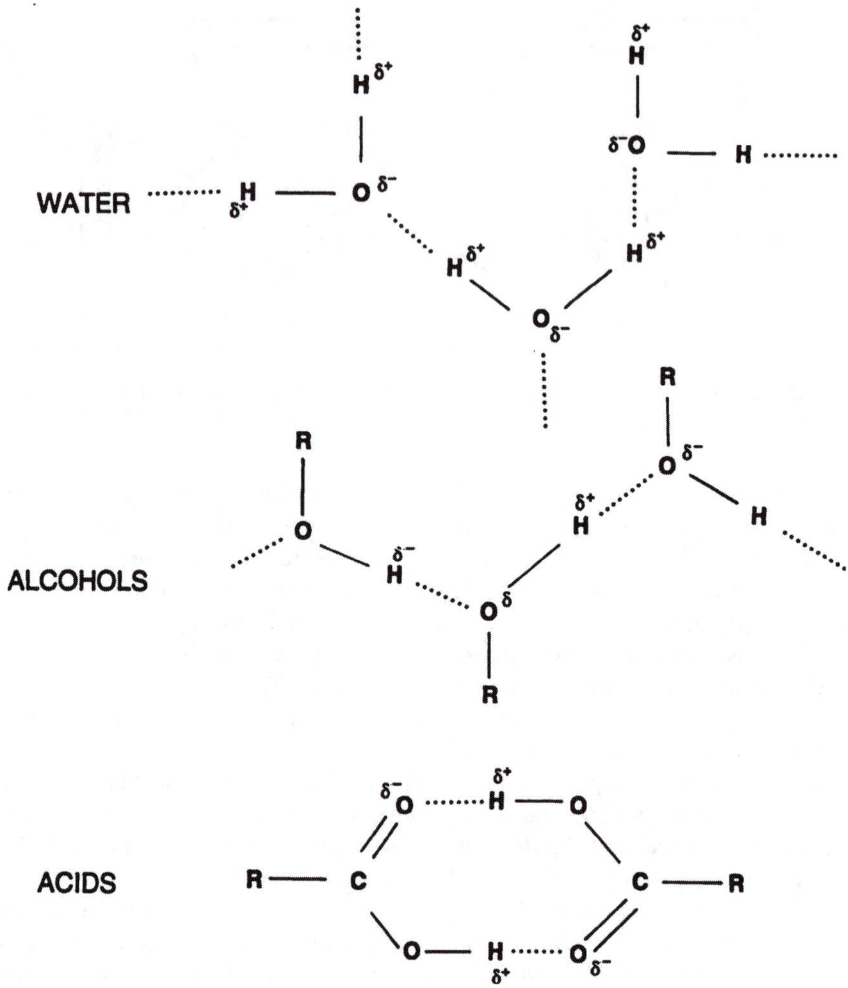

FIGURE 2.8 Hydrogen bonding in water, alcohols, and acids.

2.7 PHYSICAL-CHEMICAL PROPERTIES OF COMPOUNDS

2.7.1 Melting Point (MP)

When a substance melts, it changes from a solid to a liquid. In molecular terms, a solid has the component particles in a fixed relationship to one another. There are strong forces between chemical entities as occur, for example, with ions, high-molecular-weight substances, and highly polar substances. With ionic solids, each of the ions is held in a fixed relationship to other ions by the strong forces between the ions. This is illustrated diagrammatically with sodium chloride in Figure 2.9. This compound is held in the solid form by the fixed and close relationship between

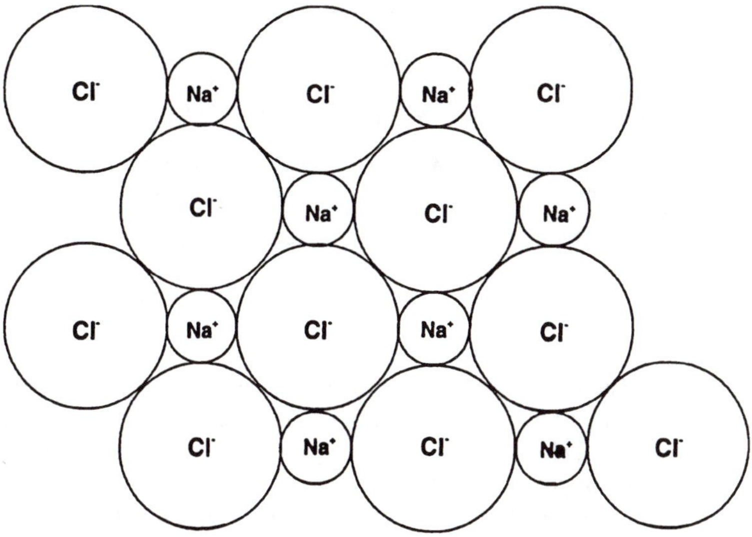

FIGURE 2.9 Diagram of the ionic structure of sodium chloride shows that this is a single layer of ions out of the solid and that other layers rest above and below this layer.

the strongly attractive forces that exist between the sodium and the chloride ions with their full positive and full negative charges. It is also noteworthy that each ion is surrounded by ions of an opposite charge, leading to an overall neutral electrostatic situation. Because of these forces, it requires a high-energy input to break this solid structure down to give a liquid. This means that this substance will exhibit a very high melting point, and in fact, sodium chloride exhibits a melting point of 801°C.

With nonpolar compounds of low molecular weight, including oxygen, nitrogen, and methane, there are no ionic or dipole forces holding separate molecules together, but there are the relatively weak London forces causing attachment between the molecules. Thus, it would be expected that these substances would exhibit a very low melting point. This is indicated by the melting point of methane, which is –183°C.

2.7.2 Boiling Point (bp)

The boiling point and related properties, such as vapor pressure (vp), have a major influence on such important environmental characteristics as evaporation and the distribution of compounds in the atmosphere. Boiling involves the conversion of a liquid with molecules in close, but random proximity, into a gas with molecules some distance apart. Gases at room temperature are substances with weak intermolecular forces. This means that generally they will be nonionic, low-molecular-weight, nonpolar substances. In a general sense, liquids can be seen as an intermediate state between gases and solids.

With ionic compounds, the ion charges hold the ions together with considerable force. Thus, we would expect that, in general, ionic compounds would have very high boiling points. For example, sodium chloride has a boiling point of 1413°C.

TABLE 2.5
Change in Boiling Point with Compounds of Similar Molecular Weight

Compound	Dipole Moment (D)	Molecular Weight	Boiling Point (°C)
H_2O	18	1.84	100
NH_3	17	1.46	–33
CH_4	16	0	–161.5

With nonionic and nonpolar compounds, the major intermolecular forces are London forces that are weak at low molecular weight. This means that this group of compounds will have a much lower boiling point than ionic compounds. For example, methane has a boiling point of –161.5°C.

With covalent compounds, the boiling point is influenced by two factors: (1) the dipole moment and (2) the size of the molecule. This latter factor is related to the strength of the London forces, which are related to molecular weight. Thus, compounds of similar molecular weight have boiling points that are related to their dipole moment. There is a comparison of the boiling points of compounds of similar molecular weight but different dipole moments in Table 2.5. It can be seen that since molecular weight is roughly constant, the boiling point in the series of compounds decreases as the dipole moment decreases.

An illustration of the action of London forces is shown in Table 2.6. In this data set, the dipole moment of the normal alkanes is constant at zero, and the molecular weight increases from methane to butane. As the molecular weight increases, the boiling point also increases in accord with the molecular weight change. This is due to the increase in the strength of the London forces with increasing molecular weight. This relationship within a series of compounds having a regular change in size, such as in a homologous series, is useful for predicting properties of members of the series that are not available. For example, there is a plot of vapor pressure against

TABLE 2.6
Change in Boiling Point with Compounds of Different Molecular Weight

Compound	Dipole Moment (D)	Molecular Weight	Boiling Point (°C)
Methane	0	16	–161
Ethane	0	30	–88
Propane	0	44	–42
Butane	0	58	0

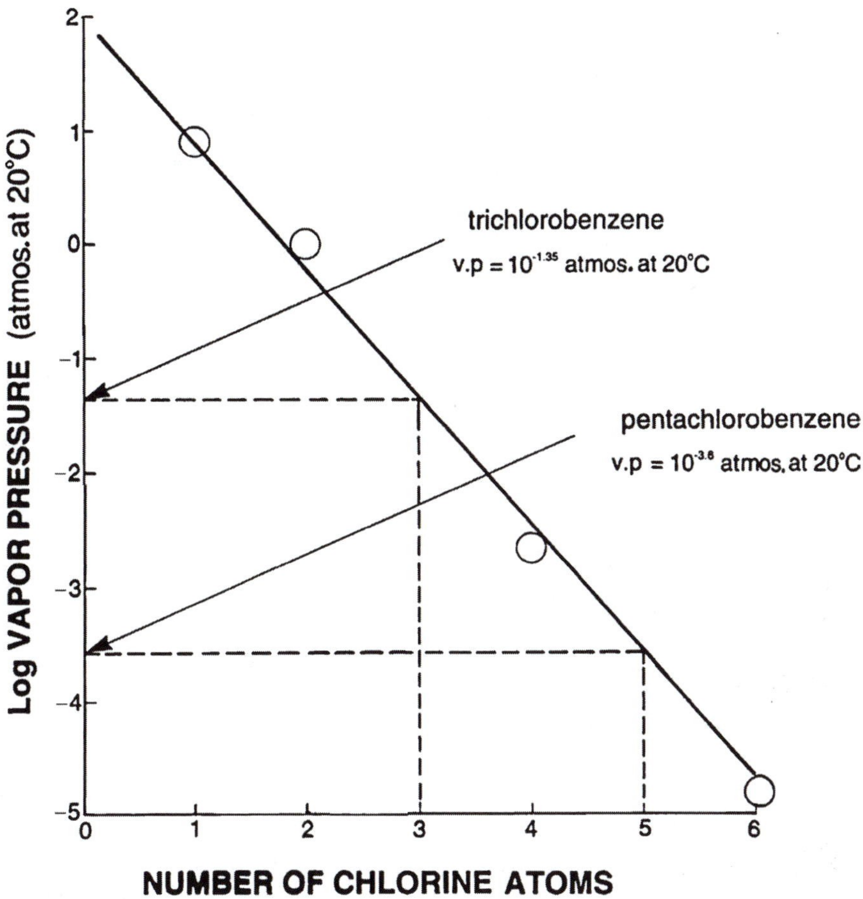

FIGURE 2.10 Plot of the relationship between log (vapor pressure) and number of chlorine atoms for the chlorobenzenes.

number of chlorine atoms in Figure 2.10 for the chlorobenzenes. The vapor pressure of some members of the series (tetra- and pentachlorobenzenes) may not be known but can be obtained by interpolation, as shown in Figure 2.10.

2.7.3 Solubility

An important environmental property of a substance is its solubility in water and in other solvents, since this property influences its dispersal in the open environment, including the oceans and aquatic systems generally; James (1986) and Reichardt (1988) have provided explanations of the theory of solubility and related properties. In addition, a range of important biological properties can be related to solubility in water and other solubility-related properties. An understanding of this property can give an understanding of many of the most important environmental properties of compounds.

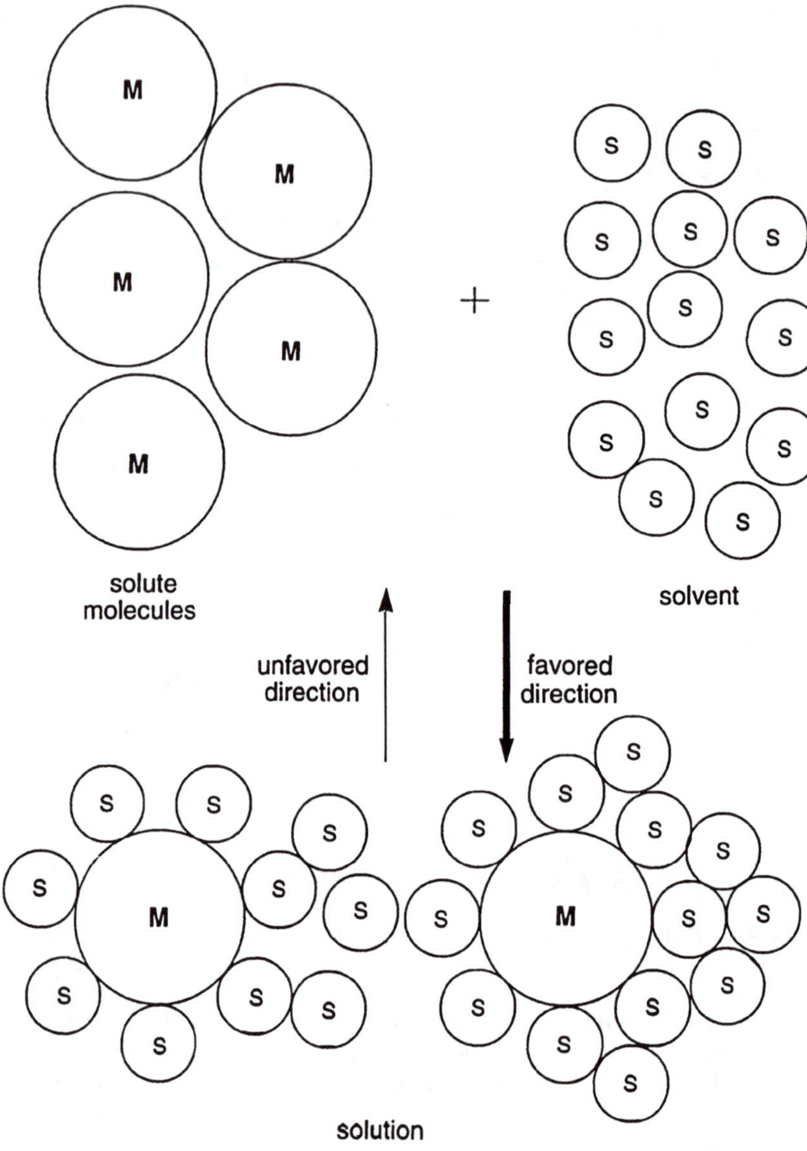

FIGURE 2.11 Interaction of solute (M) and solvent (S) molecules to produce a solution.

When a substance dissolves to form a solution, the individual molecules of the solute are separated by the molecules of the solvent, as shown in Figure 2.11. In simple terms, for a substance to dissolve, the solvent molecules (S) must form a set of bondings with the solute molecules (M), which are, in total, stronger than the solute-to-solute bonds. If this does not occur, the solvent molecules (M) will move together due to mutual attraction and the substance will come out of solution and thus be insoluble.

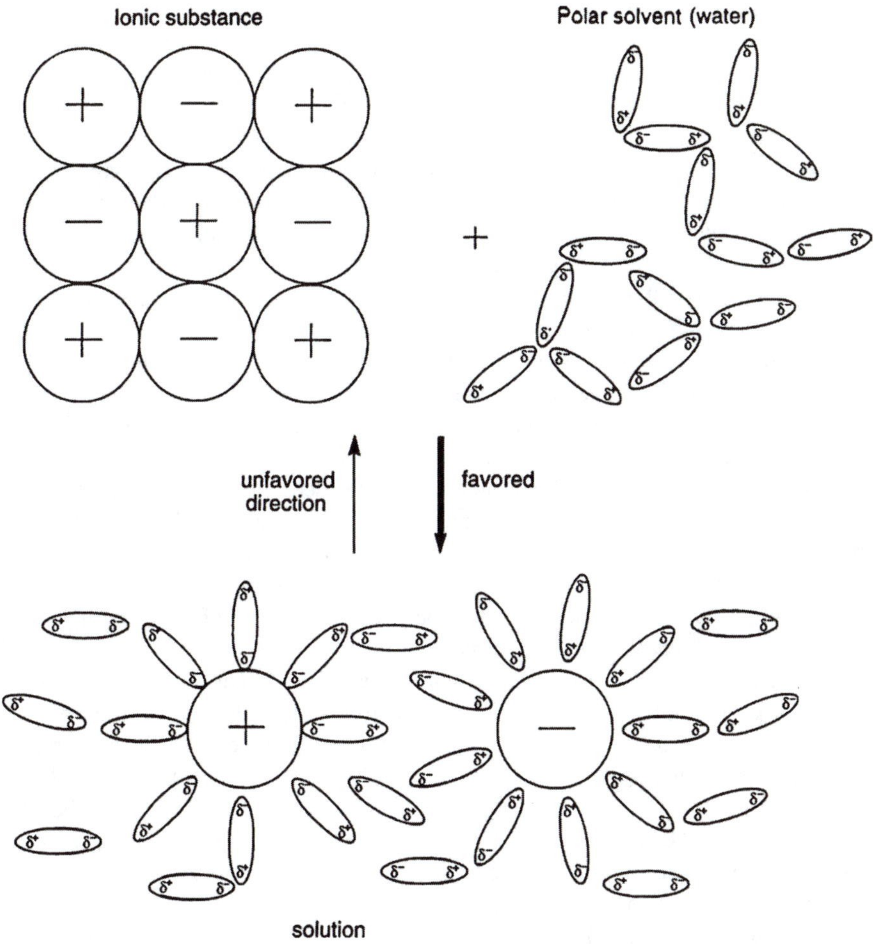

FIGURE 2.12 Solution of an ionic substance, such as sodium chloride, in a polar solvent, such as water.

The solution of an ionic compound in a polar solvent is shown diagrammatically in Figure 2.12. This is typified by the solution of sodium chloride in water. The ions in solution are each surrounded by water molecules with the oppositely charged ends attracted to the ion. The forces of attraction between the water molecules and the ions are greater than the attraction between the sodium and chloride ions. In this way, a stable solution is formed and, generally, ionic compounds are dissolved by polar and ionic solvents. A similar situation would be expected to apply to the solubility of polar covalent compounds by polar solvents.

If a nonpolar solvent was used with an ionic or polar solute, the bonds between the nonpolar solvent molecules and the ions or polar substances would be very weak and generally insufficient to form a stable solution. Thus, nonpolar solvents, such as hexane, would not be expected to dissolve polar and ionic compounds to any significant extent. For example, water, which is a polar substance, dissolves ionic and polar substances such as sodium chloride, hydrogen chloride, ammonia, and so on. On the other hand, liquid methane is a nonpolar substance and will not significantly dissolve these substances. This leads to the generalization that, in terms of polarity, like compounds dissolve and unlike compounds are insoluble in one another.

Many chemicals of environmental importance are weakly polar to nonpolar, and include compounds such as the dioxins, PCBs, and chlorohydrocarbon pesticides. The situation using a polar solvent (such as water) to dissolve these nonpolar compounds is illustrated in Figure 2.13. Since the solute is nonpolar, little bonding between water and the solute occurs. If a nonpolar solute molecule dissolves in water, its surrounding environment is quite different from that of an ion or a polar compound. The ion is surrounded by the ends of water molecules all of the same charge, so as to stabilize and neutralize the opposite charge on the ion (Figure 2.12). With the nonpolar solute, the water molecule orients to stabilize and neutralize the system by alternating the positive and negative ends facing the solute molecule, as in Figure 2.13. This creates a force on the inner surface of the cavity formed by the solute molecule that tends to contract and thus expel the solute molecule. As a result, the solubility of nonpolar substances in water decreases with increasing size of the solute molecule and the cavity it consequently forms in the solvent. This is illustrated by the plot in Figure 2.14. Here, the aqueous solubilities of the PCBs decline as the surface area of the molecules increases.

In contrast to these situations, we can evaluate the situation where a nonpolar or weakly polar solute is dissolved in a nonpolar solvent, as depicted in Figure 2.15. Common nonpolar or weakly polar solvents include *n*-hexane and *n*-octanol, but in the environment, an important nonpolar to weakly polar solvent is biota lipid. Biota lipid is a complex mixture in all organisms but always contains large numbers of nonpolar groups, leading to overall nonpolarity or weak polarity for the lipid in general. A substance must have a reasonable solubility in lipid to penetrate membranes and enter the internal system of biota. Substances that are soluble in lipid are referred to as **lipophilic** or **lipid loving**. The major components of lipids are fats, which are high-molecular-weight, nonpolar to weakly polar esters. Compounds that are lipophilic are likely to have biological effects, such as a tendency to accumulate in organisms to higher concentrations than occur in the external environment and a general toxicity to all biota. In addition, these substances are often also persistent in the environment and exhibit resistance to transformation and degradation. The general characteristics of lipophilic compounds are outlined in Table 2.7. In addition to having the property of lipid solubility, lipophilic compounds usually exhibit relatively low solubility in water. For this reason, they are often also called **hydrophobic** compounds, which means that they are **water hating**.

2.8 PARTITION BEHAVIOR

Previously, we looked at polarity and the physical-chemical properties of organic compounds. The principle of solubility discussed was that in terms of polarity: like dissolves like, and unlike substances do not dissolve. However, we have many situations that are not clear-cut, and the compounds have both polar and nonpolar

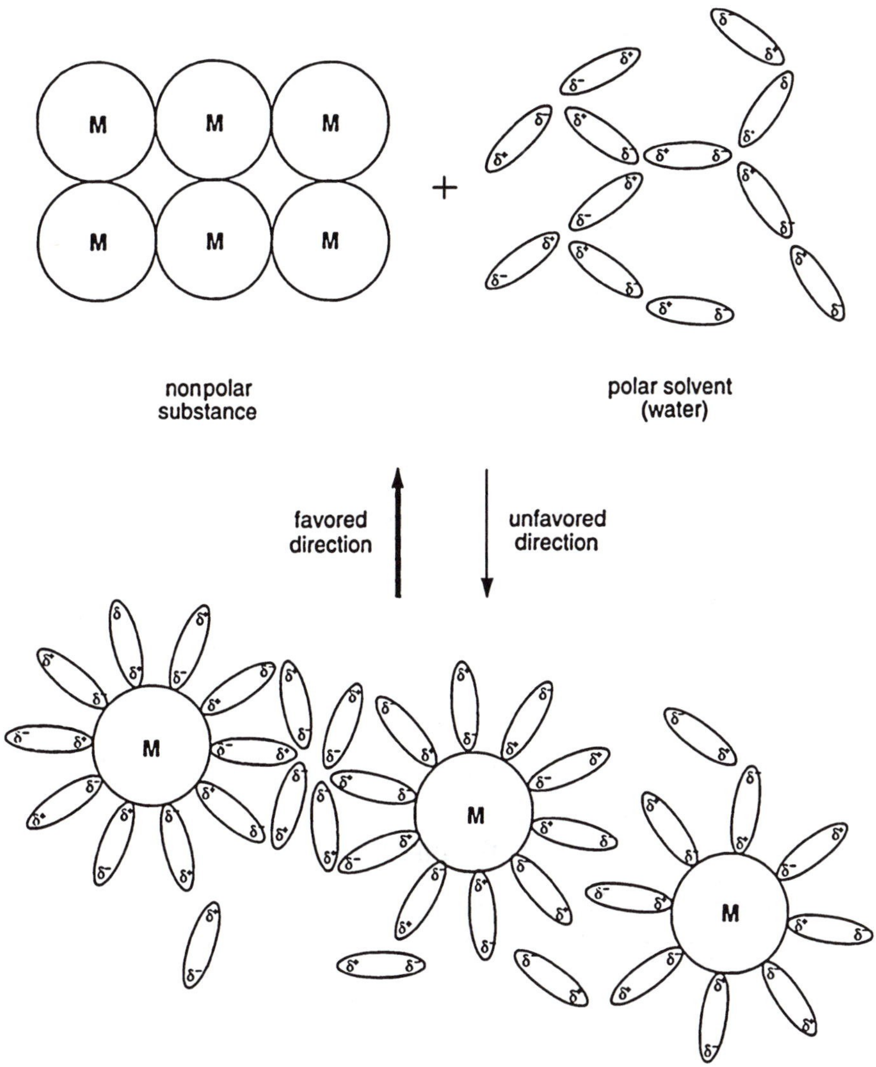

FIGURE 2.13 Solution of a nonpolar substance (M) in a polar solvent, such as water.

properties. For example, butyric acid (CH_3–$(CH_2)_2$–COOH) has polarity and in fact partially forms ions (depending on the pH of the solvent), but also has a nonpolar part, as illustrated in Figure 2.16. If placed in water, which is a polar solvent, some of the butyric acid will dissolve in the water because it is moderately soluble in water. However, if it is placed in a nonpolar solvent such as hexane, or diethyl ether, some of it will also dissolve in these nonpolar solvents. This means that it is moderately soluble in both polar and nonpolar solvents.

If one places sufficient diethyl ether and water together, they will form a two-phase system with diethyl ether forming a separate layer floating above the water, as shown in Figure 2.17. The diethyl ether (CH_3–CH_2–OH–CH_2–CH_3) is a weakly

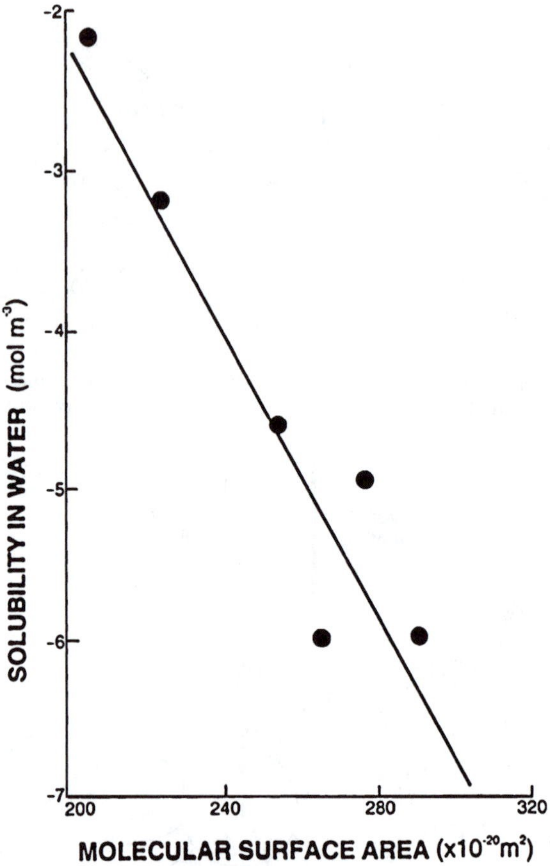

FIGURE 2.14 Relationships between the surface area of the molecule and the solubility in water for the polychlorinated biphenyls (PCBs).

polar to nonpolar liquid that has limited solubility in the polar water. If one places butyric acid in this system and shakes it thoroughly, one finds that the butyric acid has partitioned (or distributed) between the two liquids. In fact, at equilibrium, the diethyl ether acid contains a concentration of three units per volume and the water contains a concentration of one unit per volume. This partitioning can be quantitatively measured as the **partition coefficient**. The partition coefficient for butyric acid in a diethyl ether–water system is 3. The partition coefficient is constant for a given substance and two specific liquids under constant environmental conditions, such as temperature and pressure. The partition coefficient is dependent on:

1. Polarity of the substance
2. Molecular weight
3. Relationship of the polarity of the solvents used

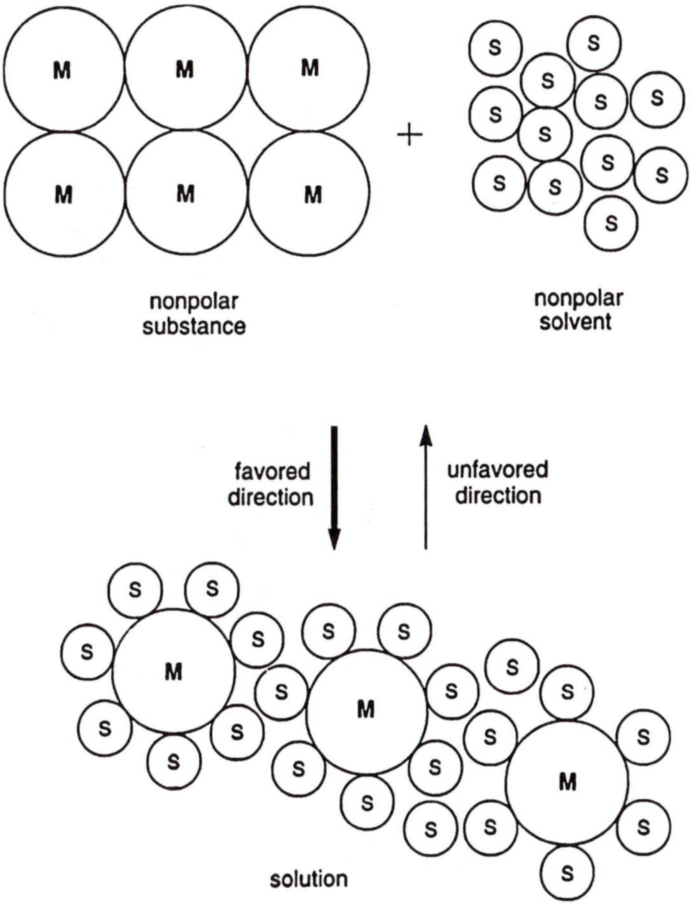

FIGURE 2.15 Solution of a nonpolar (or weakly polar) substance (M) in a nonpolar solvent such as biota lipid.

TABLE 2.7
Characteristics of Lipophilic Compounds

Characteristic	Features
Chemical structure	Nonpolar compounds such as chlorohydrocarbons, hydrocarbons, polychlorodibenzodioxins, polychlorodibenzofurans, and related compounds
Molecular weight	100–600
Aqueous solubility (mol/m^{-3})	0.002–18

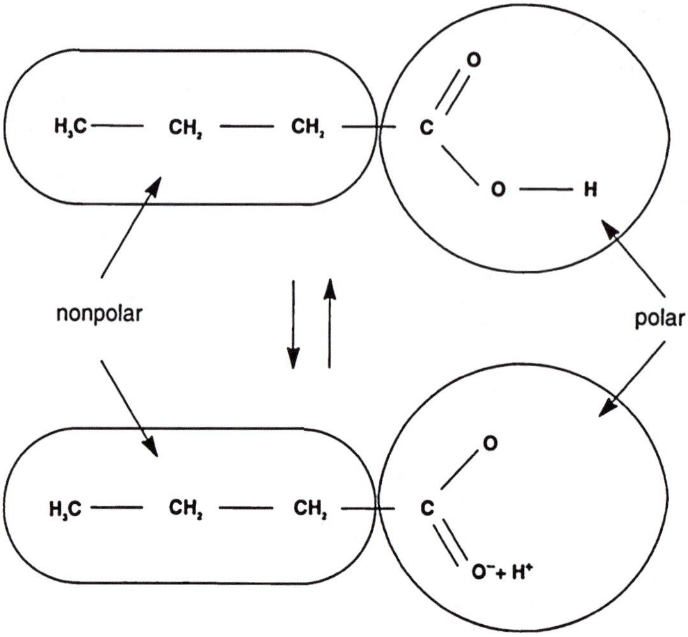

FIGURE 2.16 The butyric acid molecule, indicating polar and nonpolar sections.

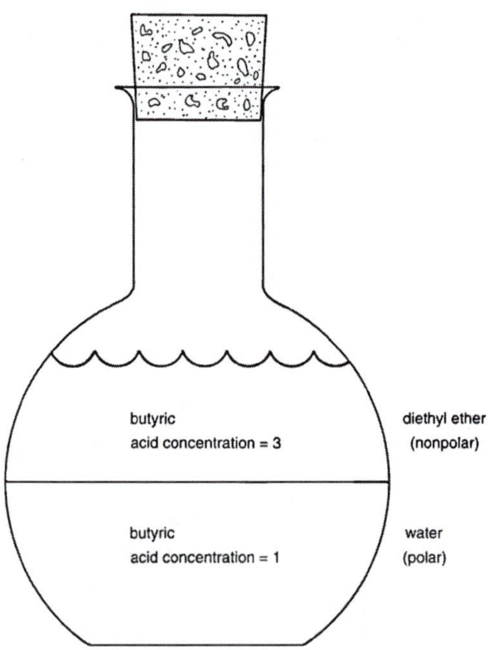

FIGURE 2.17 Partitioning of butyric acid between diethyl ether and water.

If one uses hexane instead of butyric acid in this system, since hexane is less polar than butyric acid, hexane will have a different partition coefficient. Hexane has a higher concentration in the diethyl ether phase and a lower concentration in the water phase than butyric acid. In fact, the partition coefficient of hexane in this system is approximately 100 compared to 3 for butyric acid.

One of the most widely used physical-chemical characteristics of compounds applied in environmental chemistry is the *n*-octanol–water partition coefficient. This partition coefficient is important because it imitates the biota lipid–water partition process. Octanol (CH_3–$(CH_2)_7$–OH) is in many ways similar to biota lipid since it is fat-like in many of its physical-chemical properties. The *n*-octanol–water partition coefficient (K_{OW}) is defined as

$$K_{OW} = C_O/C_W$$

where C_O is the concentration of the substance in octanol and C_W is the concentration of the substance in water at equilibrium and at a constant temperature. K_{OW} values can be measured in a **shake-flask** system, similar to the diethyl ether–water system in Figure 2.17, by adding the compound to the system and measuring the concentrations after shaking to reach equilibrium and then calculating K_{OW}. The K_{OW} values are dimensionless since both C_W and C_O are measured in the same units, i.e., units of mass per unit volume. Values have been measured for a wide variety of compounds of environmental importance, and these range from about 0.001 to over 10,000,000. Some typical illustrative values are contained in Table 2.8 (the values are expressed on a logarithmic scale for convenience).

The K_{OW} values are constant for a given compound and, apart from experimental errors, reflect the lipophilicity of a compound. As K_{OW} increases, the solubility in lipid (represented by octanol) increases relative to water. Thus, increasing K_{OW} values reflect increasing lipophilicity. An increasing K_{OW} value also usually reflects a decline in the solubility in water. A compound such as ethanol with K_{OW} = 0.49 actually has greater solubility in water than in octanol and is not a lipophilic compound.

TABLE 2.8
Some Octanol–Water Partition Coefficients (Values for Various Compounds)

Compound	K_{OW}	Log K_{OW}
Ethanol	0.49	–0.31
2,4,5-T	3.98	0.60
2,4-D	37.00	1.57
Benzene	135.0	2.13
1,4-Dichlorobenzene	3310.00	3.52
DDT	2.39×10^6	6.36

Similarly, 2,4,5-T and 2,4-D (see Figure 2.4) have K_{OW} values of 3.98 and 37.2, respectively, and these substances would be regarded as being of low lipophilicity. Compounds are usually considered lipophilic at values of >100 K_{OW}. As a general rule, lipophilic compounds have K_{OW} values in the range from 100 to 3,000,000, i.e., log K_{OW} 2 to 6.5. The lipophilic compounds in Table 2.8 are benzene, 1,4-dichlorobenzene, and DDT.

The log K_{OW} values can be predicted using relationships somewhat similar to those used to predict boiling point and vapor pressure (Figure 2.10). As outlined previously, the surface area of the cavity formed by a solute in the solvent, or the surface area of the solute itself, is a useful parameter for evaluation of environmental properties (Figure 2.14). The K_{OW} values for members of various homologous series have been found to be related to the molecular surface area of the compounds. The relationship between these characteristics for chlorobenzenes is shown in Figure 2.18, with graphical projections that can be used to provide data on members with unknown properties.

Another useful partition coefficient for describing the environmental properties of a chemical is the Henry's law constant (H), defined as

$$H = P/C_W$$

where P is the partial pressure of the compound in air and C_W is the corresponding concentration in water. This characteristic quantifies the relationship between a

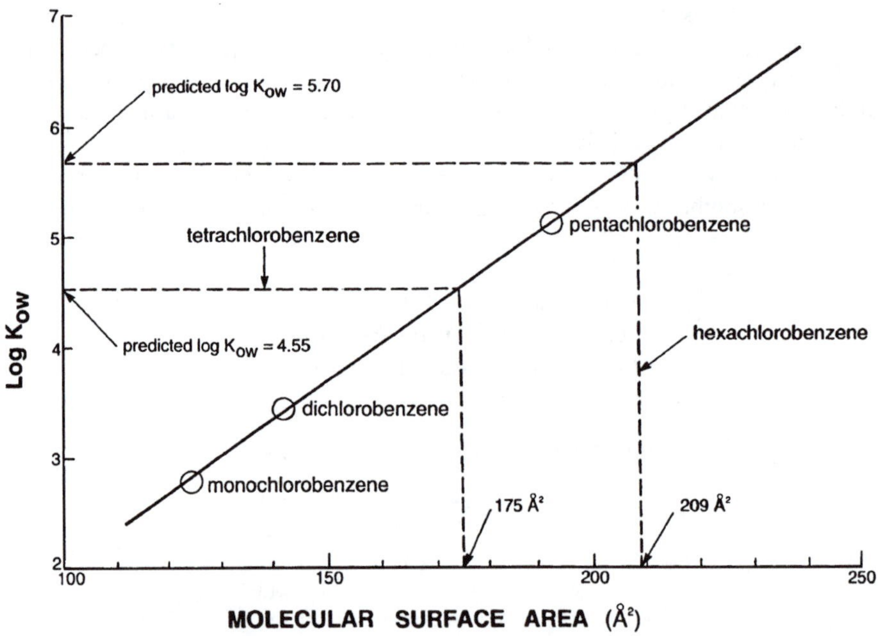

FIGURE 2.18 Relationship between log K_{OW} and molecular surface area for the chlorobenzenes with extrapolation and interpolation to estimate compounds with unknown properties.

contaminant in a water body (such as a lake or the oceans) and the atmosphere at equilibrium. Classic books by Boethling and Mackay (2000) and Lyman et al. (1990) describe a range of important environmental properties and methods to calculate values for them.

The environment can be considered to be made up of a set of phases, such as air, water, soil, atmosphere, and biota. A chemical enters the environment and partitions between these phases as the basic mechanism governing distribution. This is described in more detail in Chapter 15.

2.9 KEY POINTS

1. There are three states of matter in the environment: solids, liquids, and gases. Solids such as common salt, sand, rocks, and so on are relatively dense and consist of particles that are closely packed in a fixed relationship to one another. Liquids such a water have less closely packed particles but do not have a fixed relationship to one another. With gases, such as those in the atmosphere, the particles are in relatively rapid motion and are comparatively large distances apart.
2. Sharing of electrons by atoms to form a covalent bond can result in the polarization of the bond if the atoms are of a different element. The direction and size of the polarization of bonds in a molecule can result in compounds exhibiting overall dipole moments.
3. Ionic compounds are formed by the complete transfer of an electron from one atom in a compound to another, resulting in the formation of a fully charged entity.
4. Acids are substances that release a hydrogen ion (H^+) when they dissociate in water, whereas bases release hydroxide ions (OH^-). The equilibrium constant (K) is

$$K = [\text{products}]/[\text{reactants}]$$

5. The hydrogen ion concentration, $[H^+]$, can be conveniently expressed as the pH:

$$\text{pH} = -\log [H^+]$$

   Neutral pure water has a pH of 7, acids have pH < 7, and bases have pH > 7.
6. The intermolecular forces, generally described as van der Waals forces and measured by the dipole moment of molecules, have a major influence on properties such a melting point, boiling point, and so on. London forces also play a major role in the properties exhibited by nonpolar compounds.
7. Ionic and polar solutes will generally dissolve in water since the water molecules form stronger attachments to the solute than occur between the solute molecules and ions themselves. This is in accord with the rule "like dissolves like" in terms of polarity.

8. Nonpolar substances will not readily dissolve in water since solute and solvent molecules are not attracted and stabilization of the solubilized solute does not occur. This means that weakly polar to nonpolar chemicals important in the environment, such as dioxins, PCBs, and chlorohydrocarbon pesticides, are only sparingly soluble in water.
9. The solubility of nonpolar compounds is related to the size of the cavity formed in the water mass and is also related to the surface area of the molecule.
10. Weakly polar to nonpolar substances, such as dioxins, PCBs, and the chlorohydrocarbon pesticides, readily dissolve in the most important weakly polar to nonpolar solvent in the environment, which is biota lipid. Lipophilic (lipid-loving) substances are generally chlorohydrocarbons, hydrocarbons, and related compounds that have low solubility in water; they can also be described as hydrophobic (water-hating) compounds.
11. A substance must pass through nonpolar lipid-containing membranes to enter biota, and so lipophilic compounds are often biologically active.
12. Covalent molecules usually have a mix of polar and nonpolar sections and will partition between a nonpolar phase (such as *n*-hexane) and a polar phase (such as water). This partitioning can be quantitatively described by the partition coefficient, which is the concentration in one phase–concentration in the other phase. The octanol/water partition coefficient (K_{OW}) is a widely used environmental property and its magnitude can often be predicted using relationships with the molecular surface area and other characteristics.

REFERENCES

Boethling, R.S. and Mackay, D., *Handbook of Property Estimation for Chemicals: Environmental and Health Sciences*, Lewis Publishers, Boca Raton, FL, 2000.

James, K.C., *Solubility and Related Properties*, Marcel Dekker, New York, 1986.

Lyman, W.J., Reehl, W.F., and Rosenblatt, B.H., *Handbook of Chemical Property Estimation Methods: Environmental Behavior of Organic Compounds*, American Chemical Society, Washington, DC, 1990.

Reichardt, C., *Solvents and Solvent Effects in Organic Chemistry*, 2nd ed., VCH, Weinheim, Germany, 1988.

QUESTIONS

1. A range of different compounds and their physical-chemical properties that influence environmental behavior are listed in Table 2.9. Based on these characteristics,
   a. Draw a horizontal line dividing the polar and nonpolar substances.
   b. Draw another horizontal line between the weakly polar and polar molecules.

TABLE 2.9
Various Compounds and Some of Their Properties

Compound	bp (°C)	Water Solubility (g 100 ml^{-1})	Dipole Moment (D)
H_2	–253	2×10^{-4}	0
CH_4	–164	2×10^{-3}	0
CO_2	–78	0.15	0
HBr	–67	221	0.82
HC1	–84.9	82	1.08
NH_3	–33.5	90	1.3
H_2O	100		1.85
HF	19.5		1.82
LiF	1676	0.27	6.33

c. Can you see the reason why the ionic lithium fluoride has relatively low solubility in water?

2. There are listed in Table 2.10 the boiling points and water solubilities of a homologous series of *n*-alkanols. Based on these characteristics,
   a. Draw a horizontal line between the weakly polar and the polar compounds.
   b. How does the number of carbon atoms and molecular weight influence water solubility?
   c. Which intermolecular forces operate in this series of compounds and how does the relative importance change with the change in the number of carbon atoms?

TABLE 2.10
Some Properties of the Alcohols

Alcohol		bp (°C)	Water solubility (g 100 mL^{-1})
Methanol	CH_3OH	64	
Ethanol	CH_3CH_2OH	78	
Propanol	$CH_3(CH_2)_2OH$	97	
Butanol	$CH_3(CH_2)_3OH$	118	8.1
Pentanol	$CH_3(CH_2)_4OH$	138	2.6
Hexanol	$CH_3(CH_2)_5OH$	156	0.6
Heptanol	$CH_3(CH_2)_6OH$	176	0.1

3. The chemical structures of a number of different compounds are shown in Table 2.11. Use this (do not look up tables) to answer the following questions in general terms (e.g., gas, liquid, solid; very low, low, moderate, high).
   a. What is the physical state of these substances under normal conditions of temperature and pressure?
   b. Which of the compounds would you expect to be soluble in water?
   c. Which of the compounds would you expect to be soluble in biota lipid (lipophilic)?
4. Calculate the pH of the following solutions:
   a. Solution containing $[H^+]$ at 10^{-2} *M*
   b. 0.50 *M* HCN (hydrocyanic acid with an equilibrium constant, K, of 4.9×10^{-10})
   c. 0.50 *M* $CH_3 \cdot COOH$ (acetic acid with an equilibrium constant, K, of 1.85×10^{-5})

TABLE 2.11
Structures of Various Compounds

Number	Name	Structure	State	Water Solubility	Lipid Solubility
A	Ethane	$H_3C - CH_3$			
B	Benzene	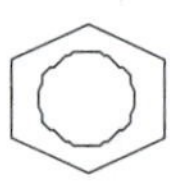			
C	Citric acid	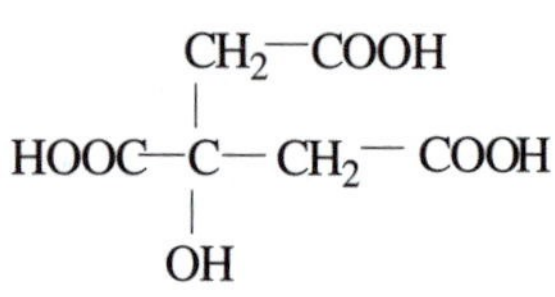			
D	Anthracene	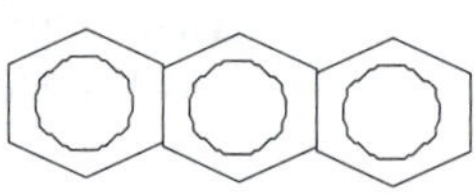			
E	Glyphosate	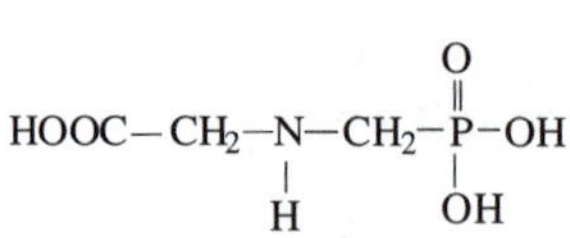			

ANSWERS

1. a. A line between CO_2 and HBr.
   b. A line between NH_3 and H_2O.
   c. The attraction between Li^+ and F^- is greater than the attraction between the water molecules and the ions; thus, Li and F ions cannot be separated by solution in water. This means that lithium fluoride will not dissolve in water.
2. a. A line between propanol and butanol.
   b. Water solubility declines with increasing numbers of carbon atoms and molecular weight.
   c. With methanol, the polar hydroxy group (OH) interacts with water and, because it is a large group in the molecule, its property of increasing water solubility dominates and methanol is soluble in water. With increasing molecular weight, the role of the hydroxy group diminishes as its dominance in the molecule declines. At the same time, London forces between molecules increase, reducing the capacity to dissolve in water and increasing the boiling point.
3.

Number	Name	State[a]	Water Solubility	Lipid Solubility
A	Ethane	Gas	Very low	High
B	Benzene	Liquid	Low	High
C	Citric acid	Solid	High	Low
D	Anthracene	Solid	Very low	High
E	Glyphosate	Solid	High	Low

[a] At normal conditions of temperature and pressure.

4. a. Hydrogen ion solution:

$$pH = -\log [H^+]$$

$$\therefore pH = -\log [10^{-2}]$$

$$= -(-2) = +2.0$$

b. Hydrocyanic acid:

$$HCN \leftrightarrow H^+ + CN^-$$

$$K = [H^+][CN^-]/[HCN]$$

$$\therefore 4.9 \times 10^{-10} = [H^+][CN^-]/0.50$$

$$\text{since } [H^+] = [CN^-] = x$$

$$\text{then } [H^+][CN^-] = x^2$$

$$\therefore 4.9 \times 10^{-10} = x^2/0.50$$

$$x^2 = 4.9 \times 10^{-10}/0.50 = 2.45 \times 10^{-10}$$

$$\therefore\ x = 1.6 \times 10^{-5}$$

$$\text{thus } [H^+] = 1.6 \times 10^{-5}$$

$$pH = -\log [H^+] - 5.20$$

c. Acetic acid:

$$CH_3\ COOH \leftrightarrow CH_3\ COO^- + H^+$$

$$K = [CH_3\ COO^-][H^+]/[CH_3\ COOH]$$

$$1.85 \times 10^{-5} = [CH_3\ COO^-][H^+]/0.50$$

$$\text{since } [CH_3COO^-] = [H^+] = x$$

$$x^2 = [CH_3\ COO^-][H^+]$$

$$\therefore 1.85 \times 10^{-5} = x^2/0.50$$

$$\therefore x^2 = 0.93 \times 10^{-5}$$

$$\therefore\ x = 0.93 \times 10^{-2.5}$$

$$\text{and } [H^+] = 0.93 \times 10^{-2.5}$$

$$pH = -\log [H^+] - 2.50$$

3 Environmental Transformation and Degradation Processes

3.1 INTRODUCTION

The environment contains many chemicals. Fox example, animals and plants themselves comprise vast arrays of organic chemicals in addition to other substances. Chemicals from animals and plants, as well as organic chemicals added to the environment by actions of human society, occur in water, soil, and the atmosphere. These chemicals may undergo **transformation**, which can be defined as any change in the chemical structure of the substance. This could be a rearrangement of the molecule into another form; alternatively, it could be the addition or loss of chemical groups by environmental processes. **Degradation** usually refers to the breakdown of the original molecule by the loss of the various component parts or by the fragmentation of the molecule into smaller substances.

Transformation and degradation processes may occur through interactions with other chemicals in the environment. This can be facilitated by the input of energy in the form of radiation or heat. Alternatively, biota may be involved, leading to the transformation and degradation of compounds through biological processes. With these processes, organic compounds are chemically acted upon by other substances in the external environment or within the biota. Both oxygen and water are substances that are reactive and available in large quantities in the environment for transformation of organic compounds. Oxygen comprises about 20% of the atmosphere, and water occurs in high proportions in biota, as well as existing in large quantities in the oceans, lakes, and rivers.

Oxygen reacts with organic compounds by oxidation processes that often involve the addition of oxygen to the molecule. This can result in the formation of a molecule that is increased in size by the addition of oxygen, or it may result in splitting the molecule into smaller oxygen-containing fragments. The oxidative degradation of an organic compound to the ultimate level results in the formation of carbon dioxide, water, ammonia, nitrate ion (NO_3^-), nitrite ion (NO_2^-), orthophosphate ion (PO_4^{3-}), hydrogen sulfide (H_2S), sulfate ion (SO_4^{2-}), and so on, depending on the conditions involved and the nature of the original compound. A simple example is the oxidation of methane by combustion:

$$CH_4 + 2O_2 \quad CO_2 + 2H_2O$$

Hydrolysis is a process that results in the addition of water to a molecule. This often results in the fragmentation of the molecule into smaller fragments that may contain additional hydrogen and oxygen. An example of this is the hydrolysis of the ester functional groups in isooctylphthalate, a substance used to make plastics pliable, as represented by the following equation:

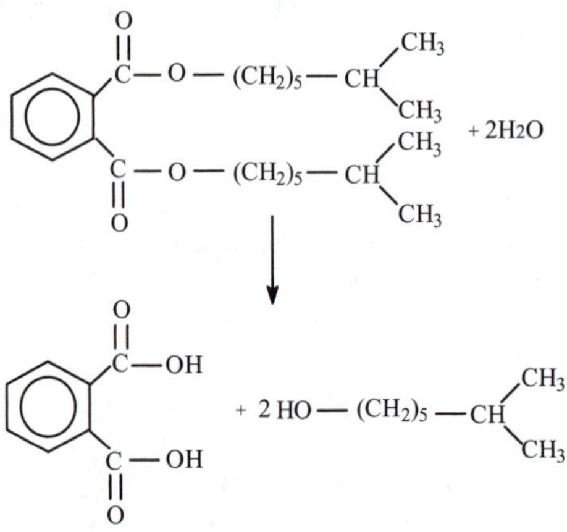

Thus, oxidation and hydrolysis may result in the production of smaller fragments that contain additional oxygen alone or both oxygen and hydrogen. Both oxidation and hydrolysis can occur when substances come in contact with oxygen or water under the appropriate conditions in air, water, soil, and biota. These processes may occur without the intervention of biota as a result of abiotic processes. On the other hand, oxidation and hydrolysis can also occur through the facilitation of biota. Reactions facilitated by animals or plants are described as biotic reactions. Both types of reactions are influenced by the prevailing temperature, the presence of oxygen and water, light, and a variety of other factors.

3.2 ABIOTIC TRANSFORMATION AND DEGRADATION

3.2.1 OXIDATION THROUGH COMBUSTION

Often spectacular oxidation of organic matter by atmospheric oxygen occurs by combustion. The burning of trees, grass, and petroleum is an example of oxidation through the combustion process. Many organic compounds can exist in the environment in the presence of the 20% of oxygen in the atmosphere without combustion occurring. However, ignition by a spark or flame initiates the occurrence of combustion. The burning or combustion of organic matter, such as wood and petroleum, is a major source of energy in human society. This has a major impact on the occurrence of oxygen and carbon dioxide in the Earth's atmosphere. The energy produced by combustion of many organic compounds can be estimated. The production of this energy is critical to the use of organic compounds as fuels. First, we will look at combustion of hydrogen as a simple example of the oxidative combustion process. An equation can be written for the combustion of one molecule of hydrogen as follows:

$$H_2 + {}^1/_2\, O_2 \rightarrow H_2O$$

This occurs with the release of energy, which is the heat of combustion. Looking at the reaction in greater detail, we can develop a concept of the reaction as occurring in a number of steps. First, the molecules can be seen as dissociating into atoms if there is sufficient energy supplied. This is illustrated by the following equation:

$$H\text{–}H \rightarrow H\cdot + \cdot H$$

The point to note is that the bond energies are the additional energy needed to break the bond or the excess energy left after the bond is formed. This is illustrated diagrammatically in Figure 3.1.

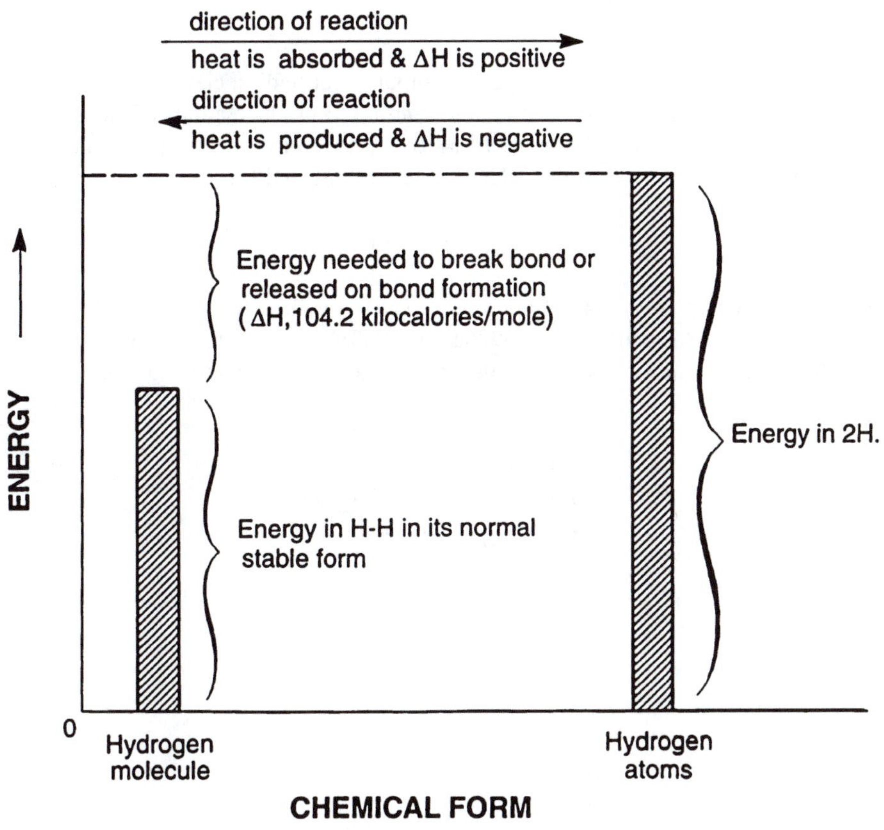

FIGURE 3.1 Diagrammatic representation of the energy present in forms of hydrogen.

The heat of reaction (ΔH) is defined as

$$\Delta H = \text{(Heat content of products)} - \text{(Heat content of reactants)}$$

If a reaction, such as the dissociation of the H_2 molecule to two H atoms, requires an input of energy to the reactants, this means that the products are of higher energy than the reactants and ΔH will be positive (Figure 3.1). This is known as an **endothermic** process. Alternatively, if a reaction, such as the reaction of two H atoms to form a H_2 molecule as the product, results in an output of energy, then the products are of lower energy than the reactants and ΔH will be negative (Figure 3.1). This is an **exothermic** process. For the H_2 dissociation, ΔH is +436 kJ mol^{-1}.

Considering the other reaction, the O_2 molecule, in the initial dissociation step, the O_2 molecule dissociates into atoms. Thus,

Input 498 kJ/mol

$$O{=}O \leftrightarrow O{:} + {:}O$$

Output 498 kJ/mol

This means that 0.5 mol O_2 needs +249 kJ input to cause dissociation to the oxygen atoms. Thus, there is a total energy input required to produce atoms of oxygen and hydrogen from water of +685 kJ mol^{-1} (436 + 249). In the final step, the atoms of hydrogen and oxygen can be considered to recombine and form water vapor. Thus,

$$2H\cdot\ (g) + {:}O\ (g) \rightarrow H_2O\ (g)$$

The amount of energy released in forming H_2O is equivalent to two O–H bonds, i.e., 2 × 463 kJ mol^{-1}, which equals –926 kJ mol^{-1} (Table 3.1). This energy is released, which on balance means that ΔH for the overall process is equal to 685 – 926, which

TABLE 3.1
Bond Dissociation Energies for a Range of Bonds

Bond	Energy (kJ mol^{-1})
H–H	436
O=O	498
C–O	351
C–H	413
O–H	463
C=O (as in CO_2)	803
C–C	347

is –241 kJ mol^{-1}. Thus, the heat of combustion is 241 kJ mol^{-1}, which is equivalent to 120.5 kJ g^{-1} (241 ÷ 2 g mol^{-1}). This calculated value compares favorably with an actual experimental value for the heat of combustion of hydrogen of 241.8 kJ mol^{-1}.

The hydrocarbons are important fuels in the form of natural gas, liquid petroleum gas (LPG), petrol, diesel, an so on. The heat of combustion is an important characteristic of these fuels and can be estimated using the characteristic bond dissociation energies shown in Table 3.1. It is important to note that these energies are average values encountered and the energy in a particular type of bond may vary with its position in a molecule. This can lead to differences between the calculated and observed heats of combustion with specific compounds.

Methane is an important component in natural gas. The oxidative combustion of methane can be represented as

$$CH_4\ (g) + 2O_2\ (g) \rightarrow CO_2\ (g) + 2H_2O\ (g)$$

The conceptual dissociation of the methane and oxygen into atoms, the recombination of these atoms to form carbon dioxide and water, and the associated energy changes are shown in Table 3.2. This indicates that the heat of combustion is

TABLE 3.2
Energy Changes and Heat of Combustion Resulting from Oxidation of Methane by Combustion

Original Molecule	Bonds Involved	Energy Involved (kJ mol^{-1})	Energy Change (ΔH) (kJ mol^{-1})
Step 1: Dissociation of Methane and Oxygen Molecules to Atoms			
H | H—C—H | H Methane	4 × C-H	4 × 413	1652
O=O oxygen	2 × O=O	2 × 498	996
		Total energy input	2648
Step 2: Recombination to Form CO_2 and H_2O			
CO_2	2 × C=C	2 × 803	–1606
H_2O	4 × O–H	4 × 463	–1852
		Total energy output	–3458
	Total energy produced	= –3458 + 2648	= –810 kJ mol^{-1}
	Heat of combustion	= +810/16	= +50.6 KJ g^{-1}

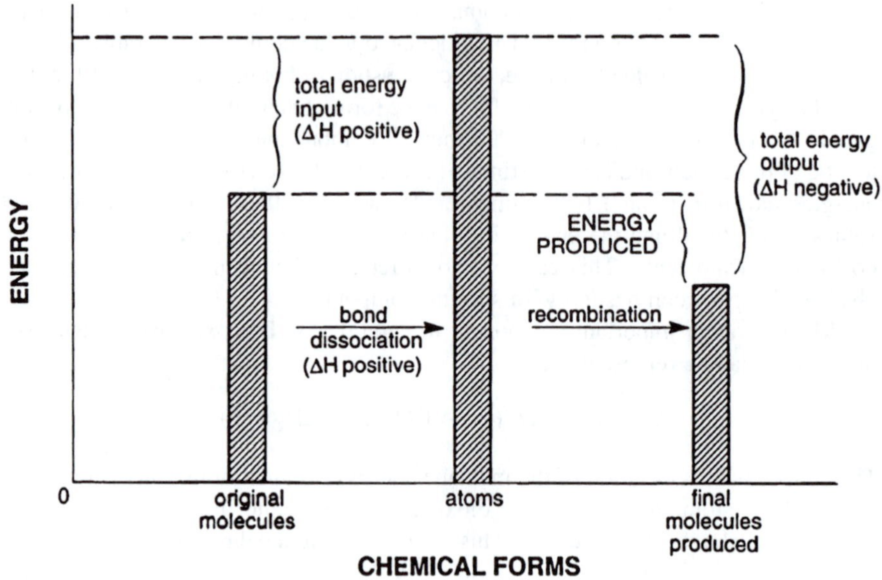

FIGURE 3.2 Diagrammatic illustration of the changes in chemical form and energy during oxidation by combustion.

calculated as –810.2 kJ mol^{-1}, which is close to the experimental measured heat of combustion of 802.5 kJ mol^{-1}. In a similar way, the heat of combustion can be estimated for many other fuel compounds.

In general terms, the basis of the estimation procedure can be illustrated as shown in Figure 3.2. Organic compounds in the form of fuels that are high in energy can be considered to be converted into their constituent atoms by the input of energy. The constituent atoms can then be seen as recombining by reforming bonds to form compounds of a lower energy content with the release of energy. The overall change in energy represents the heat of combustion, and a range of experimentally measured heats of combustion are shown in Table 3.3.

TABLE 3.3
Experimentally Measured Heats of Combustion of Some Fuels

Substance	Heat of Combustion (kJ g^{-1})
Petroleum fuels (gasoline, kerosene, diesel, etc.)	40–48
Lipids	38–40
Carbon	33
Proteins	18–23
Carbohydrates	15–18

The oxidation process with organic matter in the form of food occurs within biota, resulting in the production of carbon dioxide and water and smaller amounts of other products. Overall, this is chemically somewhat similar to oxidation through combustion and results in the release of energy that can be used by biota.

3.2.2 PHOTOTRANSFORMATION

Many organic chemicals are introduced into the environment and can absorb radiation and, as a result, undergo chemical transformation. All chemical processes require the reacting substances to have attained a certain energy. This energy can be obtained either from thermal energy (heat), as with combustion, which was considered in the previous section, or by absorption of radiation. The internal energy of a molecule is in several different forms. Thermal energy in a molecule results from being jostled by its neighbors, and it is manifested as translation, rotation, and vibration of the molecules. Absorption of electromagnetic radiation with frequencies in the infrared (IR) range (Figure 3.3 and Figure 3.4) causes increased molecular rotation and vibration. A molecule exists in specific electronic energy level states, generally the lowest energy one or the ground state. Most molecules absorb IR radiation and thermal energy, but to absorb radiation to change the electronic energy state requires ultraviolet (UV) or visible (VIS) light and specific chemical groupings in an organic molecule (chromophores).

Molecules activated by radiation absorption are described as photochemically activated molecules, and they differ in a number of important aspects from those that are thermally activated. Radiation absorption that will cause a change in the electronic state of a molecule is only possible if the energy of an incident photon corresponds to the energy difference between two internal electronic energy states of the molecule. To initiate a photochemical transformation, the photon energy must be relatively large, which corresponds to radiation in the UV/VIS region of the electromagnetic spectrum (Figure 3.4). If the light absorbed is of sufficient energy, the excited molecule produced can undergo various transformations that would essentially not occur under normal conditions. The basic characteristics of radiation can be described by relatively simple equations. Two important equations relate energy, frequency, and wavelength of radiation. First,

$$E = h\nu$$

where E is energy in Joules, h is Plank's constant (6.63×10^{-34} J), and ν is frequency (Hertz, or cycles per second, s^{-1}). The second equation is

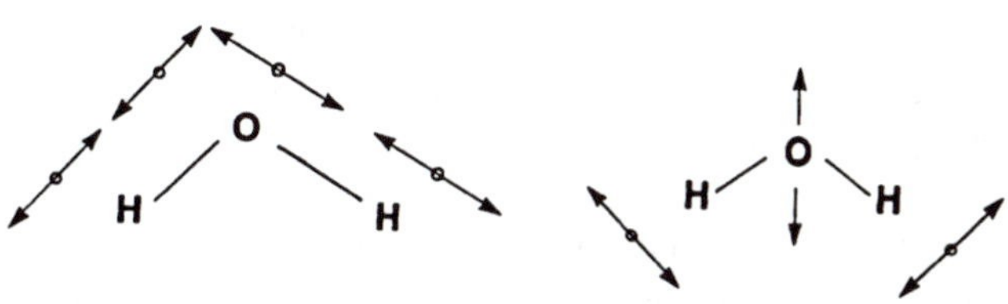

FIGURE 3.3 Some vibrational modes of a water molecule.

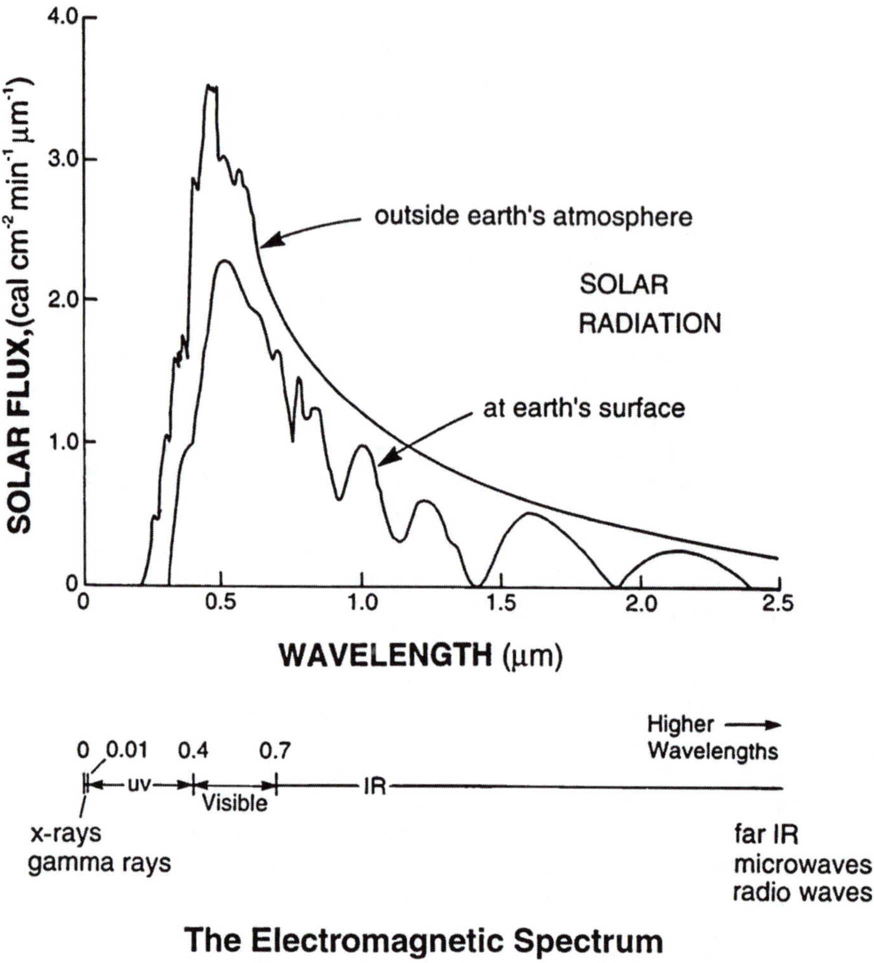

FIGURE 3.4 The spectrum of solar radiation and the electromagnetic spectrum.

$$S = \nu\lambda$$

where S is the velocity of light (assumed constant at 3×10^8 m s^{-1}) and λ is the wavelength in meters. The velocity of light is dependent on the medium that is it passing through and reaches a maximum in a vacuum. The equations above indicate that in general terms, as the wavelength (λ) increases, the frequency (ν) will correspondingly decline and the energy of a photon of radiation will also decline since energy (E) is directly dependent on frequency. Accordingly, a photon of UV radiation has greater energy than one of VIS radiation, which has greater energy than an IR photon.

Reactions induced by UV/VIS include fragmentation, oxidation, and polymerization. It is also possible that an excited molecule can return to its original state and in the process emit radiation at a different wavelength. The following are examples of photochemical reactions:

$$\begin{array}{c} \mathrm{O} \\ \| \\ \mathrm{H{-}C{-}H} + h_v \rightarrow H_2 + CO \end{array}$$
Acetone (λ $_{370}$nm)

$$O_3 + h_v \rightarrow O + O_2$$
Ozone (λ $_{320}$nm)

Photoreactions can be divided into three stages:

1. Absorption, the stage in which the absorption of photons of light gives excited molecules
2. Primary photochemical process
3. Secondary photochemical processes

Phototransformation in the environment can only occur if the UV/VIS absorption spectrum of the compound and the solar admission spectrum overlap. Solar radiation contains a substantial amount of UV/VIS radiation when it reaches the Earth's surface, as shown in Figure 3.4.

Examples of chemical groups (chromophores) that absorb UV/VIS radiation in solar radiation are given in Table 3.4. These groups all contain double bonds, in some form, aromatic rings, or series of double bonds separated by a single bond. Some structures or functional groups are poor absorbers of solar radiation; these include alcohols (–OH), ethers (R–O–R), and amines ($R{-}NH_2$). For such groups, or chemicals containing only these functional groups, phototransformation is likely to be unimportant. In addition, phototransformation can only occur when the chemical is likely to be exposed to solar radiation. For example, the atmosphere, upper layers of water bodies, and the surface of soil are likely locations in which chemicals would be exposed to solar radiation.

The Beer–Lambert law relates absorbance to concentration at a given wavelength for a chemical in solution (Figure 3.5). This law can be expressed by the following equations:

$$\log (I_o/I) = A = \varepsilon\, C\, l$$

$$I/I_o = T \text{ and } -\log T = A$$

where I_o and I are incident and transmitted light intensity (quanta or energy s^{-1} or photons $cm^2\ s^{-1}$), respectively, ε is the molar extinction coefficient ($l\ mol^{-1}\ cm^{-1}$), C is the concentration (mol^{-1}), and l is the path length (cm). This means that if 10% transmission occurs, then $I/I_0 = 10/100$, $I_0/I = 100/10 = 10$, and $\log (I_0/I) = A = 1$. Similarly, if 1% transmission occurs, then A is equal to 2.

TABLE 3.4
Groups in Molecules (Chromophores) That Absorb Ultraviolet and Visible Radiation in Solar Radiation

Group	Wavelength of Maximum Absorption (l_{MAX} in nm)	Molar Coefficient Extinction (ε) (mol^{-1} cm^{-1})
>C = O	295	10
>C = S	460	Weak
—N=—N	347	15
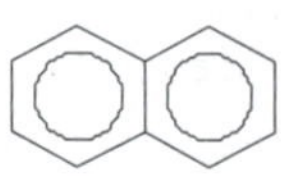	311 270	200 5000
C = C — C = O	330	20

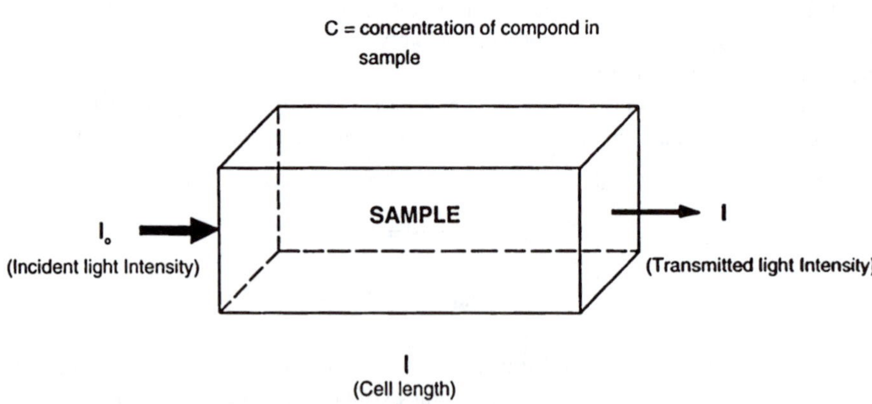

FIGURE 3.5 Passage of radiation through a sample.

3.2.3 Hydrolysis

Water is present in large quantities in the environment in oceans, rivers, and streams. It is also available in substantial quantities within all biota and in the vapor form in the atmosphere. In biota, it is the basic fluid used for the transfer of substances in biological processes. The chemical reaction of a compound with water is described as hydrolysis and was briefly outlined previously. It is one of the most important chemical processes that can act upon the many types of organic compounds occurring in the environment arising from both natural and man-made sources. The general reaction below defines the hydrolysis process:

$$R - X + H_2O \rightarrow R - OH + H - X \quad (\text{or } H^+ + X^-)$$

Organic compound

In this reaction, an organic molecule R–X reacts with water, forming a new C–O bond and cleaving a C–X bond in the original molecule. The overall effect is a displacement of the X group by a hydroxyl group.

A wide variety of functional groups and compounds are potentially susceptible to hydrolysis, including peptides, the glycosidic linkage in polysaccharides, and the ester group in fats. These reactions are shown in a generalized form in Table 3.5; it can be seen that carbohydrates, proteins, and fats, which are the major components of living tissue, are all susceptible to hydrolysis. Hydrolysis also occurs with a wide range of synthetic compounds, including many pesticides and other substances. These reactions can be mediated by biota or can occur without the need for biological assistance. However, when they occur abiotically, the rates of reaction can be very slow. Some chemical structures and groups tend to be resistant to hydrolysis, including alkanes, polycyclic aromatic hydrocarbons, alcohols, aldehydes, and ketones.

The importance of hydrolysis from an environmental fate view is that the reaction introduces a hydroxyl group into the parent molecule and may fragment the molecule into smaller groups. The hydroxyl group is a polar group (see Chapter 2) and usually tends to increase the polarity of the molecule, but this depends on the nature of the group that is removed. The products of hydrolysis are usually more susceptible to biotransformation, and the hydroxyl group makes the chemical more water soluble, as does the smaller size of molecular fragments that may be produced. With compounds having high K_{OW} values (see Chapter 2), the K_{OW} values of the products will be less and the biological activity altered. Furthermore, the product is usually less toxic than the initial starting material, but there are some exceptions to this. For example, the butoxy ethanol ester of 2,4-dichlorophenoxyacetic acid (2,4-D) exhibits an EC_{50} of >100 mg l^{-1} to daphnia, whereas the hydrolysis product (which is 2,4-D itself) exhibits greater toxicity at 47 mg l^{-1}. This is illustrated in Figure 3.6. Hydrolysis reactions are commonly catalyzed by H^+ or OH^- ions. This results in a strong dependence of the rate of hydrolysis on the pH of the water in which the reaction occurs.

TABLE 3.5
Groups That Are Susceptible to Hydrolysis

Group	Products
R_1–C(=O)–N(H)–R_2 →	R_1–C(=O)–OH + H–N(H)–R_2 (acid, amine)
R_3–O–R_4 (disaccharides), oligosaccharides, polysaccharides →	R_3–OH + HO–R_4
R4–C(=O)–O–R5 (fats)	R4–C(=O)–OH + HO–R5 (fatty acids, glycerol)
R–X (pesticides which are alkyl halides, amides, carbamates, and esters) →	R–OH + H–X

Example:

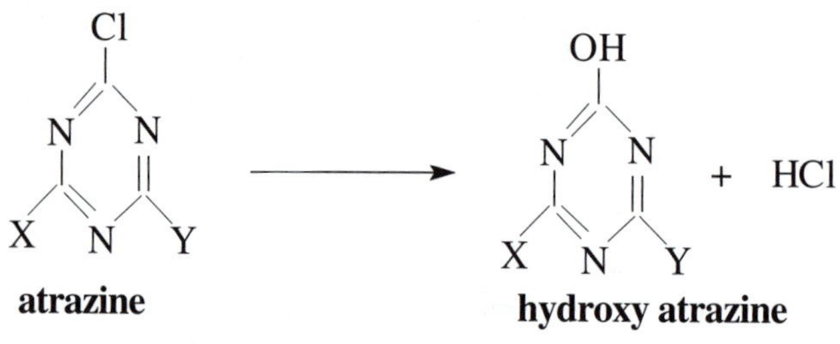

3.3 BIOTRANSFORMATION AND BIODEGRADATION

Biotransformation and biodegradation of chemical compounds by the action of living organisms is one of the major processes that determines the fate of organic chemicals in aquatic and terrestrial environments. These processes can be divided into two broad categories: (1) microbial transformations and (2) transformation by higher organisms. These two different groups are described below.

3.3.1 Microbial Transformation

Microorganisms are ubiquitous in the environment, as is shown by the populations of bacteria in water bodies illustrated in Table 3.6. Generally, as the amount of organic matter increases in a water body, the microbial population also increases. Microorganisms play a major role in the biogeochemical cycles of various elements that occur in the environment. Frequently, microbial transformation is the most important and possibly the only significant process that can decompose an organic

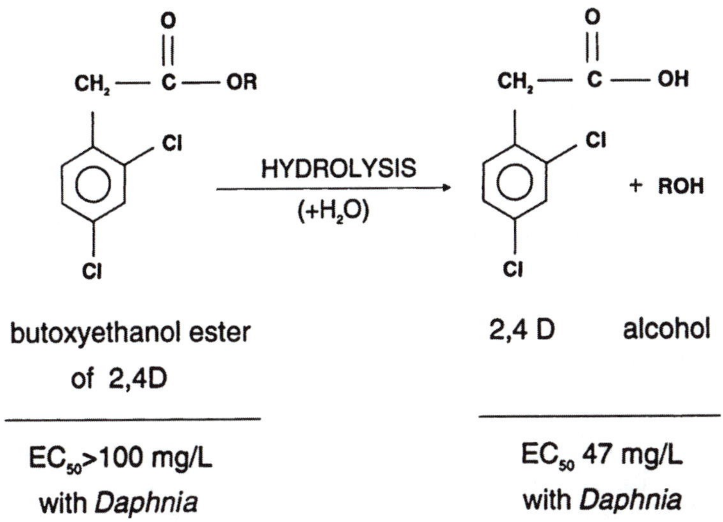

FIGURE 3.6 Toxicity of butoxyethanol ester of 2,4-D compared to its hydrolysis product 2,4-D.

TABLE 3.6
Size of Typical Bacterial Populations in Natural Waters

Environment	Bacteria Numbers (ml^{-1})
Clear mountain lake	50–300
Turbid, nutrient-rich lake	2000–12,000
Lake sediments	$8 \times 10^{9-5} \times 10^{10}$
Stream sediments	10^7–10^8

xenobiotic chemical in the environment. Microorganisms include bacteria, small, single-celled organisms; fungi, nonphotosynthetic organisms; algae, photosynthetic organisms; protozoans, unicellular, eukaryotic microorganisms; and viruses, parasitic microorganisms unable to multiply outside the host tissues.

Most bacteria fall into the size range 0.5 to 3.0 μm. In general, it is assumed that a filter with a 0.45-μm-pore size will remove all bacteria in water passing through it. The majority of bacteria in aquatic environments are nutritionally heterotrophic. Fungi are aerobic organisms (require atmospheric oxygen) that can be uni- or multicellular and generally can thrive in more acidic media than bacteria. Perhaps the most important function of fungi in the environment is the breakdown of cellulose in wood and other plant materials. To accomplish this, fungal cells secrete an enzyme (cellulase) that breaks insoluble cellulose down to soluble carbohydrates that can be absorbed by the fungal cell. Although fungi do not grow well in water, they play an important role in determining the composition of natural water because of the large amounts of their decomposition products that enter natural water bodies. A particularly important example of these decomposition products is the humic substances (discussed in Chapter 17) that occur in soil and runoff water entering natural water bodies. Algae are photosynthetic and are abundant in both fresh and saline waters, soil, and other sectors of the environment.

3.3.2 Types of Microbial Degradation

The rate at which a compound is biotransformed and biodegraded by microorganisms depends upon its role in microbial metabolism and a variety of other factors. Heterotrophic bacteria, which are capable of using complex carbon compounds as their principal source of energy, can degrade organic compounds to provide the energy and carbon required for growth. This is known as metabolism of growth substances; these substances are identifiable by their ability to serve as the sole carbon force for a bacterial culture. Many toxic and synthetic substances function as growth substrates for bacteria in a manner similar to naturally occurring organic compounds. Metabolism of growth substances usually results in relatively complete degradation or mineralization to carbon dioxide, water, and inorganic salts.

3.3.3 Patterns of Growth

Before the degradation of a compound can begin, the microbial community must adapt itself to the chemical, in many cases, which results in a lag phase initially when little growth occurs (Figure 3.7). With growth substances, both laboratory and field investigations have shown that this adaptation results in a lag time of 2 to 50 days before the microbial community adjusts. Frequently, specific organisms within a community with specific enzymes systems are required for degradation to occur. The adjustment period can involve species selection and numbers increase, as well as the production of specific enzymes systems to match the substrate. This production of enzymes is referred to as enzyme induction. These factors are outlined below:

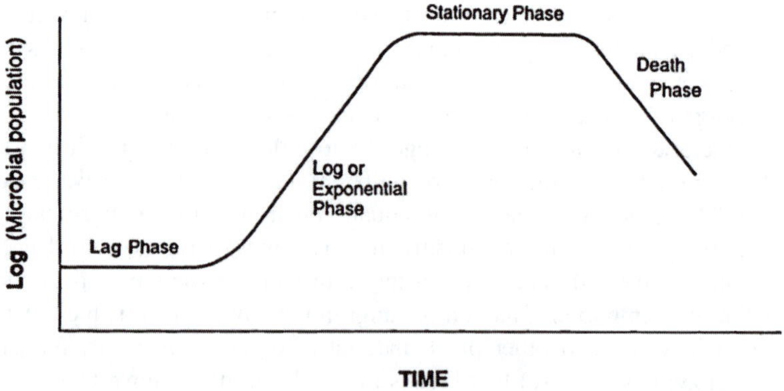

FIGURE 3.7 Typical pattern of growth of a microbial population.

1. **Prior exposure to the organic compound**: Prior exposure to the organic compound substrate reduces the adaption or lag time. Thus, lag times in pristine environments should generally be much longer than in locations that have been previously exposed.
2. **Initial numbers of suitable species**: Areas with larger microbial communities should require relatively short lag times to develop a viable population of degrading microorganisms (Table 3.6).
3. **The presence of more easily degraded carbon sources**: The presence of more easily degraded carbon sources may delay the adaptation of the microbial community to more persistent contaminants. For example, it has been found that microorganisms degraded added glucose completely before degrading hydrocarbons in lake water.
4. **Concentration of the organic compound**: There may be concentration thresholds below which adaptation does not occur. On the other hand, too high a concentration of the organic compound may be toxic to the microorganisms.

When the lag phase is completed, population growth occurs rapidly and usually increases at an exponential rate. At the completion of this phase, the microorganism population effectively establishes an equilibrium with growth substances available, and a stationary phase in terms of population occurs. Finally, the growth substances are exhausted and wastes accumulate, leading to a decline in the population number, as illustrated in Figure 3.7.

3.3.4 Co-metabolism

Compounds that co-metabolize usually degrade only in the presence of another carbon source. This is in contrast to growth substances that are able to serve as the sole carbon source for a microbial community. Microorganisms can degrade compounds that they apparently cannot use for growth or energy via co-metabolism. Co-metabolism is believed to occur when enzymes of low specificity alter or degrade a compound to form products that other enzymes in the organism cannot degrade. Substances that undergo co-metabolism are usually similar in structure to natural substrates, but are altered or degraded without necessarily providing significant amounts of energy to the microorganism. Often, this is an important mechanism for the degradation of pesticides in soil.

The kinetics of co-metabolism differ significantly from that of growth substances. Often, no lag period occurs before co-metabolism begins, and accumulation of intermediate products resulting from partial degradation is likely. Generally, slower rates of degradation are observed compared to metabolism of growth substances.

3.3.5 Reduction-Oxidation (Redox) Processes in Natural Waters

In aerobic aquatic environments, microorganisms utilize molecular oxygen (O_2) as a major reactant in the degradation of organic compounds. The concentration of oxygen dissolved in water depends on factors such as temperature, salinity, biological activity, and reaeration rate. Usually, the dissolved oxygen concentration in natural waters ranges from 0 to about 10 mg l^{-1} due to the low solubility of oxygen in water. With sufficient oxygen, the complete aerobic degradation of carbohydrates, and reduction of oxygen to form water, can be simply expressed as follows:

$$6CH_2O + 6O_2 \rightarrow 6CO_2 + 6H_2O + \text{Energy utilized by organism}$$

Carbohydrate

This process is a redox process with each molecule of oxygen having an oxidation state of zero, being reduced by addition of hydrogen and two electrons to an oxidation state of –2, as it is in water. Thus, the availability of electrons can be seen as a controlling factor for redox processes in aquatic systems.

If the organic compound contains nitrogen, the normal products of carbon dioxide and water are formed, but the nitrogen present is converted into ammonia (NH_3) and nitrate ion (NO_3^-). It is noteworthy that nitrate is not produced directly, but rather, with aerobic, heterotrophic microorganisms, ammonia is the first product formed and nitrate is formed by oxidation of this ammonia by different microorganisms. From this general type of reaction, bacteria and other microorganisms extract the energy needed to carry out their metabolic processes, to synthesize new cell material, for reproduction and for movement.

As the concentration of dissolved oxygen (DO) is depleted, microbial degradation pathways change and degradation reactions are modified. When DO is removed, anaerobic degradation without atmospheric oxygen occurs, having generally lower energy yields and microbial growth rates. Most organic substances are biodegraded more slowly under anaerobic conditions, although there are a number of exceptions to this general rule. Rate constants derived for degradation in oxygenated systems do not apply to anaerobic systems. The dissolved oxygen content of waters in various parts of the environment can vary considerably. In zones such as sediments and bottom waters that lack a mechanism for aeration, DO can be in short supply and organic compounds are typically microbially degraded utilizing a sequence of available oxidizing agents. Initially, aerobic degradation occurs, but when the DO falls to 0.5 to 1.0 mg/l or less, nitrate begins to substitute for molecular oxygen in the breakdown process as an oxidizing agent. This process can be simply expressed by the following equation:

$$2CH_2O + 2NO_3^- \rightarrow 2CO_3^{2-} + 2H_2O + N_2$$

Carbohydrate Nitrate

If nitrogen was present in the original organic compound undergoing oxidation, then ammonia could also be produced. As the nitrate is consumed, other oxidizing agents such as SO_4^{2-} are utilized, with the production of carbon dioxide, water, and hydrogen sulfide (H_2S). With nitrogen-containing compounds, ammonia (NH_3) is produced as well. Generally, different microorganisms are responsible for, or associated with, the use of the various oxidizing agents.

In the complete absence of dissolved oxygen, the degradation of organic compounds can still proceed. This can be represented for carbohydrates by the following equations:

$$6CH_2O \rightarrow 3CH_4 + CO_2$$

Carbohydrate Methane

$$6CH_2O \rightarrow 2C_2H_5\,OH + 2CO_2$$

Carbohydrate Ethanol

These processes release a relatively low amount of energy, as can be seen by the production of methane and ethanol, which are capable of further oxidation with the release of relatively large amounts of energy.

Microorganisms require nutrients, such as nitrogen and phosphorus, in order to metabolize organic substances. The availability of nutrients is one of the most important factors controlling the activity of heterotrophic microorganisms in aquatic environments. Nutrient limitation has been reported to retard the biodegradation of organic compounds, particularly in marine environments, which are generally relatively deficient in both nitrogen and phosphorus.

As well as the major nutrients (C, N, and P), microbial growth and activity can be affected by essential micronutrients (growth factors, trace metals). For example, the degradation of petroleum oil in seawater can be enhanced by the addition of iron, which is consistent with the role of iron as a cofactor in some of the enzymes responsible for hydrocarbon oxidation.

The pH scale, previously described in Chapter 2, is used to conveniently characterize acidity as the concentration of the hydrogen ions in a system. It can be seen from the discussion above that reduction-oxidation (redox) conditions also lie on a somewhat similar continuum in aquatic systems. Redox can be characterized as the effective concentration of electrons; even though electrons do not exist in solution, they are available from substances in the water. Thus, the pE scale is the negative of the log of the effective concentration of electrons in water. Low pE values indicate that there are relatively high concentrations of electrons available and that reducing conditions exist, whereas high pE values indicate that few electrons are available and oxidizing conditions are prevalent.

3.3.6 Terrestrial Systems

Soils are heterogeneous systems composed of varying proportions of organic and mineral matter, and are the result of many complex processes, such as the disintegration and weathering of rocks and the decomposition of plant and animal material. In an average soil, mineral matter accounts for about 50% of the soil volume, with organic material typically representing less than 10%. Air and water occupy the remaining volume of the soil in a complex system of pores and channels.

Well-drained soils are generally aerobic in nature, with a fairly ready supply of oxygen from the atmosphere.

Microorganisms are generally considered to be aquatic in nature, but soil microorganisms tend to exist in a sorbed state attached to the soil solid matter. Decreasing soil moisture content usually decreases the rate of degradation of organic compounds such as pesticides. This is particularly true at lower soil moisture levels approaching air-dried soil conditions. At excessive soil moisture levels, the soil may change from an aerobic to an anaerobic condition with generally reduced microbial degradation rates.

3.3.7 Degradation by Higher Organisms

Higher organisms often tend to degrade endogenous (i.e., arising from within the body) chemicals such as bile acids, fatty acids, steroids, and other hormones, as well as xenobiotic compounds such as petroleum hydrocarbons. One of the major enzyme systems for facilitating degradation of xenobiotic lipophilic compounds is the **mixed-function oxidase** (MFO) system. The name *mixed-function oxidase* is used because the system acts as an oxygenase and an oxidase. The key enzyme in the system is cytochrome P_{450}, which is based on a heme structure. **Cytochrome P_{450}** is not one or two enzymes, but a family of closely related enzymes called **isozymes**. This enzyme system facilitates the insertion of oxygen into C–H bonds of substrates containing aliphatic groups. This is illustrated by the reaction below:

$$\underset{\text{Substrate}}{R-H}+O_2 \xrightarrow{\text{Cytochrome } P_{450}} R-OH+H_2O$$

With aromatic compounds, epoxidation occurs, but overall an oxygen atom is inserted into substrates.

Most enzymatic reactions are highly specific for only one substrate or a small series of closely related substrates. However, the MFO enzyme series is distinguished by the fact that there is a lack of specificity. These enzymes are able to act on xenobiotic compounds to which an organism has never been exposed before. The enzyme system can be generated as a result of exposure; this is described as **enzyme induction**. Cytochrome P_{450} is found most abundantly in the liver of vertebrates. This reflects the liver's role as the body's primary site for degradation of xenobiotic lipophilic compounds. The occurrence of this enzyme system is not restricted to mammalian tissues, but also occurs in fish, birds, yeast, some plants, and some bacteria. Induction is triggered by exposure to certain xenobiotic lipophilic compounds described as **inducers**. Not all xenobiotic compounds are equally effective as inducers, but the best inducers are chemicals that are lipid soluble and have a relatively long half-life in the organism.

The most potent inducing agent yet discovered is dioxin, more accurately described as 2,3,7,8-tetrachlorodibenzodioxin (see Chapter 6). This substance is often described by the abbreviation TCDD. Other inducers include DDT-type insecticides, barbiturates, PAHs such as benzo-(a)-pyrene, and ethanol. It is interesting to note that cigarette smokers generally have higher cytochrome P_{450} levels than nonsmokers. In fact, elevated levels of cytochrome P_{450} have been proposed as an indicator of previous exposure to xenobiotic substrates such as petroleum hydrocarbons.

The degradation or biotransformation process often occurs in two steps. The activity of the cytochrome P_{450} isozymes is referred to as phase I reactions. Phase I reactions typically result in the insertion or introduction of an OH group in place of a simple hydrogen. This results in the insertion of a relatively reactive, polar functional group into a compound. This makes the product more soluble in water, with a lower K_{OW} value, and also places a reactive site on the molecule. Phase II reactions are known as conjugation reactions since they conjugate or join together the product mentioned above with another substance. These reactions usually involve the coupling of a phase I product with a naturally occurring derivative of a carbohydrate, peptide, or sulfate. The product of this reaction also is of higher polarity, greater water solubility, and thus is more easily eliminated than the original xenobiotic compound. Neither humans nor animals are born with a full complement of P_{450} isozymes. In addition, it may take several years for human infants to acquire adult levels of biotransformation activity.

3.4 KINETICS OF TRANSFORMATION AND DEGRADATION

It is frequently important to determine how rapidly transformation and degradation occur in the environment so that the extent and period of contamination can be evaluated. As a result, numerous methods have been developed to simulate transformation of organic compounds in the various compartments of the environment. One of most widely used evaluations in environmental management is a test of water and wastewater that measures the capacity for biodegradation of organic matter present and consequent demand for oxygen. This is the **biochemical oxygen demand** (BOD); it is a measure of the amount of dissolved oxygen consumed by microorganisms in degrading the organic matter in a water sample over a 5-day period. Not only does it measure the degradation of organic matter present, but at the same time, it evaluates the possible dissolved oxygen effects of organic wastes present when discharged to a water body.

3.4.1 Kinetic Expressions

Environmental transformation and degradation involve reaction of an organic compound with another substance. As discussed previously, the other substance usually involved is either water or oxygen or, more often, both of these substances. A kinetic expression that can be used is as follows:

$$-\frac{dc}{dt} \alpha \ CX$$

where C is the concentration of organic compound present, t is the elapsed time, and X is the concentration of oxygen or water. This is a second-order kinetic expression where the rate of degradation is proportional to the concentration of two reactants. In most environmental situations, the amount of oxygen or water available is very large and effectively constant. This means that the concentration of organic matter is relatively low and the following expression applies:

$$-\frac{dc}{dt} \alpha\ C$$

This expression describes first-order kinetics, and the rate of reaction is proportional to the concentration of the organic reactant present. Assuming this to be the normal situation in the environment, kinetic expressions for degradation of growth substances by microorganisms can be relatively complex, but the following expression for degradation is often used:

$$-\frac{dC}{dt} \alpha\ BC$$

where B is the size of the bacterial population. This expression is a second-order rate expression, similar to the expression above, in which the population of bacteria was not considered. In this case, the second-order rate of reaction is proportional to the population of microorganisms and the concentration of the organic reactant present. Both B, the viable microbial population, and C, the concentration of the organic compound, will change with time. Most environmental chemical transformations and degradation facilitated by microorganisms occur in an open water, soil, air, or other phase. This situation often mitigates against the increase in population of microorganisms in a specific area and tends to keep the population relatively constant. This means that the first-order expression indicated previously can generally be applied to most environmental degradation processes; that is,

$$-\frac{dC}{dt} \alpha\ C$$

and

$$-\frac{dC}{dt} = kC$$

where k is the first-order transformation or degradation rate constant with units of time^{-1}. On integration and rearrangement, the following equation can be derived:

$$\ln (C_o/C_t) = kt \text{ or } C_t = C_o e^{-kt}$$

where C_o is the concentration of the organic substance at time zero and C_t is the concentration of the substance after a time t. This can be plotted as shown in Figure

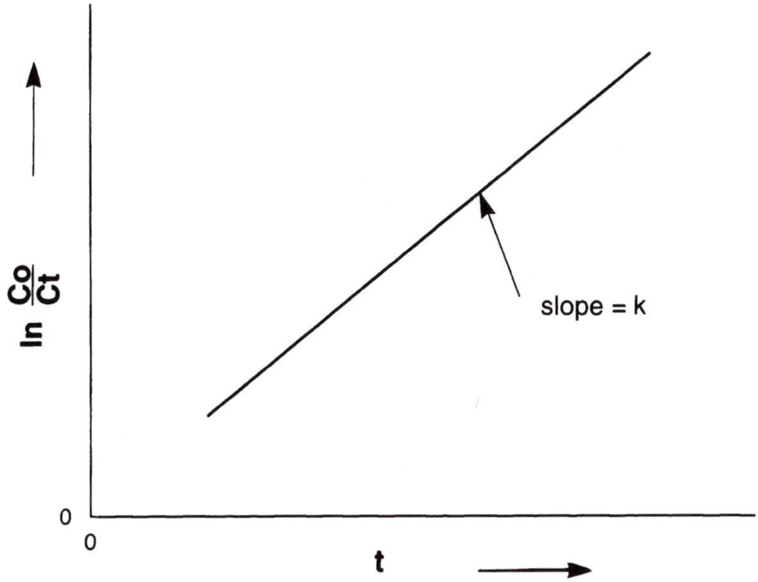

FIGURE 3.8 Theoretical plot of ln (C_o/C_t).

3.8 with ln (C_o/C_t) against the elapsed time t. The slope of this plot represents the rate constant k. From the equation above, it can be shown that

$$\ln C_t = \ln C_o - kt$$

Thus,

$$\ln C_t = \text{constant} - kt$$

This relationship can be plotted as shown in Figure 3.9, and a measure of persistence of the compound is the rate of loss that can be determined as the rate loss constant, k, which is the reverse slope of the plot in Figure 3.9.

A more convenient measure of the persistence of a chemical in the environment is the half-life, $t_{1/2}$. An expression for this can be derived as follows. It can be shown that

$$t = \frac{\ln(C_o/C_t)}{k}$$

Starting at any point in time when half of the original substance has been degraded, then

$$\frac{C_o}{C_t} = \frac{1}{0.5} = 2$$

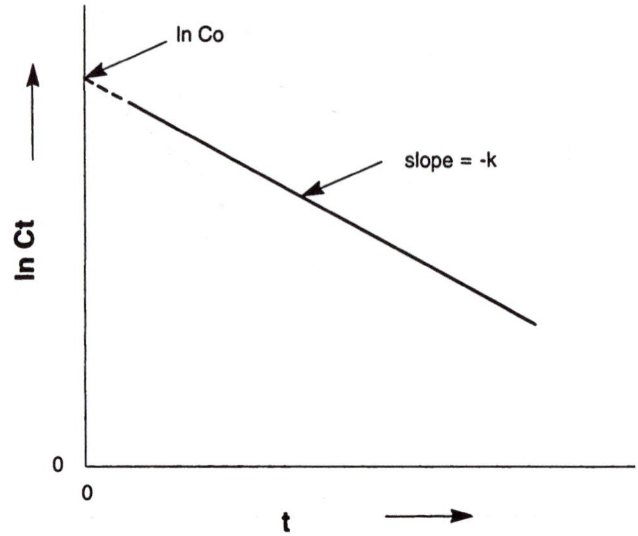

FIGURE 3.9 Theoretical plot of ln C_t against the elapsed time.

Thus,

$$t_{1/2} = \frac{\ln 2}{k} = \frac{0.693}{k} = \text{constant}$$

This means that the half-life ($t_{1/2}$) is constant for a given compound and a given environmental degradation process that occurs under specific conditions. This has been found to be generally true for most environmental processes since these obey approximate pseudo-first-order kinetics. The actual degradation rate constants measured in the environment show a high level of variability. The pH, temperature, availability of water, availability of oxygen, and other factors all influence persistence in a particular situation. These factors vary within different phases of the environment, causing this high degree of variation. Examples of some half-lives that have been measured are shown in Table 3.7. Howard et al. (1991) have assembled a comprehensive set of data on the degradation of compounds in the environment.

3.4.2 Volatilization

In some situations, disappearance from an environmental compartment involves loss to another compartment rather than degradation. These processes also often follow first-order kinetics. For example, volatilization of pollutants from water to the atmosphere is a very important physical loss process. Even substances with very low water solubility, very high boiling points, and low volatility, such as the PCBs and DDT, volatilize at a significant rate over relatively short periods. The processes involved in volatilization are:

1. Diffusion to the air–water interface from within the water mass

TABLE 3.7
Half-Lives of Some Pesticides in Soil

Compound	Half-Life (years)
Chlorohydrocarbons	3–10
DDT	1–7
Dieldrin	10
Toxaphene	10
Organophosphorus	
Dyfonate	0.2
Chlorfenfos	0.2
Carbophenothion	0.5
Carbamates	
Carbofuran	0.05–1

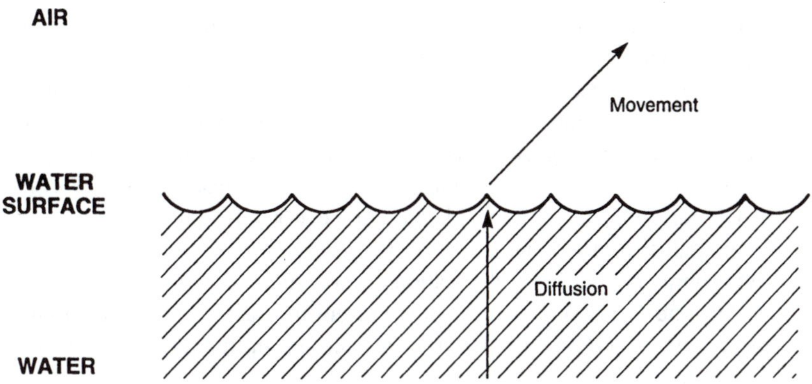

FIGURE 3.10 Processes involved in the volatilization of a compound from water.

2. Movement of the substance away from the air–water interface into the atmosphere

These processes are illustrated diagrammatically in Figure 3.10. Volatilization generally follows first-order kinetics and it can be shown that:

$$C_t = C_o\, e^{-(Kt/D)}$$

where C_t is the concentration after an elapsed time t, C_0 is the concentration at time zero, K is the liquid exchange constant, and D is the depth of an individual compound molecule taken from the surface. The half-life in water is a useful environmental characteristic and, from the expression above, can be shown to be represented by the following equation:

$$t_{1/2} = \frac{\ln 2D}{k}$$

TABLE 3.8
Examples of the Half-Lives of Some Compounds Due to Volatilization from Water

Compound	Half-Life (in hours, depth of 1 m)
n-Octane	5.6
DDT	73.9
Mercury	7.5

Examples of half-lives for volatilization of various compounds in water are shown in Table 3.8.

3.5 KEY POINTS

1. Substances in the environment, both synthetic and natural, are transformed and degraded by abiotic and biotic processes.
2. There are two major substances in the environment that participate in degradation and transformation processes. These are oxygen in the atmosphere, or originating from this source, leading to oxidation processes; and water, in water bodies, the atmosphere, and biota, leading to hydrolysis.
3. The energy produced by oxidative combustion of hydrocarbons can be estimated from the bond dissociation energies.
4. Environmental phototransformation of molecules occurs as a result of the absorption of UV/VIS wavelengths present in solar radiation. Absorption occurs when certain chromophores are present in the molecular structure of the organic compound.
5. Two important equations relating energy to frequency and wavelength of radiation are

$$E = h\nu$$

$$S = \nu\lambda$$

6. The Beer–Lambert law relates absorbance to concentration of a chemical and takes the form

$$\log\left(\frac{I_o}{I}\right) = A = \varepsilon\, C\, l$$

7. The following general equation describes hydrolysis reactions with organic compounds in the environment.

$$R - X + H_2O \rightarrow R - OH + H - X \text{ (or } H\cdot + X^-)$$

Organic compound

8. Many biotransformation and biodegradation reactions are facilitated by microorganisms that are common in most sectors of the natural environment.
9. Oxidation reactions of carbohydrates take different forms depending on the availability of oxygen and can be expressed by the following equations:

Aerobic conditions:

$$6CH_2O + 6O_2 \rightarrow 6CO_2 + 6H_2O$$

Carbohydrate

Oxygen in water at about 0.5 to 1.0 mg l^{-1} or less:

$$2CH_2O + 2NO_3^- \rightarrow 2CO_3^{2-} + 2H_2O + N_2$$

Carbohydrate Nitrate

Absence of oxygen:

$$6CH_2O \rightarrow 3CH_4 + 3CO_2$$

Carbohydrate Methane

10. The redox conditions in an aquatic system can be evaluated by measuring the effective concentration of electrons present using a scale of the negative of the log of the effective electron concentration, the pE scale.
11. The mixed-function oxidase enzyme system (cytochrome P_{450}) is induced in organisms due to exposure to xenobiotic lipophilic compounds. This system facilitates oxidation by insertion of oxygen into C–H bonds as follows:

$$\underset{\substack{\text{Lipophilic}\\ \text{compound}}}{R - H} + O_2 \xrightarrow[P_{450}]{\text{Cytochrome}} R - OH + H_2O$$

12. Most biotransformations and biodegradation processes in the environment follow pseudo-first-order kinetics according to the following equation:

$$\ln = \frac{C_o}{C_t} = kt$$

13. For first-order processes, the persistence of a chemical in the environment is best expressed by the half-life ($t_{1/2}$):

$$t_{1/2} = \frac{0.693}{k}$$

REFERENCES

Connell, D.W. and Miller, G.J., *Chemistry and Ecotoxicology of Pollution*, John Wiley & Sons, New York, 1984.

Howard, P.H., Boethling, R.S., Jarvis, W.F., Meylan, W.M., and Michalenko, E.M., *Handbook of Environmental Degradation Rates*, Lewis Publishers, Boca Raton, FL, 1991.

QUESTIONS

1. When hydrocarbons undergo combustion in air, small quantities of nitrogen dioxide (NO_2) and nitric oxide (NO) are produced and can play an important role in atmospheric pollution. Hydrocarbons do not contain any nitrogen at all, so why are nitrogen dioxide and nitric oxide produced? Write equations for reactions that may be involved.
2. Ethane and ethanol are both commonly used fuels, but ethane is usually in petroleum-based fuels, whereas ethanol is obtained by fermentation of plant matter. In addition, these two substances are closely related chemically since ethanol has the same structure as ethane, but has an additional oxygen atom in the molecule to form an alcohol group.

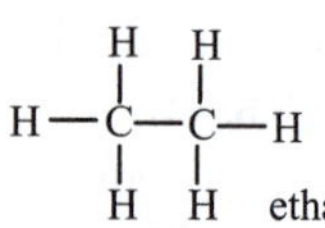
ethane

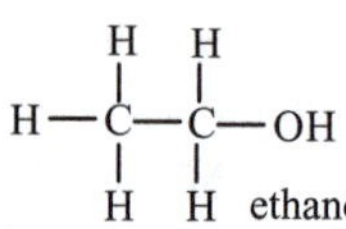
ethanol

Compare these two compounds as fuels on the basis of calculated heats of combustion and production of carbon dioxide from complete combustion.

3. Explain the fundamental reasons for the difference in the heats of combustion of ethane and ethanol in terms of their bond dissociation energies.
4. An antibiotic, Microbitol, has the following properties:

Molecular weight	350
Maximum UV/VIS absorption	347 nm
Molar extinction coefficient ε, mol l^{-1}	5000
Half-life in water	15 days

It has been identified as a possible contaminant in a water body and a sample (1 l) has been collected and extracted with solvent (1 ml), which was expected to achieve 100% efficiency. The Absorbance at 347 nm was measured and found to be 15 in a 1-cm-length cell. Calculate the concentration of Microbitol in the original water sample and estimate the time involved until the Microbitol is effectively zero (0.001 mg l^{-1}).

ANSWERS

1. Air involved in the combustion of hydrocarbons contains about 78% nitrogen and 21% oxygen. Under the conditions produced by combustion, N_2 and O_2 react to produce NO and NO_2. The reactions involved can be expressed as follows:

$$N_2 + O_2 \rightarrow 2NO$$

$$N_2 + 2O_2 \rightarrow 2NO_2$$

2. In evaluating the complete combustion of ethane, the following equation can be used:

$$C_2H_6 + 3^1/_2\, O_2 \rightarrow 2CO_2 + 3H_2O$$

$$(MW = 30) \quad (MW = 44)$$

The heat of combustion can be calculated as follows:

a. Ethane has 6 × (C–H) and oxygen $3^1/_2$ × (O=O).
b. (6 × 413) = 2478 kJ mol^{-1} + ($3^1/_2$ × 498) = 4221 kJ mol^{-1}.
c. Total energy input = 4221 kJ mol^{-1} ethane.
d. The products comprise 4 × (C=O) to form CO_2 + 6 × (O–H) to form 3 H_2O.
e. (4 × 1803) = –3212 kJ mol^{-1} ethane + (6 × 463) = –2778 kJ mol^{-1} ethane.
f. Total energy release = –5990 kJ mol^{-1} ethane.

Overall energy change = –1769 kJ mol^{-1} ethane.
Thus, heat of combustion/unit mass = 1769/30 kJ g^{-1}.

$$\text{Heat of combustion} = 59.0 \text{ kJ g}^{-1}$$

The amount of CO_2 produced is 88 g mol^{-1} 88/30 g/g = 2.9 g/g ethane.

$$CO_2 \text{ produced is 2.9 g/g ethane}$$

Considering the complete combustion of ethanol, the following equation is obtained:

$$C_2H_5OH + 3O_2 \rightarrow 2CO_2 + 3H_2O$$

$$(MW = 46) \quad (MW = 44)$$

The heat of combustion can be calculated as follows:

a. Ethanol has 5 × (C–H), 1 × (C–O), 1 × (O–H), 1 × (C–C), and oxygen 3 × (O=O).
b. (5 × 413) + (1 × 351) + (1 × 463) + (1 × 347) + (3 × 498) = +720 kJ mol^{-1} ethanol input.
c. The products comprise 4 × (C=O) to form $2CO_2$ and 6 × (O–H) to form $3H_2O$.
d. Early release is –(4 × 803) – (6 × 463) = –5990 kJ moe^{-1} ethanol.

Overall energy change = –1270 kJ mol^{-1} ethanol.

$$\text{Heat of combustion} = 1270/46 = 27.6 \text{ kJ g}^{-1}$$

The amount of CO_2 produced is 88 g mol^{-1} or 88/46 g/g ethanol = 1.9 g/g ethanol.

$$CO_2 \text{ produced is } 1.9 \text{ g/g ethanol}$$

Thus, in summary:

1. Ethane gives 59.0 kJ/g of heat on combustion and 2.9 g CO_2/g ethane.
2. Ethanol gives 27.6 kJ/g of heat on combustion and 1.9 g CO_2/g ethanol.

These calculations indicate ethane could be a much better fuel than ethanol, but CO_2 production is much higher.

3. Diagrammatically, the energy changes occurring in ethane and ethanol on combustion can be represented as shown in Figure 3.11.

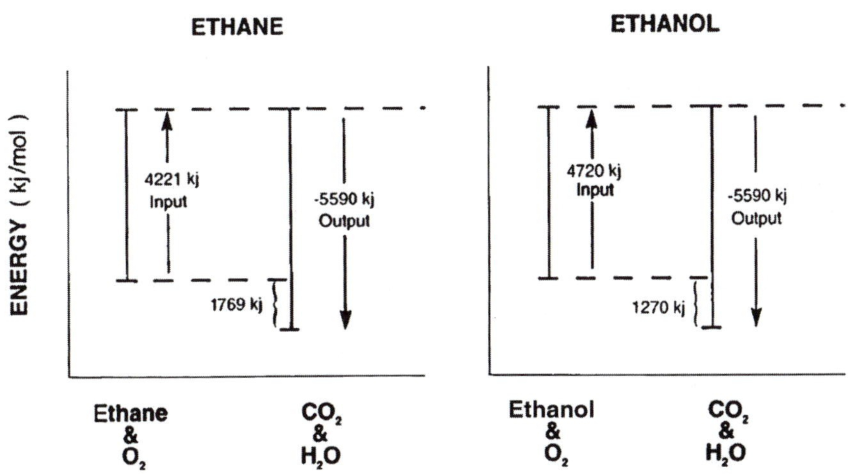

FIGURE 3.11 Energy changes in ethane and ethanol on combustion.

This indicates that for each mole of ethane and ethanol, there is the production of the same amount of energy output on the formation of the products, but there is a difference in the energy input to each initial set of reactants. Ethane and O_2 require 4221 kJ mol^{-1} to form atoms from their components, whereas ethanol and O_2 require 4720 kJ mol^{-1} due to the presence of extra bonds resulting from the additional O atom.

4. Concentration in water:
   The Beer–Lambert law states that

$$A = \varepsilon.C.l.$$

Substituting into this equation, then

$$15 = 5000.C.1$$

$$\therefore \text{C in solvent} = 15/5000 \text{ mol } l^{-1}$$

$$\text{C in water} = (15/5000)(1/1000) \text{ mol } l^{-1}$$

$$= (15/5000)(1/1000)\ 350 \text{ g } l^{-1}$$

$$= (15/5000)\ 350 \text{ mg } l^{-1}$$

$$= 1.05 \text{ mg } l^{-1}$$

Time to reach zero concentration:
The first-order rate equation states that $\ln (C_o/Ct) = kt$.
Information available: $C_o = 1.05$ mg l^{-1}, $Ct = 0.001$ mg l^{-1}.
k can be obtained from the half-life.

$$t_{1/2} = 0.693/k$$

$$\therefore k = 0.693/t_{1/2} = 0.693/15 = 0.046 \text{ day}^{-1}$$

Substituting into the first-order equation:

$$\ln (1.05/0.001) = 0.046.t$$

$$\ln 1050 = 0.046.t$$

$$\therefore t = 6.96/0.046 = 151 \text{ days}$$

4 Environmental Toxicology

4.1 INTRODUCTION

The history of toxicology dates from ancient times and is a major theme in the development of chemistry. The ancient papyrus document titled "Ebers" dates from 1500 B.C. and contains the earliest known reference to the beneficial and toxic effects of a range of medicinal products. Other early references to toxic substances were by Hippocrates and Aristotle. It is interesting to note that the famous Roman historian Pliny, the Greek physician Nikander, and the Roman architect Vitravius all commented on the harmful effects of lead ingestion. Thus, this can be considered one of the earliest observations of harmful effects of chemicals in the environment.

The next major breakthrough in toxicology was around 1500 by the Swiss chemist Phillipus von Hohenheim Paracelsus, commonly called Paracelsus. He is credited with moving alchemy away from a prime concern with the transmutation of elements into gold, and toward chemicals in medicine. Paracelsus is often quoted in modern times with his statement that there is a relationship between the dose of the toxicant and the response (Randy, 1995). He expressed it as follows:

> All substances are poisons; there is none that is not a poison. The right dose differentiates a poison and a remedy.

Other key breakthroughs occurred in 1755 when Percival Pott discovered the link between soot and scrotal cancer in chimney sweeps and in 1815 with the publication of *General System of Toxicology* by Mattieu Orfila. In 1962, Rachel Carson's book *Silent Spring* first alerted society to the potential danger of many organic chemicals, particularly DDT and related pesticides, when discharged to the environment. Since then the field of environmental toxicology has undergone quite dramatic growth.

Environmental toxicology is the study of the toxic or poisonous effects that chemicals in the environment exert when individual organisms or relatively small numbers of individuals are exposed. The effects of chemicals on whole ecosystems are a related, but different topic described as **ecotoxicology** and outlined in Chapter 18. Strictly speaking, environmental toxicology is only concerned with the effects of chemicals, although in a broader sense other forms of pollution, such as thermal pollution, are often included. Toxic effects are by definition not beneficial but deleterious, and the type and severity of toxic effects that can occur are very broad, ranging from temporary sublethal effects to lethality (Landis and Ming Hu, 2004; Wright and Welbourn, 2002).

Toxicology, including environmental toxicology, is principally concerned with effects on animals, although toxic effects on plants can be significant in some situations. In broad terms, toxicology can be considered to consist of three areas:

Clinical toxicology: The investigation of the effects of pharmaceutical chemicals on human health.
Forensic toxicology: Concerned with the illegal use of chemicals, such as illicit drugs.
Environmental toxicology: Covers the effects of chemicals in the environment on organisms.

Environmental toxicology presents many new challenges to the field of toxicology. Toxicology, up to recent times, has been principally concerned with relatively large doses of chemicals with short exposure periods, usually of the order of days, leading to lethal effects, for example, possible lethal effects of medicinal compounds on human beings. Sublethal effects are also of interest in environmental toxicology. A typical investigation might include an area being studied for the lethal effects of agricultural pesticides on fish. However, there are additional aspects that relate to low exposures over long periods with more subtle, but still very important sublethal effects. Some areas of interest here are the influence of low levels of chlorine in drinking water on human health and effects of trace petroleum contamination in aquatic systems. To cope with these aspects, environmental toxicology has been extended by a range of new approaches to evaluate the deleterious effects of chemicals. Of particular importance are **ecotoxicology** and **risk assessment**, covered in Chapter 18, and **genotoxicology**, covered in Chapter 16.

A toxicant in the environment could occur in water, air, soil, or food as a result of direct discharges, transport from elsewhere, or biotransformation of another substance. To exert a toxic effect on an organism, it must be transferred to the site of action within the organism by the processes generalized in Figure 4.1. This means the chemical is taken up by the organism through the stomach, gills, lungs, etc., and then distributed throughout the body, where it may be biotransformed and excreted. Some of the original compound, or its biotransformed product, may be transferred to the site of action. With carbon monoxide poisoning, the site of action is the hemoglobin in blood to which the carbon monoxide attaches, leading to a loss of its oxygen-carrying capacity and subsequent toxic effects. When the toxicant is attached to the site of action, the observed toxic effects occur.

4.2 ROUTES AND MECHANISMS OF TOXICANT ENTRY TO ORGANISMS

4.2.1 Mechanisms of Entry to Organisms

In order for a chemical to exert a toxic effect, it must move from the ambient environment into the organism. For any compound to enter an organism, regardless of the route of entry, it must cross cell membranes. There are three principal mechanisms by which toxicants cross cell membranes: passive diffusion, facilitated

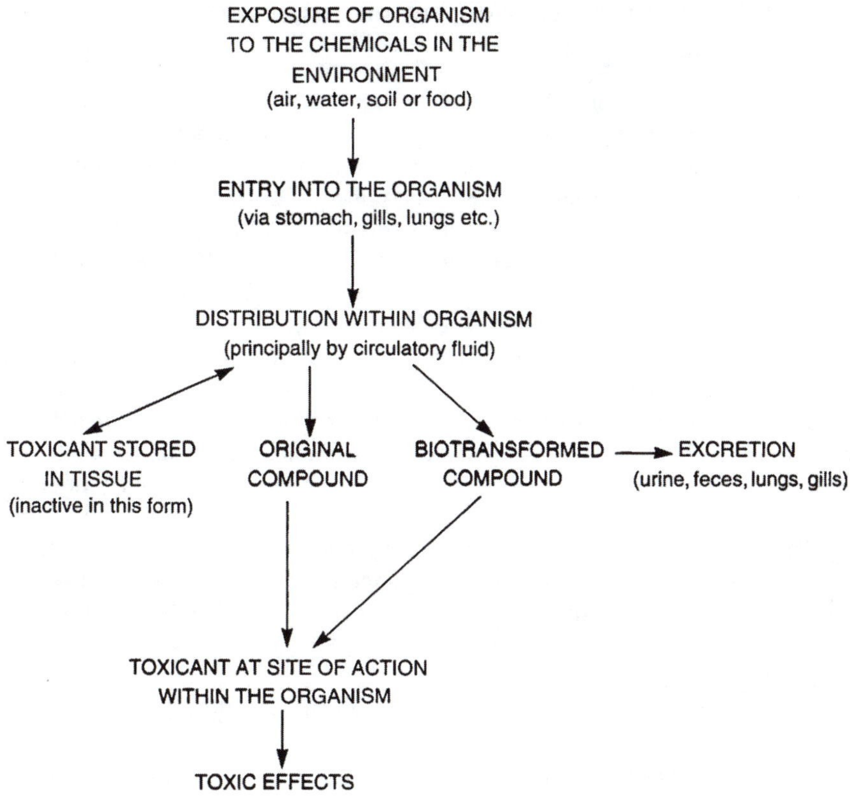

FIGURE 4.1 Generalized transfer processes for a chemical from the environment to the site of toxic action.

diffusion, and active transport. Of these, passive diffusion appears to be the predominant mechanism for toxicants.

Passive diffusion is largely governed by the difference in toxicant concentration on either side of the membrane, i.e., the concentration gradient. The basic structure of cell membranes consists of a lipid bilayer; thus, where passive diffusion is the mechanism of entry, a toxicant will cross the membrane only when the compound is lipid soluble and the concentration of the toxicant is lower on the inside of the membrane than on the outside.

Facilitated diffusion is also a diffusion process, so a favorable concentration gradient is required for the compound to cross the membrane. The diffusion is facilitated by chemical processes that aid in the transport, and thus the rate of transfer across the membrane is higher than for simple diffusion. In active transport, the compound is transported via a carrier; however, this mechanism requires an energy input. Because of the expenditure of energy, toxicants can transfer across a membrane regardless of the direction of the concentration gradient or any other gradient.

4.2.2 Chemical Properties of Compounds That Cross Cell Membranes

There are a number of physical and chemical properties that affect the penetration of chemicals through lipid membranes. Some of the most important properties are ionic state, molecular size, and lipophilicity, commonly measured by the octanol–water partition coefficient (K_{OW}, described in Chapter 2). The influence of these properties on cross-membrane transport is shown in Table 4.1. Lipophilic (fat-loving) compounds have K_{OW} values in the range of 10^2 to $10^{6.5}$, with lipophilicity increasing with increases in the K_{OW} value. For example, DDT is a highly lipophilic compound with a K_{OW} value of $10^{6.2}$ and can readily cross through cell membranes.

The ionic state can have a major effect on the ability of toxicants to cross cell membranes. Ions attract a surrounding shell of water molecules (see Chapter 2), effectively preventing their passage through the lipophilic layer in cell membranes, which is hydrophobic (water-hating) in nature. Less ionized compounds pass through slowly, whereas nonionized compounds, with their greater lipid solubility, cross readily.

The pH of the solution is a variable that influences the extent of ionization of chemicals and thus their ability to cross membranes. The relationship between pH and the extent of ionization has been quantified by the Henderson–Hasselbach equation:

$$pH = pK_a + \log [\text{ionized form/nonionized form}]$$

where pH is the negative logarithm of the hydrogen ion concentration ($-\log [H^+]$) and pK_a is the negative logarithm of the acid ionization constant ($-\log K_a$).

When the pK_a of a chemical is numerically equal to the pH, then, according to the Henderson–Hasselbach equation, there is an equal number of ionized and non-ionized forms of the compound (i.e., log [ionized form/nonionized form] = 0 and ionized form/nonionized form = 1). Any change in pH will change the extent of ionization. A variation of one pH unit will cause a 10-fold change in the ratio of ionized to nonionized forms. This will, in turn, have significant effects on the ability of chemicals to be transferred across lipid membranes (Table 4.1).

TABLE 4.1
Physical and Chemical Properties That Affect Toxicant Transfer across Membranes

Parameter	Effect on Cross-Membrane Transfer
Ionic state	Dissociated molecules in the ionic state are not transferred readily
Molecular size	Smaller compounds are transferred more readily
Lipophilicity	Lipophilic compounds are transferred more readily ($K_{OW} = 10^2$ to $10^{6.5}$)
Concentration	Compounds at higher concentrations are transferred more readily

4.2.3 Routes of Entry to Organisms

There are three principal routes of entry for toxicants into an organism:

1. Inhalation through the respiratory system (lungs or gills)
2. Absorption through the skin
3. Ingestion through the gastrointestinal tract (GIT)

If plants are under consideration, a different set of entry routes would be relevant.

The gills of aquatic organisms are essentially the aquatic equivalent of the lungs of terrestrial animals. They have similar structures and use similar features to maximize the uptake of oxygen, which makes them excellent sites for toxicant uptake. The mechanism of toxicant transfer from gills or lungs to blood is diffusion. In order to maximize the rate of diffusion, gills have a very large surface area, typically 2 to 10 times the body surface area with fish. There is a very small distance through which the toxicant must diffuse (i.e., 2 to 4 μm) and blood pass through the gills in the opposite direction to the water flow over the gills. The countercurrent flow of blood and water maximizes the difference in toxicant concentration between the two phases, and thus increases the rate of diffusion. In aquatic organisms, this is the predominant form of toxicant entry, except perhaps in some cases, for food.

The lungs have a similar purpose to the gills and are structured to maximize contact between atmospheric oxygen and blood to facilitate exchange. This provides an excellent medium to also exchange toxicants in the vapor or gaseous form in the atmosphere. Toxicants can also be attached to particles in high concentrations. The respiratory system of terrestrial mammals has several mechanisms to remove or prevent the entry of particulate pollutants into the body. The defense mechanisms are impedance to movement through the nasal passage, the inertia of large particulates, and the lungs' self-cleaning system. Many of the cells that line the respiratory system secrete mucus and have cilia. This mucus is thick and sticky and entraps particles that strike it. The cilia all beat in one direction, pushing the particle-contaminated mucus up to the mouth, where it is then generally swallowed and enters the gastrointestinal system prior to excretion. There are no mechanisms to prevent the entry of vapors or gases. However, many gases and vapors are respiratory irritants or have an odor that can offer a degree of protection if avoidance behavior results.

Skin acts as a selective barrier to chemicals; some can penetrate it while others cannot. The barrier properties of skin are primarily due to only one of the seven layers of cells that toxicants must cross to enter the body, which is described as the **stratum corneum**. This is composed of densely packed, biologically inactive cells and is one of the cell layers that comprise the epidermis, the outermost part of the skin. The barrier qualities of skin vary with the location of the skin due to changes in the thickness and quality of the stratum corneum and the abundance of either sweat glands or hair follicles.

The gastrointestinal tract consists of the mouth, esophagus, stomach, intestines, and rectum (Figure 4.2). A toxicant inside the GIT has not yet entered the human body. To enter the body, it must cross the lining of the GIT and enter the bloodstream. The mouth, esophagus, and rectum play a minor role in the absorption of food or toxicants. In

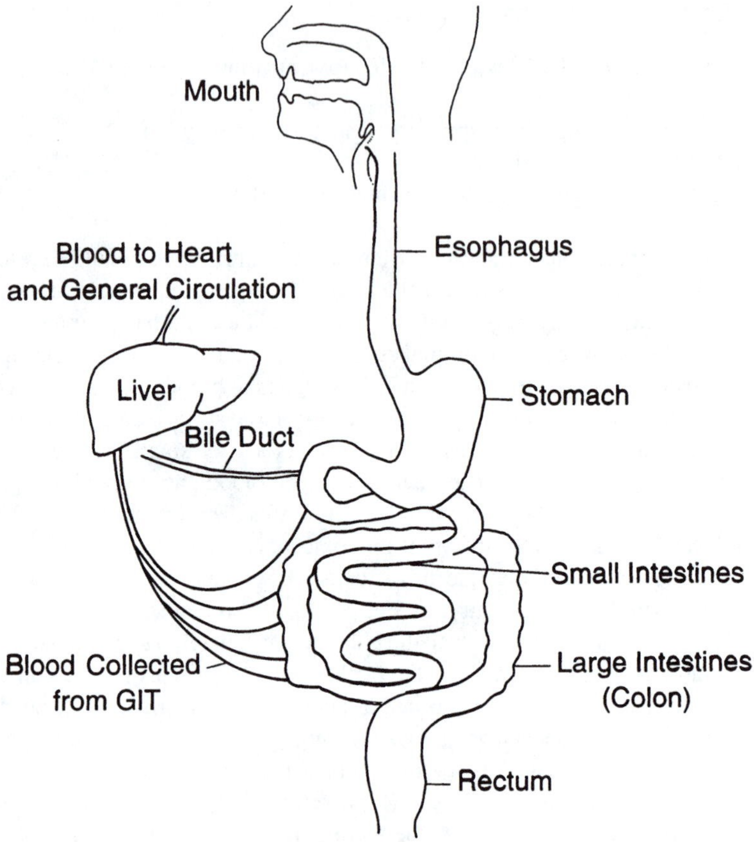

FIGURE 4.2 The anatomy of the gastrointestinal tract.

contrast, the stomach and intestine walls are designed to maximize the absorption of food substances and thus are ideal sites for absorption of toxicants. However, the characteristics of the absorption in the GIT may well be different from those for the same compound if it were absorbed elsewhere. The stomach is acidic with a pH of approximately 1.5. If the pH of the stomach is different from the pH of the toxicant carrier, often food, the chemical form of the toxicant (i.e., the ratio of ionized to nonionized form) will change in accordance with the Henderson–Hasselbach equation (see Section 4.2.2), which in turn will affect the extent of absorption across the GIT. The length of time that a toxicant stays in the GIT is also very important, as increased exposure can lead to increased concentrations of the toxicant in the body.

4.3 DISTRIBUTION OF TOXICANTS WITHIN THE ORGANISM

Once a toxicant crosses a cell membrane, it has entered the body, but it does not stay in those cells at the entry point; it is distributed throughout the body (Figure

4.1). The principal distribution mechanism is through the blood circulatory system. Blood is not homogeneous; its components belong to three different fractions: water, lipids, and proteins. A toxicant will partition into each of these three phases according to the physical and chemical properties of the toxicant and the different blood fractions. If the toxicant is water soluble, the bulk of it will be found dissolved in the aqueous phase of blood. However, if the toxicant has a high affinity for either lipids or proteins, it will bind reversibly to that fraction. The toxicant in the aqueous phase is not bound to water but is simply dissolved and is hence described as free. It is this free toxicant that can exert toxic effects, whereas bound forms are biologically inactive (i.e., they cannot exert a toxic effect). Examples of the distribution of chemicals between free and bound forms are presented in Table 4.2. DDT has very low aqueous solubility and a high K_{OW} value, and thus is highly lipophilic and expected to be located principally in the lipid fraction. Parathion is somewhat similar, but less lipophilic, and so has a higher proportion in the aqueous fraction. Nicotine is very water soluble and correspondingly occurs mainly in the aqueous fraction.

The reversible binding of toxicants to either lipid or protein biomolecules can be represented by the following equation:

$$T_f + \text{Biomolecule} \leftrightarrow T_b$$

where T_f is the free (unbound) toxicant and T_b is the bound toxicant. The term *reversible* means that toxicant molecules are continuously moving from being free to bound and can become unbound (or free) at some point in time. As the formation of bound toxicants is reversible, an equilibrium can be established between the bound and free forms of the toxicant. The position of the equilibrium is governed by the relative affinity of the toxicant to the various blood components or tissue. In most cases, the toxicant forms an attachment with the tissue in which it is stored and thus becomes biologically inactive. Such attachments are also reversible, so equilibria are established between compartments of the body that store toxicants and the blood (Figure 4.3), and hence all the compartments are in equilibrium.

TABLE 4.2
The Relative Distribution of Several Chemicals between the Free and Bound Forms in Blood

Chemical	Log K_{ow}	Aqueous Solubility (mg l^{-1})	% Free	% Bound
DDT (an organochlorine pesticide)	6.2	0.003	0.1	99.9
Parathion (an organophosphate pesticide)	3.8	6.5	1.3	98.7
Nicotine	1.3	Very soluble	75.0	25.0

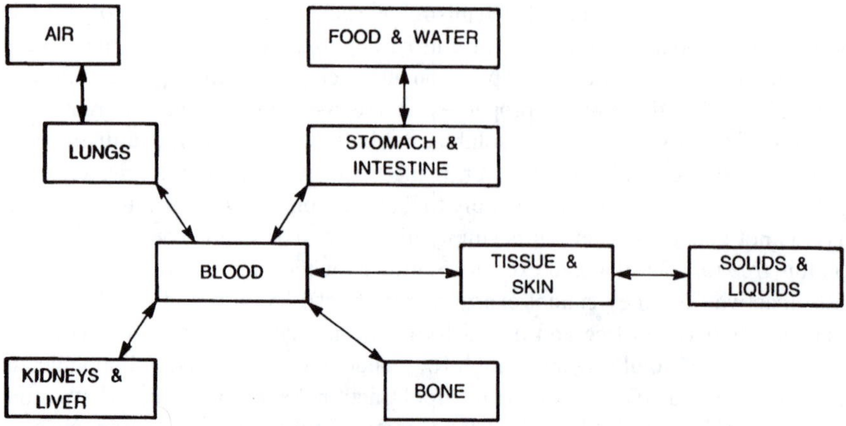

FIGURE 4.3 The routes of entry and distribution of toxicants between the various compartments in the body.

Due to the equilibria, a change in the concentration in any compartment will lead to concentration changes in the other compartments. For example, if a person has been routinely exposed to a toxicant so that a steady concentration was always in the blood and then the exposure source was removed, the toxicant level in the blood will drop. However, the toxicant will move out of the other compartments where it was stored and into the blood until equilibrium is once again established.

There is evidence to indicate that with some toxicants the concentration of toxicant that has accumulated at the site of action is a critical factor infuencing toxicity. This can be considered to equate to the concentration of residual chemical in the body of the organism and is often referred to as the **critical body burden** or residue. With persistent lipophilic chemicals, this concentration is reasonably consistent at 1 to 12 mmol kg^{-1} for lethal effects in a range of organisms, including fish, earthworms, crabs, daphnids, and mussels.

4.4 BIOTRANSFORMATION OF TOXICANTS*

4.4.1 Principles of Biotransformation

Many environmental toxicants are substances that are resistant to chemical transformation and degradation. Often, these substances are lipophilic with K_{OW} values in the range of 10^2 to $10^{6.5}$, or log K_{OW} from 2 to 6.5. Some examples are DDT, dieldrin, and polychlorinated biphenyls (PCBs).

The very property that allows these lipophilic toxicants to pass through lipid membranes and thus enter an organism is also a characteristic that limits their excretion. The normal means by which organisms rid themselves of dangerous chemicals is by solubilization in urine and feces, and here the water solubility of the chemicals has a strong influence on their capacity to be removed. As many

* Also see Chapter 3, "Environmental Transformation and Degradation Processes."

environmental organic toxicants are reasonably lipophilic, and therefore have low aqueous solubilities (see Chapter 2), they cannot be readily excreted via the urinary or fecal routes. Thus, organisms have developed means to increase the aqueous solubility of toxicants by various biotransformation processes. Biotransformation normally occurs in two stages: phase I and phase II. However, compounds with sufficiently high aqueous solubilities and reactivities may skip or eliminate phase I altogether. All biotransformation reactions are facilitated by enzyme systems.

4.4.2 Phase I Transformations

Phase I reactions generally tend to involve such enzymes as mixed-function oxidases (MFOs), reductases, and hydrolases, which oxidize, reduce, or add either the hydroxyl (OH–) or hydronium (H_3O^+) ions to reactants, respectively. This increases the polarity, water solubility, and reactivity of the toxicant and lowers the K_{OW} value, reflecting decreased lipophilicity. Of the MFO enzymes, the cytochrome P_{450}-dependent monooxygenase system is the most important.

4.4.3 Phase II Transformations

The reactions that occur in phase II are called **synthetic** or **conjugation reactions**, as metabolites from phase I are combined (conjugated) with naturally occurring metabolic intermediates. Examples of such metabolic intermediates include UDP-glucuronic acid, amino acids such as glycine and glutamine, and other substances. This process further increases the aqueous solubility, lowers the K_{OW} value, and greatly facilitates the excretion of toxicants via the urinary system.

4.4.4 Sites of Biotransformation

The most important organ associated with biotransformation is the liver (hepatic tissue), followed in order of decreasing importance by the kidneys, lungs, and intestine. In nonliver tissue (extrahepatic tissue), only certain types of cells contain the necessary enzymes for biotransformation. Such cells have a limited range of enzymes, and those enzymes present generally have lower activities than liver tissue.

4.5 EXCRETION OF TOXICANTS

There are three principal routes for the excretion of toxicants: urinary, fecal, and elimination through the lungs or gills. The relative importance of each of these routes depends on both the type of organism and the chemical involved. However, it is important to realize that all bodily secretions (e.g., saliva, breast milk, semen, sweat, and tears) and all cells lost from the body (e.g., dead skin cells, hair, fingernails, toenails, and the molted skin or exoskeleton) lead to the removal of toxicants. This occurs, as mentioned previously, because toxicants are distributed widely in an organism.

If a toxicant is biotransformed to a water-soluble compound, the bulk of it will move into the aqueous component of the blood. Eventually, the blood passes through the kidneys, where the aqueous component and the associated biotransformation product are removed and excreted in the urine. The fecal route of excretion receives toxicants from two main sources. First, toxicants that are not absorbed from food and liquids that enter the gastrointestinal tract are automatically present in the feces. The other source is bile. When toxicants reach the liver, they are biotransformed and removed by either entering urine or bile. The principal properties that determine whether a toxicant or transformation product enters the bile or urine are the aqueous solubility and molecular size. The larger and less water soluble compounds enter the bile, which is then released into the intestine to mix with the feces.

Elimination by the lungs is only important for chemicals that are gases or are highly volatile, i.e., have high vapor pressures. The means by which such compounds are transported from the blood to air in the respiratory system appears to be passive diffusion. An example of a compound eliminated by this mechanism is ethanol. With humans, approximately 2 to 4% of ethanol consumed is excreted in exhaled breath. Police take advantage of this and use instruments (breathalyzers) that measure the concentration of alcohol in exhaled air in order to estimate the blood alcohol concentration, assuming an equilibrium is reached in the partition process between air and blood.

The gills can be a very important route of elimination of nonpolar, lipophilic compounds for aquatic organisms, particularly fish. For example, half of the pentachlorophenol in fish is eliminated by the gills. The mechanism of exchange from blood to water is passive diffusion. Thus, passive diffusion is the predominant mechanism for the uptake and elimination of lipophilic toxicants across the gills.

4.6 CLASSES OF POISONS BASED ON EFFECT

It is important to recognize that the type of toxic effect exerted varies with the type of chemical. Chemicals can be classified into different groups based on the type of deleterious effect they exert:

Toxicants: Any substance that causes a deleterious biological effect when living organisms are exposed to it. This is the largest class of compounds, as all compounds can exhibit toxic effects, depending on the dose as postulated by Paracelus (see Section 4.1). In fact, the other categories of chemicals discussed below are really subsets of this group.

Teratogens: Any substance that causes defects in the reproduction process by either reducing productivity or leading to the birth of offspring with defects.

Mutagens: Any substance that leads to inheritable changes in the DNA of sperm or ovum cells.

Carcinogens: Any substance that causes a cell to lose its sensitivity to factors that normally regulate cell growth and replication. Such cells replicate without restriction to form a growing mass called a tumor.

In this chapter, only toxic properties without teratogenic, mutagenic, or carcinogenic aspects are discussed. Further information on the other classes of toxic properties can be found in Chapter 16.

4.7 QUANTITATIVE PRINCIPLES OF TOXICOLOGY

4.7.1 Background

For most chemicals, there are relationships between the biological effect observed and either the toxicant concentration in the ambient environment (e.g., water with fish) or dose (e.g., occurrence of lead in food with humans). These relationships are collectively termed *dose–response relationships.* An aspect of environmental toxicology that is of particular importance is that some environmental toxicants are in fact essential to growth and development at low concentrations, but become toxic at higher concentrations. These substances are mainly metals, such as iron, magnesium, zinc, copper, and a variety of other substances. The general nature of the relationship between dose (or concentration) and biological response is different for chemicals essential to the organism's metabolism and those that are nonessential (as illustrated in Figure 4.4). For essential substances, the relationship is parabolic. When present at concentrations well below those required by the organism for normal growth and development, deleterious effects and possibly death occur due to their absence. At increased concentrations, the effects become less harmful; i.e., first irreversible and then reversible effects occur, until the toxicant is present at a concentration range suitable for normal metabolism, at which no deleterious effect

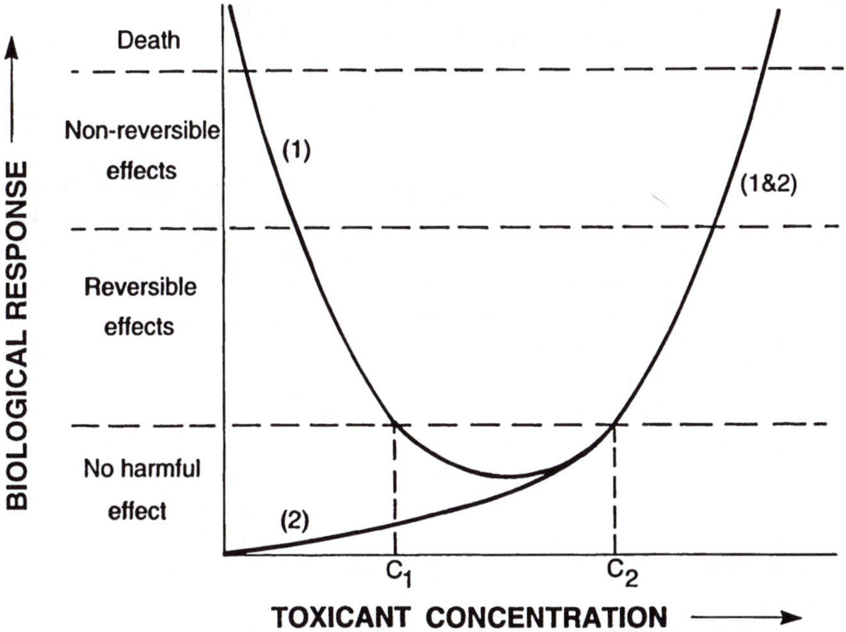

FIGURE 4.4 Biological responses to differing concentrations of (1) essential chemicals and (2) nonessential chemicals. Normal metabolic activity with an essential chemical occurs between concentrations C_1 and C_2, which below C_1 the chemical is in deficiency and above C_2 is in excess.

are exerted. As the concentration of an essential substance increases beyond this range, toxic effects become apparent and become increasingly deleterious, leading eventually to death.

The relationship for nonessential compounds is somewhat similar to an exponential growth curve (Figure 4.4). At zero and low concentrations, no measurable toxic effects are apparent. However, as the concentration increases, the severity of the deleterious effects increases, with initially reversible effects occurring, then irreversible effects, and finally death. An example of a nonessential chemical is toluene (C_6H_5–CH_3). This substance is a neurotoxin (i.e., it affects nerves) and has been implicated in central nervous system depression, dizziness, nausea and vomiting, respiratory depression, and death. From general observations of doses and the responses that occurred, the well-known principle "the right dose differentiates a poison and a remedy" was formulated by Paracelsus. Even such innocuous compounds as water, oxygen, and sodium chloride can exert toxic effects and even be fatal at high enough doses.

The chemical form and route of entry of toxicants affects the toxicity. For example, pure metallic mercury (Hg) is much less toxic than methylated mercury [$(Hg(CH_3)_2$]. Curare (the poison used by some South American Indians to kill prey) is a very rapid acting poison if introduced directly into the blood, whereas consumption in food has no effect. Another example is amygdalin, which is found in the pits of peaches and apricots. If consumed orally, it is approximately 40 times more toxic than by intravenous injection; this is because the intestines contain bacteria with enzymes that convert the amygdalin to more harmful products.

4.7.2 Measures of Toxicity

Individuals from a population of the same species have different susceptibilities (or tolerances) to toxicants. The distribution of tolerance is generally accepted to be a normal or Gaussian distribution when plotted against the logarithm of the toxicant concentration or dose (Figure 4.5). Those individuals affected at low levels are termed *intolerant* or *susceptible*, while those that are only affected at much higher levels are termed *tolerant* or *nonsusceptible*. The majority of individuals have a tolerance between these two extremes. Examples of the variation in tolerance include the different amounts of alcohol that can be consumed by various individuals before the same effects appear, and the observation that some people can smoke heavily and never develop lung cancer while others who smoke lightly may develop cancer. This variation in tolerance is due to such differences as innate tolerance, rates of metabolism, and body composition.

All measures of toxicity refer to either a concentration or a dose (amount) that causes a toxic effect, and it is important to clearly differentiate between these. **Concentration-based** measures of toxicity state the concentration in the surrounding environment (e.g., water for aquatic toxicity tests, air in inhalation tests, and soil for terrestrial toxicity tests) needed to exert the particular toxic effect and are expressed in units such as mg/l, mol/l, or mmol/l. On the other hand, **dose-based** measures of toxicity are expressed on a basis of mass per mass of organism tested, e.g., grams of toxicant per kilogram of tissue. Dose-based measures of toxicity are

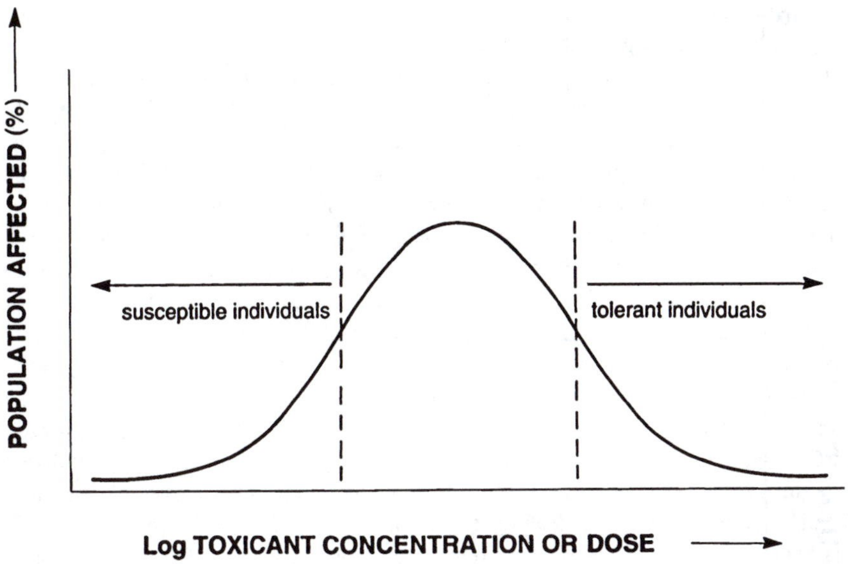

FIGURE 4.5 The susceptibility, or tolerance, of a population of individuals of a single species to different concentrations of a toxicant, plotted as percentage responding against log [toxicant concentration].

used when a known mass of a toxicant is administered directly to the test organism. This is done by exposing the animals to the chemical by injection, direct introduction to the stomach (gavage), or application to the skin. However, concentration-based measures of toxicity are used when the toxicant is dissolved in water. Thus, concentration is nearly always used in aquatic and inhalation toxicity tests, where the toxicant is dissolved in ambient environment or air, respectively.

The data in Figure 4.5 can be replotted, instead of the proportion (as percent) affected at each concentration range, as the total proportion that has responded up to that concentration. Thus, this new curve (Figure 4.6) should start at zero cumulative percentage of the population affected at zero concentration and increase to a maximum of 100% cumulative percentage for some higher toxicant concentration. This is described as a cumulative distribution curve, commonly used in toxicology and can be used in human health and ecological risk assessment (see Chapter 18).

The principal measure of the toxic effects used in toxicity studies is the 50% effect level, where 50% of the individuals are more tolerant and 50% are less tolerant. This represents the average organism in the population and exhibits the greatest consistency in experimental measurements. Measures of toxicity with different magnitudes, for example, 90 and 10% of the test organisms affected, are used, but less frequently. The concentration, or dose, corresponding to the percent affected is referred to as the **effective concentration** (EC), or **effective dose** (ED), for that percent affected; for example, EC_{90}, EC_{50}, and EC_{10}, with ED as the acronym with dose. With lethal effects the abbreviations LC and LD are used.

The biological effects measured as percent of population affected in a toxicity test are called **biological endpoints**. A large number of different endpoints have

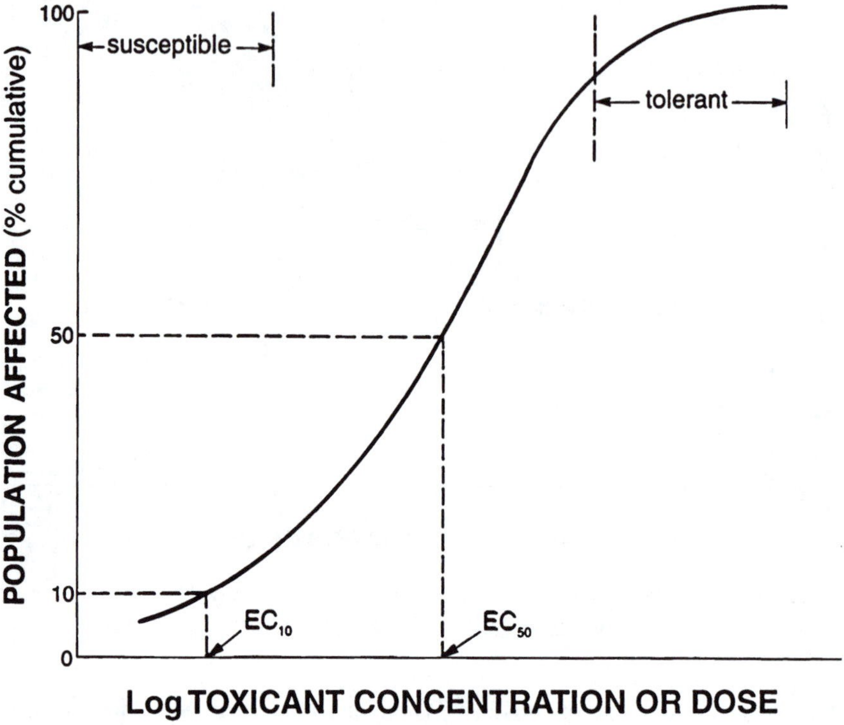

FIGURE 4.6 The susceptibility, or tolerance, of a population of individuals plotted as the cumulative percentage responding against log [toxicant concentration].

been used. They can be subdivided into two principal categories: lethal and sublethal. Lethality is simply the quantification of the number of test animals that have died or are still alive. There is a much greater number of sublethal endpoints, which include changes in the rate of respiration or the activity of an enzyme; inhibition of growth, mobility, metabolic rate, bioluminescence, and reproductive capability; and a stimulation of deformities. Some common toxicity measures are shown in Table 4.3.

As mentioned previously, measures of lethality are useful in environmental toxicology only where lethal levels are likely. In most situations in the environment, sublethal levels are present, and so techniques are needed to address this. Some of these utilize other measures of toxicity based upon statistically significant differences between treatments and the controls. Such measures are termed **no observed effect level** (NOEL) or **no observable adverse effect level** (NOAEL), and **lowest observable effect level** (LOEL) or the **lowest observable adverse effect level** (LOAEL). These values can be expressed in terms of concentrations or dose. The relationship between the values is indicated in Figure 4.7. The NOAEL is a maximum value, as indicated, and is not represented by the level corresponding to point A. Similarly, the LOAEL is a minimum level, as indicated in Figure 4.7, and is not represented by the level at point B or those at higher concentrations.

TABLE 4.3
Abbreviations of Some Common Toxicity Measures

EC_{50}	The concentration of a chemical, in the ambient environment, that exerts a 50% change of the measured effect on the test organisms when compared to the control under standard conditions and over a specified exposure period.
IGC_{50}	The concentration of a chemical, in the ambient environment, that inhibits growth of the test organisms by 50% when compared to the growth of the control under standard conditions and over a specified exposure period.
LC_{50}	The concentration of a chemical, in the ambient environment, that is lethal to 50% of the test organisms.
LOEL	The lowest concentration of toxicant used in a toxicity test that has a statistically significant effect on the biological response when compared to the control.
NOEL	The maximum concentration of a toxicant used in a toxicity test that has no statistically significant effect on the biological response when compared to the control.

Note: The above measures have equivalent measures expressed in terms of dose, e.g., LD_{50}.

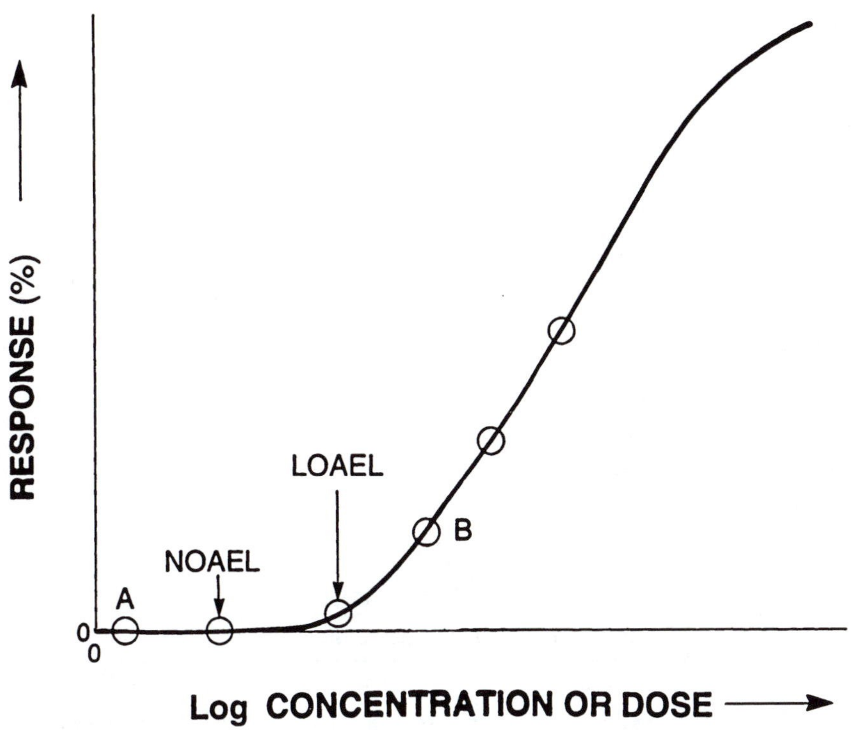

FIGURE 4.7 Diagrammatic illustration of the lowest observed adverse effect level (LOAEL) and no observed adverse effect level (NOAEL).

In evaluating mechanisms of toxicity, it is often useful to compare the toxicity of different chemicals in terms of the number of molecules (e.g., mol l^{-1} or mmol l^{-1}). For example, if two compounds A and B both had toxicities of 45 mg/l, they may not necessarily have the same toxicity in molecular or molar terms. Assuming these compounds had molecular weights of 126 and 50 amu, respectively, the toxicity expressed in mol l^{-1} would be 0.37 and 0.9 mol l^{-1}, respectively, which means that compound A is more toxic.

4.7.3 Factors Influencing Toxicity

The toxicity of a compound varies with the period of exposure. An experiment could be carried out to determine the concentration needed to exert a particular toxic effect, e.g., LC_{50} at different exposure times. A typical plot of the toxicity values (e.g., LC_{50} against the exposure time) would resemble Figure 4.8. This clearly illustrates that as the period of exposure increases, the amount of toxicant required to exert the required effect decreases. The minimum concentration that will exert the toxic effect, regardless of the period of exposure, is called the **incipient** or **threshold toxicity** of a chemical. To obtain the **threshold toxicity**, a vertical line is drawn from the

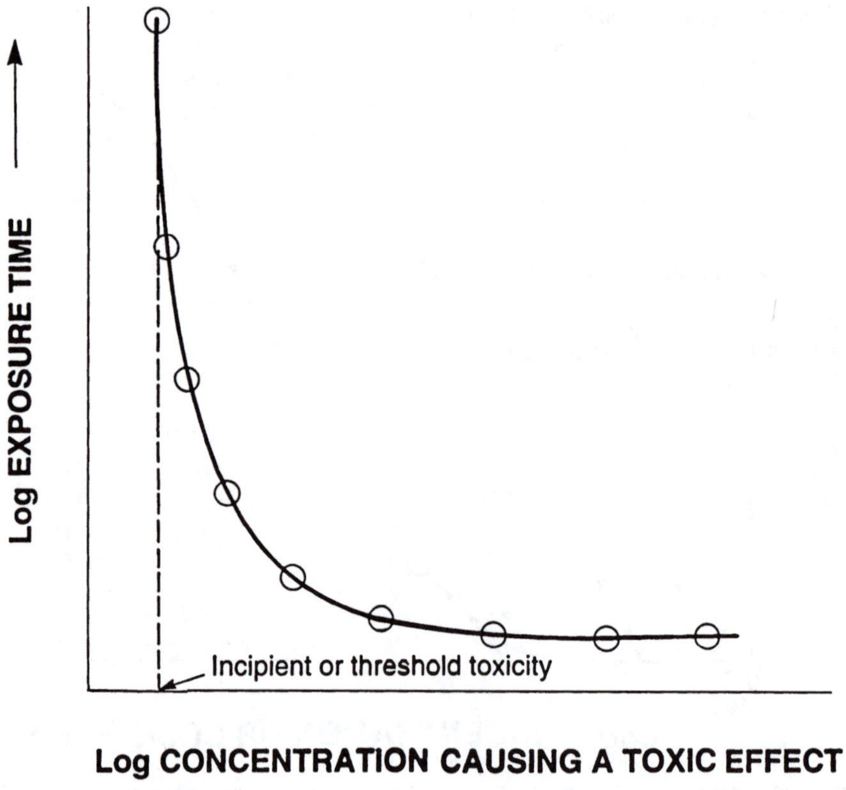

FIGURE 4.8 The variation in toxicant concentration to cause lethal toxicity with exposure time.

point where the toxicity curve becomes effectively parallel with the y-axis (time) down to the x-axis (Figure 4.8). If the toxicity curve does not become effectively parallel to the y-axis, it should be extrapolated until it does, and then the incipient toxicity can be determined.

Different species and even different strains of the same species can have markedly different susceptibilities to the same compound. For example, the amounts of dioxin (i.e., 2,3,7,8-tetrachlorodibenzo-*p*-dioxin) required to kill 50% of exposed guinea pigs and hamsters are 1 and 5000 μg/kg of body weight, respectively, and there is only 65% agreement between the results of carcinogenic tests for mice and rats. This difference in susceptibility can cause problems, as most toxicity tests are not conducted on the organisms of concern, since they may not be suitable or available; rather, they are conducted on surrogate species. For example, rainbow trout (*Salmo gairdnerii*) and fathead minnow (*Pimephales promelas*) are commonly used to represent all freshwater fish.

The sex of an organism can also affect its tolerance to chemicals. This is largely related to the rate of metabolism and body composition. The rates of metabolism of chemicals between the sexes are often different. These differences in metabolism appear at puberty and are usually maintained throughout adult life. It is not possible to generalize which sex metabolizes chemicals faster than the other. The effect of body composition on tolerance is often a result of the partitioning of compounds between the lipid and water phases of organisms. For example, females generally have more fat than males of the same mass, so they will have higher concentrations of fat-soluble compounds and lower concentrations of water-soluble compounds on a gram-per-kilogram (g kg^{-1}) basis.

There is a wide range of other factors that affect or modify the toxicity of chemicals. Some of these include:

1. Composition of diet
2. Age and general health of the test organisms
3. Hormone levels (e.g., pregnancy)
4. Experimental conditions under which the toxicity was determined

An important point concerning the stated toxicity of a chemical to an organism is that it is not a constant. Under different conditions, the toxicity may change. The stated value is specific for that particular species, with that particular body composition based on sex, age, health, etc., under the stated experimental conditions for the given exposure period and route of exposure.

4.8 EXPERIMENTAL TESTING FOR TOXICITY: BIOASSAYS

4.8.1 Types of Bioassays

Bioassays are experiments, usually conducted in the laboratory, where toxic effects to specific toxicants are measured under controlled conditions. Apart from being able to vary the test organism, there are many other variables that can be modified to produce different bioassays. Two of the variables that can be modified are the duration of exposure to the toxicant and the constancy of toxicant concentration. Variation in the exposure period produces four main types of bioassays: acute, subacute, multigenerational, and chronic. **Acute** toxicity experiments are conducted for short periods, i.e., a maximum of 96 hours. In contrast, **chronic** tests are conducted so that the organism is exposed for a significant proportion of its normal expected lifetime. In between these two are **subacute** tests, in which the exposure to the toxicant should not exceed 10% of the normal life expectancy of the species. Other bioassays that are designed to determine if toxicants affect reproduction or are teratogens must be conducted over several generations and are thus called **multigenerational** bioassays. The length of chronic, subacute, and multigenerational bioassays varies extensively, depending on the test organism. For instance, a chronic bioassay for rats lasts for 6 months to 2 years, but only 28 days for shrimp; and a multigenerational (three generations) bioassay for cladocerans (e.g., *Daphnia magna*) takes 7 to 21 days, while for rodent species, it would take approximately 20 months.

Experiments with aquatic organisms are usually conducted in sets of aquaria and the toxicant is introduced into the water. The toxicant concentration is maintained using three types of experimental systems: static, semistatic and flow-through. In **static** experimental conditions, the test organisms are exposed initially to the toxicant at the required concentration. Thus, with time, the chemical is absorbed by the animal, adsorbed by the test vessel, and biodegraded or volatilized, and the concentration in the ambient environment will decrease (Figure 4.9). In **semistatic** bioassays, the toxicant is introduced at regular intervals throughout the bioassay by replacing the depleted aquarium water with water at the specified concentration. The toxicant concentration will therefore tend to oscillate due to the removal and replenishment of the toxicant (Figure 4.9). **Flow-through** tests are carried out where the toxicant is continuously introduced to the test animals with a corresponding outflow. Under such a scenario, the toxicant concentration should remain essentially constant (Figure 4.9).

Flow-through bioassays are generally viewed as the best form of bioassay since the test organisms will exhibit a toxic effect, resulting from exposure to a constant known concentration. Such experiments are more reproducible, and thus the results are more reliable.

4.8.2 Calculating Toxicological Data

The simplest way to evaluate toxicity is to plot the logarithm of concentration or dose of the toxicant on the x-axis against the biological response, as cumulative percentage of test organisms affected, on the y-axis, as shown in Figure 4.10. In the case of aquatic organisms, where a set of aquaria are used with each at a different concentration, the response in each aquarium is observed as a cumulative percentage.

To estimate the effective concentration for a toxicological effect, for example, the EC_{50}, from this graph, a horizontal line is drawn from the point of 50% effect

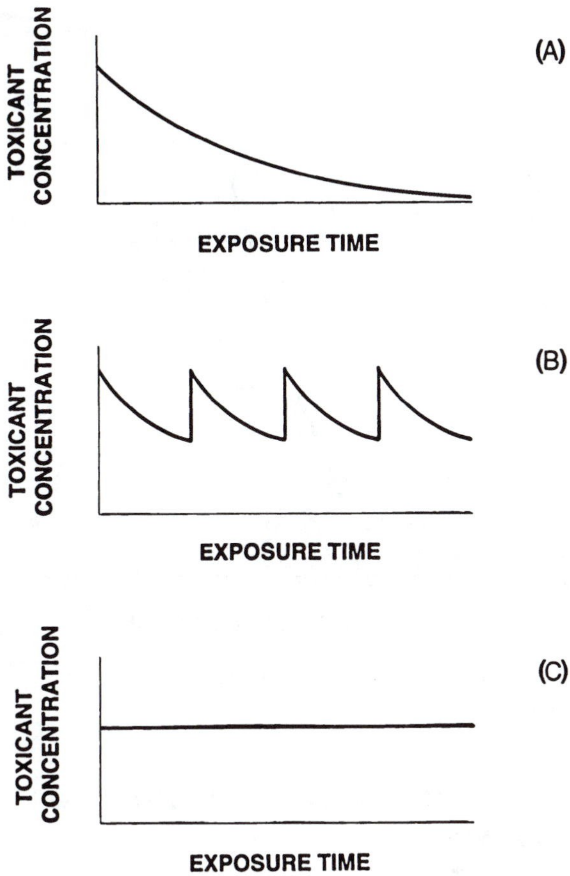

FIGURE 4.9 The change in toxicant concentration with time for different aquaria systems: (a) static, (b) semistatic, and (c) flow-through.

to where it intersects the toxicity curve (point A, Figure 4.10). From this intersection, a vertical line is drawn down until it intersects the x-axis. This point, point B, is the concentration of toxicant that affects 50% of the test organisms exposed to the toxicant under the stated experimental conditions (i.e., EC_{50}). Lethality is a common endpoint, and so the LC_{50} or LD_{50} can be estimated by this technique.

Other information, besides an estimate of toxicity, can be derived from dose–response plots. The gradient of the linear portion, shown in Figure 4.10, indicates the potency of the toxicant or the sensitivity of the organism to the toxicant. For a toxicant with a steep gradient, small changes in the concentration of toxicant may cause quite large changes in the magnitude of the biological response. For toxicants with small gradients, small changes in the concentration may cause very little change in the biological response. This information can be utilized in deriving safe exposure levels for toxicants.

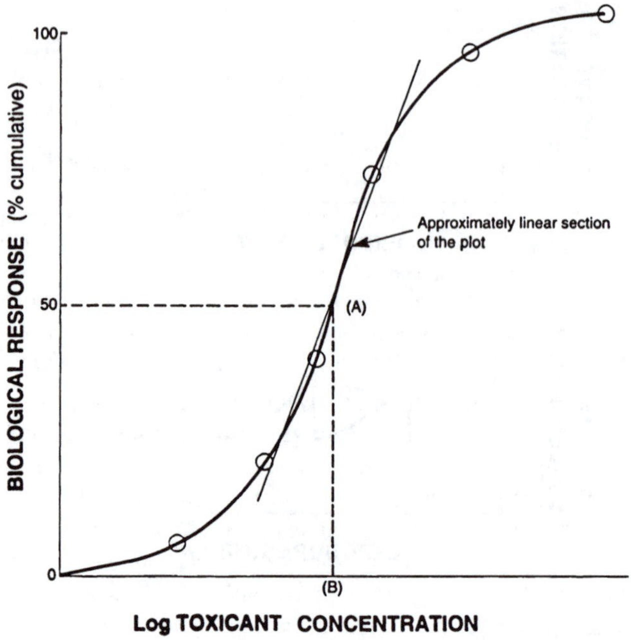

FIGURE 4.10 Plot of cumulative percent of organisms giving a response against the log [toxicant concentration].

4.9 ALTERNATIVE, MORE HUMANE METHODS FOR TOXICITY ASSESSMENT

Over the last two decades there has been increasing pressure from the public and animal liberationists to cease all animal testing, or at least to minimize the suffering involved and the number of test organisms used. As a result, many countries have passed legislation banning certain toxicological tests, such as the Draize test, where chemicals are added to the eyes of animals. In some countries, government approval must be obtained for every lethal toxicity test using vertebrate animals, the rationale being that such animals can feel pain. Great progress has been made in the development of new, more humane methodologies. For example, new range-finding methods have been developed that use less organisms and yet provide sufficiently accurate information.

Other advances have been the rapid increase in the development and use of:

1. *In vitro* bioassays, where cell cultures are used to assess the toxicity of chemicals to particular organs.
2. Bioassays using microorganisms or other simple organisms and relating these to the toxicities of higher animals.

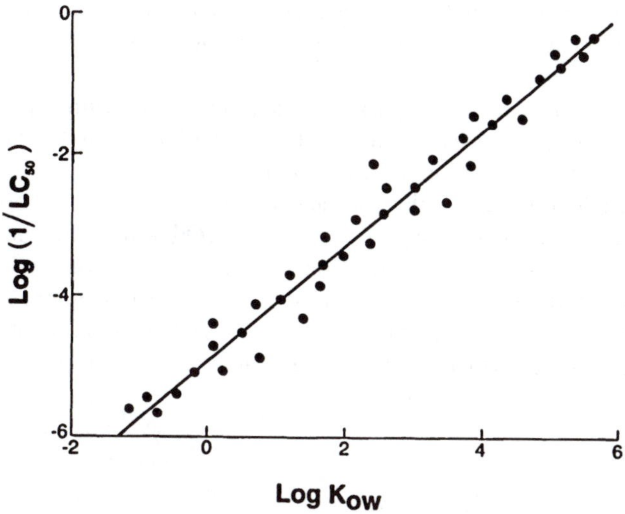

FIGURE 4.11 Plot of log K_{OW} against log [1/LC_{50}] for fish for a range of chemicals.

3. Relationships that attempt to predict the toxicity of compounds based on physicochemical properties and descriptors of the compounds. These are called quantitative structure-activity relationships (QSARs) or structure-activity relationships (SARs).

An example of a QSAR is shown in Figure 4.11, where the relationship between lethal toxicity, as log [1/LC_{50}], and the log K_{OW} value are plotted for a range of chemicals. This relationship can be used to estimate the toxicity of other compounds of unknown toxicity.

The use and development of these more humane and less animal intensive bioassay tests will increase in the future. Despite this, most toxicologists would say that these new methods cannot fully replace animal studies; rather, they are a complement.

4.10 KEY POINTS

1. All substances will exert toxic effects as long as the dose is high enough.
2. To exert a toxic effect, the toxicant must enter the body and be distributed to the site of action (target organ).
3. The toxicity of a compound depends on several factors: (a) the rate of entry and excretion; (b) the physical-chemical properties of the chemical; (c) the duration of exposure; (d) the compound's susceptibility to biotransformation; and (e) the species, age, sex, and health of the test animal.
4. Biotransformation of lipophilic chemicals is generally due to induction of the mixed-function oxidase (MFO) enzyme system in biota. Biotransformation products generally have increased aqueous solubility and a decreased K_{OW} value, and therefore this process decreases the time the chemical stays in the body.

5. Elimination of chemicals is a critical measure in minimizing toxic effects. Chemicals are eliminated from the body by urine and by fecal and respiratory excretion.
6. Toxicity is measured by exposing animals or cells to chemicals. Numerous endpoints can be measured in toxicity tests, such as death, fertility, movement, growth, photosynthesis, rate of activity, and mutations.
7. The main measure of toxicity is the concentration that exerts a 50% effect on the test population, e.g., LD_{50} and LC_{50}. Other measures include the no observed effect level (NOEL) and lowest observed effect level (LOEL).
8. The methods used to determine the toxicity of compounds are undergoing radical change in an attempt to obtain the necessary information more humanely. QSARs can be used to estimate toxicity from physical-chemical properties, such as the K_{OW} value.

REFERENCES

Landis, W.G. and Ming-Ho, Y., *Introduction to Environmental Toxicology*, 3rd ed., Lewis Publishers, Boca Raton, FL, 2004.

Rand, G.M., Ed., *Fundamentals of Toxicology: Effects, Environmental Fate and Risk Assessment*, 2nd ed., Taylor and Francis, Washington DC, 1995.

Wright, D.A. and Welbourn, P., *Environmental Toxicology*, Cambridge University Press, Cambridge, U.K., 2002.

QUESTIONS

1. What is the scientific rationale for the 50% effect level being the most common measure of toxicity?
2. What is the difference between the data analysis methods used to determine the toxicity measures described as the LC_{50} and NOEL?
3. What is the key physicochemical property of toxicants that will affect their uptake across the skin and the stomach?
4. What properties of toxicants are modified by biotransformation reactions to facilitate their excretion, and how is this achieved?
5. A bioassay has been carried out on shrimp in aquaria and the following data attained after 24 hours:

Number of Shrimp Surviving (of 200/aquarium)	Concentration in Aquarium Water (μg/l)
200	11.0
164	14.5
124	19.1
72	21.9
22	30.2
0	57.5

Plot this data out and graphically estimate the LC_{50} and LC_{80}.

6. Using the plot in Figure 4.11, where the LC is measured in units of mg l^{-1}, estimate the 24-hour LC_{50} for fish for compound A with K_{OW} = 3500 and compound B with K_{OW} = 135,000.

ANSWERS

1. This represents the level of toxic effect that occurs with the average organism, with 50% of the individuals being more tolerant and 50% less tolerant. Also, the error in measurement is least at this level, and therefore this estimate of toxicity is most accurate.
2. Measures such as the LC_{50} are determined using linear relationships or similar methods that are then used to calculate the concentration that causes the toxic effect. NOEL values are determined by evaluating if there is a significant difference between the toxicity of the control and different concentrations of the toxicant.
3. The key factor is lipid solubility, measured as the K_{OW} value. Toxicants in the stomach are subject to quite acidic conditions, which, in accordance with the Henderson–Hasselbach equation, can lead to quite large changes in the extent of ionization. As ionized compounds have lower K_{OW} values, the uptake will decrease.
4. The biotransformation reactions increase the polarity of the toxicants and hence the aqueous solubility. This is done by adding hydroxyl functional groups or combining with normal body metabolic products.
5. The data can be plotted out as in Figure 4.12, allowing the estimation of the log LC_{50} as 1.30 and the log LC_{80} as 1.43, thereby giving LC_{50} and LC_{80} values of 19.99 and 26.9 μg l^{-1}, respectively.
6. Compound A: K_{OW} = 3500
   Thus, log K_{OW} = 3.54
   Estimating from Figure 4.11,
   log $1/LC_{50} = -2$
   and $1/LC_{50} = 0.01$
   Compound A: LC_{50} = 100 mg l^{-1}
   Compound B: K_{OW} = 135,000
   Thus, log K_{OW} = 5.13
   Estimating from Figure 4.11,
   log $1/LC_{50} = -0.7$
   and $1/LC_{50} = 0.20$
   Compound B: LC_{50} = 5.0 mg l^{-1}

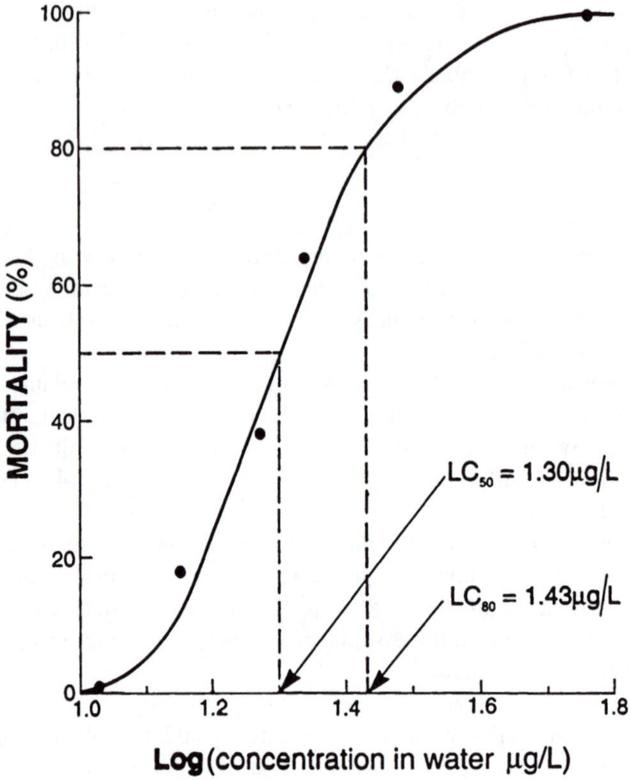

FIGURE 4.12 Plot of data on lethality to shrimp from question 5.

Part III

Contaminants in the Environment

5 Petroleum Hydrocarbons

5.1 INTRODUCTION

Petroleum is a major fuel in contemporary society and is widely used in motor cars, cargo ships, planes, and transport vehicles. Petroleum is also a source for a range of chemical products, such as plastics, lubricating oils, and bitumen. Crude and processed petroleum is transported on sea and land in large quantities. Discharges to the oceans are generally of the order of several million tonnes per year, while discharges due to incomplete combustion are generally placed at tens of million tonnes per year (Anon., 1976). A qualitative indication of sources of petroleum to the environment is shown in Table 5.1.

The combustion of petroleum fuels is usually incomplete, and unchanged petroleum as well as partly combusted products are discharged to the atmospheric environment. This comprises the major environmental discharges, which are not due to the spectacular accidents resulting in spills but the low concentrations present in municipal wastes or sewage and urban runoff. Sewage contains low concentrations due to the activities of the many urban industries that use petroleum. The petroleum in urban runoff is principally due to petroleum on road surfaces. As a result of these discharges, petroleum hydrocarbons are distributed throughout the world's oceans and the atmosphere as major global contaminants (Jordan & Payne, 1980). On the other hand, petroleum from natural sources (Table 5.1) has been entering the environment over long spans of geological times. This means that some of the contam-

TABLE 5.1
Sources of Petroleum to the Environment

Source	Quantity
Marine Environment	
Municipal waste (sewage)	+++
Urban runoff	++
Industrial waste	++
Accidents	++
Natural marine seeps	++
Terrestrial Environment	
Incomplete combustion	+++
Evaporation	++
Industrial operations	++

ination is natural, and some generated from human sources. There is a tendency for petroleum contamination to occur to the greatest extent adjacent to urban complexes.

5.2 CHEMICAL NATURE OF PETROLEUM

Liquid petroleum is a naturally occurring, oily and flammable liquid comprised principally of hydrocarbons. Unrefined liquid petroleum is generally termed *crude oil*. The name *petroleum* literally means "rock oil" and gives some indication of its origin. Petroleum is commonly found in sedimentary rocks where it is generated from the deposition of organic matter derived from microorganisms, plants, and animals at the bottom of ancient water bodies. Burial and compaction occur, destroying microorganisms and providing modest temperature (generally <200°C) and pressure. Over geological time spans of millions of years, these conditions convert the decayed organic matter into petroleum.

5.2.1 Hydrocarbons

The chemical composition of crude oil or petroleum is quite complex and variable. This reflects the variety of organic matter from which it was originally formed and the complex biochemical and chemical processes that occur during the transformation to petroleum. Typically hydrocarbons (i.e., compounds of carbon and hydrogen only) constitute 50 to 90% of petroleum. The types of hydrocarbons found are *n*-alkanes, branched alkanes, cycloalkanes, and aromatics. The *n*-alkanes are also known as normal paraffins and are found in all crude oils. Depending on the sample, the *n*-alkanes range in size (and boiling point) from C_1 (methane) up to C_{40} or more.

The branched alkane content of petroleum is at least as great as the *n*-alkane content, and usually larger. The branches are typically methyl (CH_3) groups. This branching inhibits microbial transformation. The mass percentage of normal and branched alkanes in particular samples of petroleum from Southern Louisiana and Kuwait are shown in Table 5.2. Structures of typical alkanes found in petroleum are shown in Figure 5.1.

TABLE 5.2
Mass Percentage of Chemical Components in Particular Samples of Petroleum from Southern Louisiana and Kuwait

Chemical Component	Southern Louisiana Crude	Kuwait Crude
Alkanes (normal and branched)	28.0	34.1
Cycloalkanes[a]	44.8	20.3
Aromatics[a]	18.6	24.2
Polar (O-, N-, S-containing) components	8.4	17.9
High boiling, high-molecular-weight material	0.2	3.5

[a] Includes compounds with alkyl side chains.

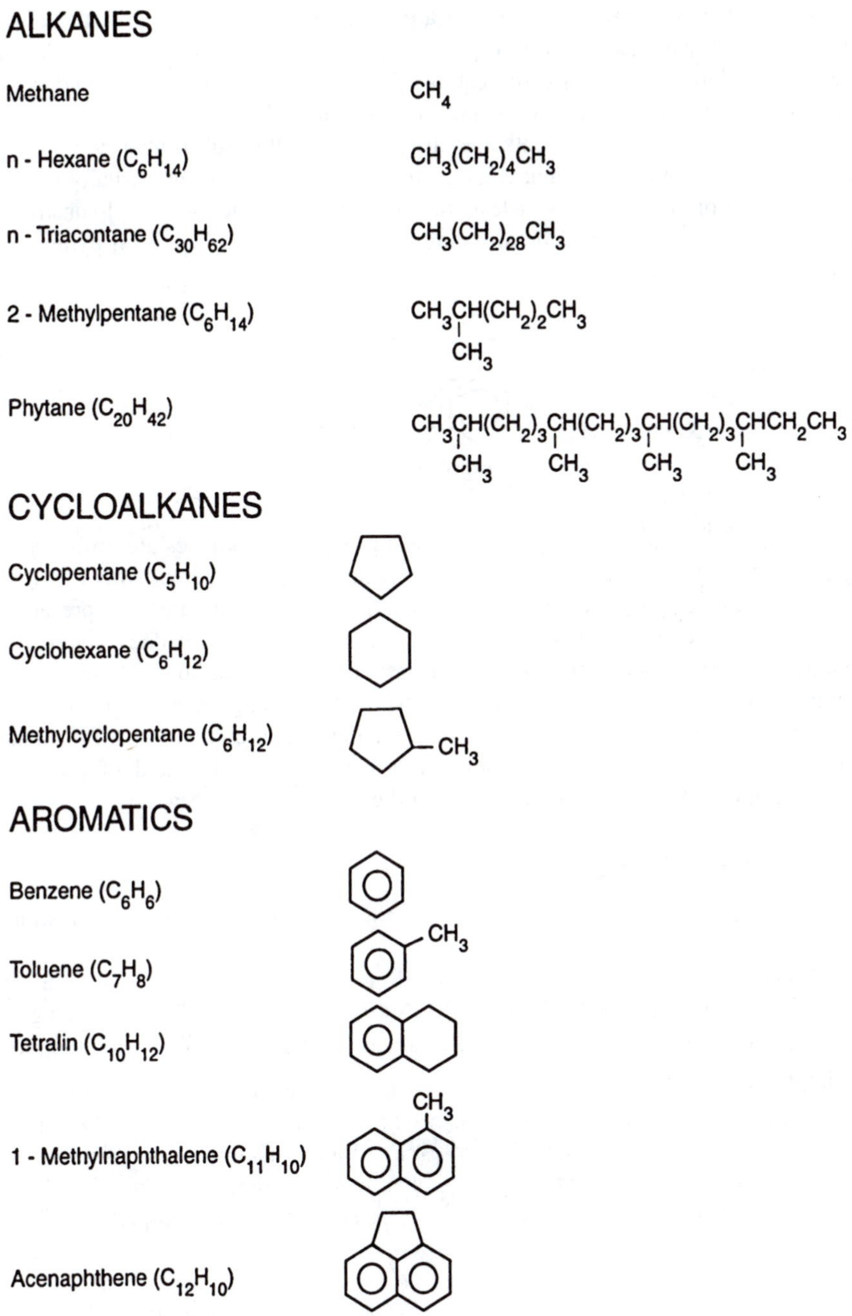

FIGURE 5.1 Structures of typical hydrocarbons found in petroleum samples.

Cycloalkanes (or naphthenes as they are also known) usually comprise the largest group of hydrocarbons by mass in petroleum. The cycloalkanes most abundant in petroleum contain five or six carbon atoms in the ring, i.e., cyclopentanes and cyclohexanes. Compounds with seven- and eight-membered rings also occur in small amounts. This group of cycloalkanes also includes alkyl-substituted compounds, e.g., methyl cyclopentane (Figure 5.1). Aromatic hydrocarbons are usually present to the extent of 20% by mass or less. In addition to benzene- and alkyl-substituted benzenes, fused-ring polycyclic aromatic hydrocarbons, e.g., benzo [a] pyrene, are characteristic components.

Benzo [a] pyrene

Some of these are carcinogenic, and as a group, the aromatics are probably the group that is of greatest environmental significance of the different sorts of hydrocarbons present. Alkyl derivatives of aromatic hydrocarbons are also present in petroleum. It is important to note that crude oil or petroleum does not usually naturally contain alkenes or other similar unsaturated hydrocarbons. These occur in refined petroleum products, often as a result of catalytic cracking processes in refineries. These processes change the size and, in some instances, the hydrogen/carbon ratio of molecules with the aim of producing a higher yield of profitable transportation fuels (gasoline, jet fuel, and diesel).

5.2.2 Nonhydrocarbons

Sulfur occurs in petroleum in various forms; some chemical structures are illustrated in Figure 5.2.

Some petroleum contains elemental sulfur, and it can also occur as hydrogen sulfide (H_2S) and carbonyl sulfide (COS). Sulfur is also present in a wide range of hydrocarbons, largely in the form of mercaptans (or thiols), sulfides, and thiophene derivatives.

Mercaptan R-SH

Sulfide R-S-R

Thiophene

Sulfur compounds tend to be concentrated in the higher boiling fractions of petroleum. Sulfur-containing compounds represent unwanted impurities in

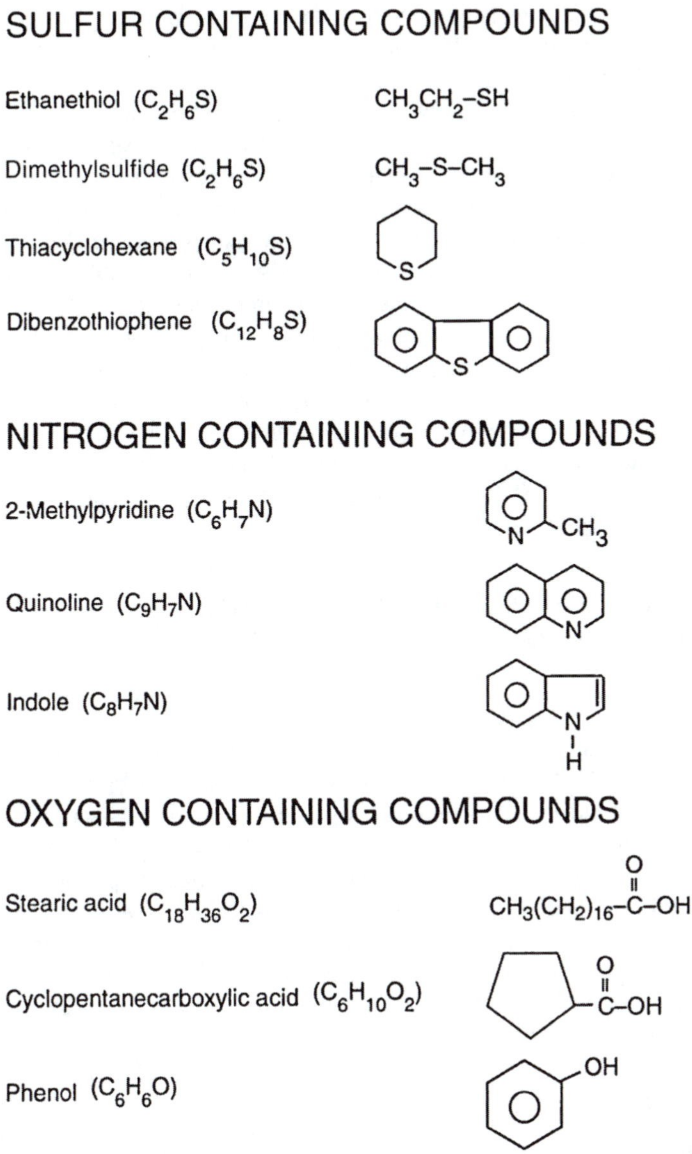

FIGURE 5.2 Some sulfur-, nitrogen-, and oxygen-containing compounds typically found in petroleum samples.

petroleum since they are generally corrosive to metals, malodorous, and may poison various catalysts used in refineries. In addition, on combustion the sulfur present in many compounds is converted to sulfur dioxide (SO_2), an unwanted air pollutant.

Nitrogen-containing compounds are present in most petroleum and, like the sulfur-containing compounds, tend to be concentrated in the high boiling fractions. The presence of oxygen-containing compounds in most petroleum is relatively minor. The oxygen is usually present in the form of various functional groups, particularly carboxylic acids. Typical examples of nitrogen- and oxygen-containing compounds found in petroleum are shown in Figure 5.2.

Most samples of petroleum are dark in appearance, the color ranging from black to yellow-brown. The origin of this color is relatively large particles of a complex nature that are colloidally dispersed in the petroleum hydrocarbons. As well as having a high sulfur, nitrogen, and oxygen content, these particles contain metals, particularly vanadium, nickel, cobalt, and iron. With few exceptions, the chemical form in which these metals occur is unknown. Some exist as water-soluble salts, e.g., NaCl enclosed within the colloidal particles. Despite probable contact between the petroleum and water underground for extended periods, extraction of these salts into water has not occurred. Petroleum from Venezuela in particular, and also Canada, contains relatively high levels of vanadium (up to 1000 ppm) in the form of vanadyl cations (VO^{2+}) complexed to a porphyrin molecule, which acts as a tetradentale group. On high-temperature combustion, the vanadium is likely to be discharged in the form of vanadium pentoxide (V_2O_5) and associated with particulate material in the atmosphere or any solid residue.

5.3 PETROLEUM HYDROCARBONS IN THE ENVIRONMENT

5.3.1 Dispersion

Petroleum hydrocarbons are emitted or discharged into all phases of the environment: air, water, and soil. Most petroleum has a density less than 1 g cm^{-3}, and so material discharged to the oceans will initially float on the surface of marine waters as a slick, since typical seawater at 298 K (25°C) has a density of 1.03 g cm^{-3}. Dispersion or spreading is probably the most significant process affecting petroleum hydrocarbon slicks in the first 6 to 10 hours following discharge. Eventually, most slicks form a layer about 0.1 mm thick or less on the water. Drift is governed by wind direction and speed, together with wave action and currents. If drift is likely to result in landfall, then coral and mangrove communities, for example, can be impacted and intertidal benthic strata fouled.

5.3.2 Evaporation

Evaporation is the process in which molecules escape from the surface of a liquid into the vapor phase or atmosphere. This can be a major fate of material on the surface of marine waters, particularly in the first few days following spillage.

Evaporation removes the lower-molecular-weight, more volatile components of the petroleum mixture. The rate of evaporation depends on the concentration of the compound in the slick and its vapor pressure. The vapor pressures and boiling points of *n*-alkanes as related to carbon number are plotted in Figure 5.3. For those *n*-alkanes that are gaseous at room temperature, their vapor pressures at that temperature are greater than 1 atm or 101.625 kPa. As the molecules become larger, their vapor pressures decrease and boiling points increase. Increasing boiling points

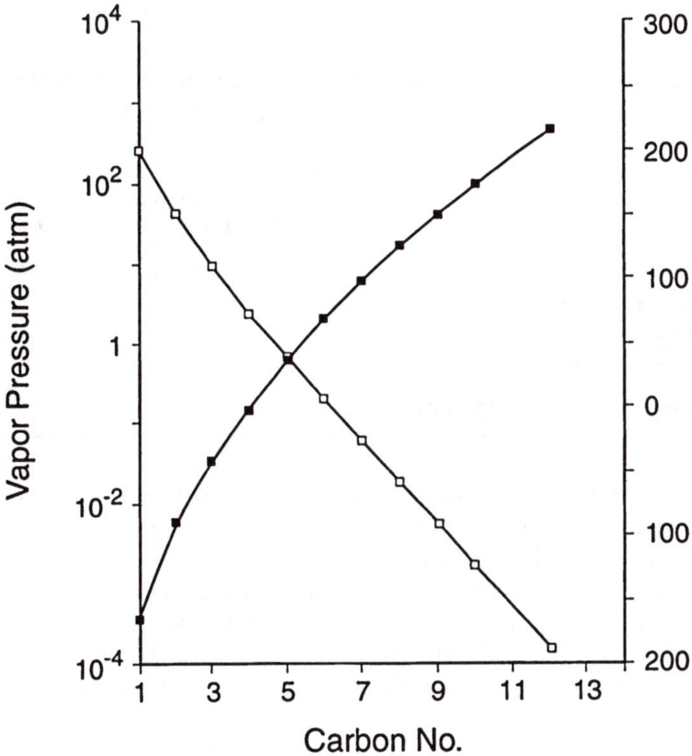

FIGURE 5.3 A plot of pure component vapor pressure (atm) and atmospheric pressure boiling points (°C) vs. carbon number for a series of *n*-alkanes.

reflect stronger intermolecular forces in the liquid phase, which must be overcome in order to vaporize. Since the *n*-alkanes are all nonpolar, the intermolecular forces must be dispersion or London forces. The overall strength of these attractions is related to molecular size. It is not surprising then that as the size of the *n*-alkanes increases in a regular manner (by increments of CH_2 groups), so does the boiling point.

Hydrocarbons with vapor pressures equal to that of *n*-octane (0.019 atm at 298 K) or greater will be lost quickly via evaporation. Pentadecane ($C_{15}H_{32}$) is the smallest *n*-alkane commonly found in weathered petroleum. Table 5.3 contains details of the fate of unrecovered petroleum from the *Exxon Valdez*, which grounded in Prince William Sound in March 1989. It can be seen that a significant amount of petroleum was subject to evaporative loss.

5.3.3 Dissolution

As well as being the most volatile, lower-molecular-weight hydrocarbons also tend to be the most water soluble. For example, lower-molecular-weight aromatic hydrocarbons are more water soluble than larger ones. Similarly, smaller *n*-alkanes have a greater aqueous solubility than larger ones. Generally, for lower-molecular-weight

TABLE 5.3
Overall Fate of Nonrecovered Petroleum from the *Exxon Valdez* Approximately 3 Years after the Spill,[a] Expressed as a Mass Percentage

Fate	Mass Percentage of Nonrecovered Spilled Petroleum
Associated with subtidal sediments	15
Associated with intertidal strata	2
Evaporation and subsequent photolysis	24
Photolysis and microbial transformation in water	59

[a] Estimated mass of spill is 3.55×10^4 tonnes.

hydrocarbons, evaporation is more important than dissolution. In comparing the aqueous solubility of different types of hydrocarbons, we should compare compounds of a similar size. Table 5.4 shows that among C_6 hydrocarbons, the order of aqueous solubility is aromatic > cycloalkane > branched alkane > *n*-alkane. In fact, benzene is some 200 times more soluble than *n*-hexane. There are several reasons for these differences in the closely related compounds, and some of these are discussed in Chapter 2. The size of the surface of the molecule and the dipole moment are important factors involved.

TABLE 5.4
Aqueous Solubilities of Various C_6 Hydrocarbons

Compound	Structure	Aqueous Solubility (g/m³)	Aqueous Solubility (mol/m³)
n-Hexane	$CH_3(CH_2)_4CH_3$	9.5	0.11
2-Methyl pentane	$CH_3CH(CH_3)(CH_2)_2CH_3$	13.8	0.16
Cyclohexane	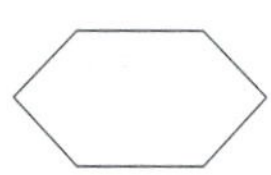	55	0.65
Benzene	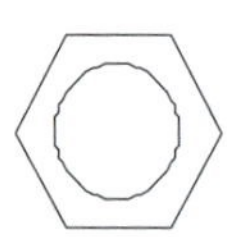	1780	22.8

Aqueous solubility values are usually for solution in pure water. Seawater contains dissolved ions and is a phase that hydrocarbons have even less affinity for than water. As a general rule of thumb, hydrocarbons are about 75% as soluble in seawater as pure water. Molecules containing N,S, or O atoms are generally more polar than hydrocarbons. Such molecules are likely to be among the more soluble components of slicks, particularly if the molecules are small.

5.3.4 Emulsion Formation

Emulsions are a colloidal state where fine droplets of one liquid are dispersed in another. With petroleum and water, two kinds of emulsion are possible: an oil-in-water (O/W) emulsion and a water-in-oil (W/O) emulsion. With an O/W emulsion, fine droplets or globules of petroleum (typically 0.1 to 1 μm in diameter) are dispersed in water. These fine droplets can have, in total, a relatively large surface area. Because of this, the rates of processes such as microbial transformation and photodegradation are often maximal. These O/W emulsions are often inherently unstable and tend to coalesce back to a slick.

5.3.5 Photooxidation

Of the processes affecting petroleum hydrocarbons we have encountered thus far, none have actually chemically transformed them, or removed them from the environment as a whole. In the presence of oxygen, sunlight can transform or degrade components of petroleum mixtures (Payne & Phillips, 1985). Photooxidation is oxidation initiated by light, particularly solar ultraviolet light. On exposure to this light, a number of physical and chemical changes occur in petroleum. Compounds with oxygen-containing functional groups, e.g., alcohols, ketones, aldehydes, carboxylic acids, and peroxides, are usually formed. Petroleum invariably contains compounds with benzylic methylene groups. These are CH_2 groups that are adjacent to aromatic rings, as shown below.

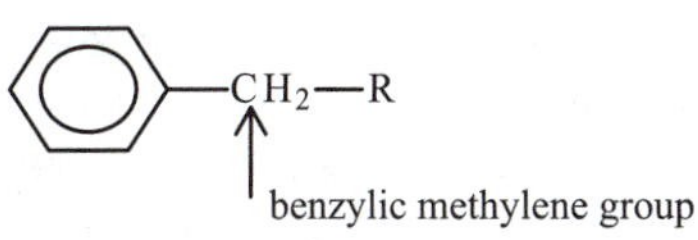

Benzylic methylene groups that are part of a ring are particularly prone to hydroperoxide formation. This can occur with common petroleum components, such as tetrahydronaphthalene (or tetralin) and indan, for example, as shown below.

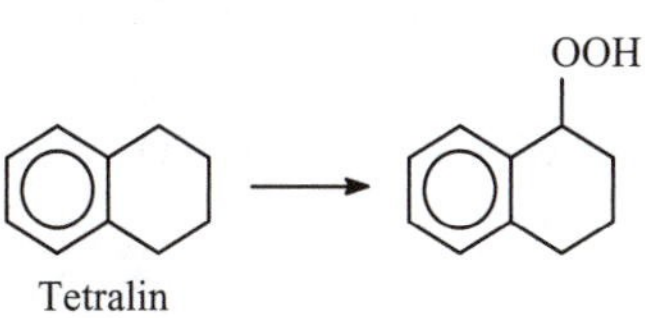

Hydroperoxides are more water soluble than the original parent hydrocarbon and, in addition, are relatively toxic to a variety of marine organisms. Whatever the mechanism of their formation, the O–O bond of hydroperoxides is quite weak. It can be split homolytically, thermally, or by ultraviolet light, as shown below.

$$ROOH \xrightarrow{h\nu} RO\bullet + \bullet OH$$

These radicals can attack other hydrocarbons, ultimately forming a variety of oxygenated derivatives with increased polarity due to the presence of oxygen, leading to increased aqueous solubility. The turbidity and dark color of petroleum often increase on exposure to sunlight. This is due to the accompanying formation of polar oxidized higher-molecular-weight material formed via radical recombination or reactions between various oxygen-containing compounds.

5.3.6 Microbial Transformation

As shown in Table 5.3, a significant fate for spilled *Exxon Valdez* oil was aqueous microbial transformation. In principle, the majority of the components of petroleum are able to be transformed by a variety of microorganisms. Whether this occurs in practice in the marine environment depends on factors such as the number and type of microorganisms present as well as the availability of oxygen and nutrients. The term *bioremediation* is now widely used and, in the context of petroleum spills, refers to the use of microorganisms to degrade the hydrocarbons to CO_2 and H_2O. Petroleum is a natural and long-standing part of our environment, and it should come as no surprise that hydrocarbon-oxidizing bacteria, fungi, algae, and yeasts exist.

The microbial transformation of hydrocarbons occurs either within the aqueous phase on dissolved or dispersed hydrocarbon molecules or at the hydrocarbon–water interface. Microorganisms do not exist within the hydrocarbon phase. Generally, *n*-alkanes, particularly smaller ones, are most rapidly transformed by microorganisms under aerobic conditions. In anoxic environments, e.g., sediments, saturated hydrocarbons are effectively resistant to microbial transformation. Most petroleum is less dense than seawater, so such a situation is unlikely.

The first step in the bacterial transformation of an *n*-alkane, such as hexane, is the insertion of an oxygen atom into the molecule, usually on a terminal carbon atom. Because only one of the two oxygen atoms of molecular oxygen is inserted into the hydrocarbon, the enzyme that catalyzes this process is known as a monooxygenase. Also necessary is NADH, which acts as a source of electrons and reducing power. NADH is a hydrogen carrier in the overall respiration process in which organisms transform and degrade organic compounds. In general, for *n*-alkanes,

$$R\text{–}CH_3 + O_2 + NADH + H^+ \rightarrow R\text{–}CH_2OH + NAD^+ + H_2O$$

The enzyme involved in the oxidation contains iron, but the iron is not associated with a heme molecule. Figure 5.4 shows a likely pathway for the bacterial

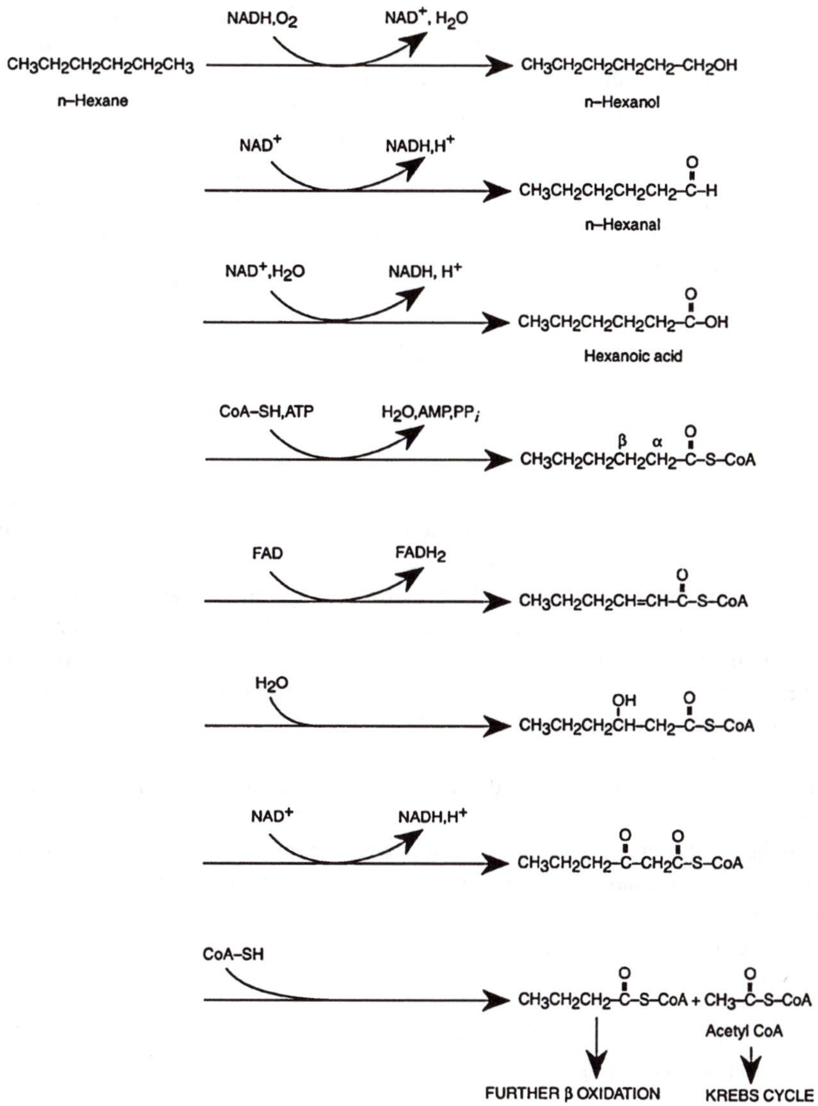

FIGURE 5.4 A pathway for the microbial transformation of an *n*-alkane.

oxidation of *n*-hexane. Eventually, *n*-hexanoic acid is formed, which can then undergo β oxidation, which is a normal degradation process for fatty acids derived from triglycerides (or fats). Essentially, β oxidation involves the successive cleavage of two carbon fragments from the main chain. The two carbon fragments are in the form of acetyl-coenzyme A (CoA), which can be fed into the Krebs (tricarboxylic acid) cycle of respiration and used for energy. The carbon ultimately appears as CO_2.

Branch alkanes and cyclic alkanes tend to be less readily oxidized by microorganisms. For some branched hydrocarbons, the reason can be seen in Figure 5.4. If an alkyl group is attached to the β carbon of hexanoic acid, a ketone cannot be formed at this position later in the sequence. If branched hydrocarbons such as these are to be transformed by microorganisms, other pathways, which are probably relatively slow, must be followed. The extent of microbial transformation of discharged or spilt petroleum, and hence its possible age, is often inferred from the ratio of a relatively degradable component, such as an alkane, and a more resistant component, such as a branched alkane (e.g., pristane and phytane).

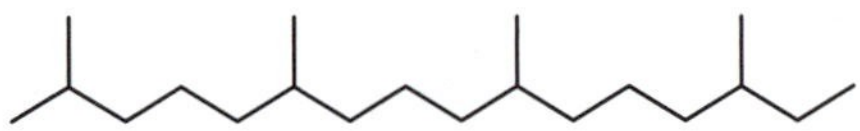

Phytane ($C_{20}H_{42}$)

Eventually, however, even compounds such as pristane and phytane are biotransformed, and even more resistant petroleum components are used to quantify the extent of microbial transformation.

Aromatic compounds can also be transformed by microorganisms (under both aerobic and anoxic conditions). Each step is catalyzed by a particular enzyme or group of enzymes. The first step involves a dioxygenase-type enzyme inserting both oxygen atoms from molecular oxygen into the benzene molecule, subsequently forming in the next step, a *cis*-diol. Although there are exceptions, bacteria and higher organisms transform aromatic hydrocarbons by different mechanisms. Bacteria tend to produce *cis*-diols and use the hydrocarbon as a carbon source and for energy.

Higher organisms (including humans) transform aromatic hydrocarbons initially using a monooxygenase enzyme, subsequently forming *trans*-diols. Higher organisms excrete the *trans*-diols or their derivatives, as the original aromatic hydrocarbon is not used as a carbon source. These processes are illustrated below.

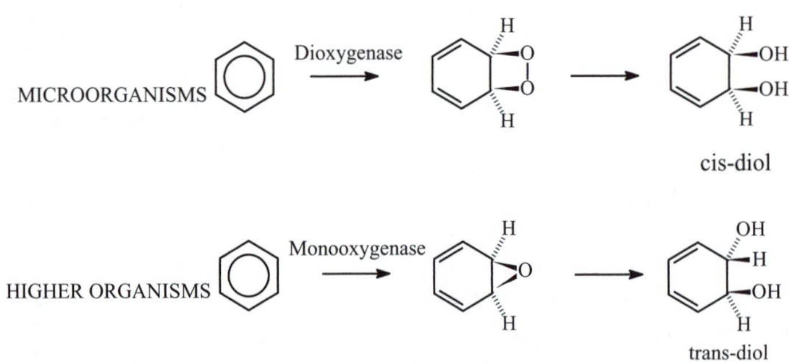

Higher-molecular-weight aromatic compounds may also be transformed. For example,

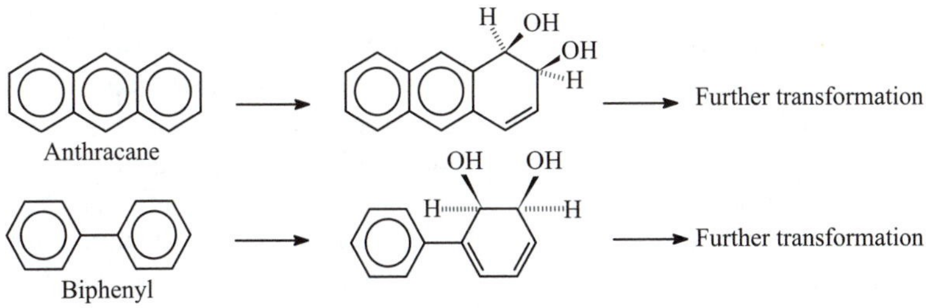

It must be reemphasized, though, that for the processes mentioned to actually occur at a reasonable rate, microorganisms, in particular bacteria, need to be present in sufficient numbers. This is not always the case in marine systems.

5.4 PETROLEUM AND AQUATIC ORGANISMS

The toxicity of petroleum varies according to its composition. As a chemical group, aromatic hydrocarbons are generally more toxic than other types of hydrocarbons. A summation of toxicity of the water-soluble aromatic fraction from various petroleum products is shown in Table 5.5. As a general rule, larvae and juveniles of all species are more sensitive than adult life-forms. Levels of 0.1 to 10 mg/l have been found to exert lethal effects on these groups. The adults are less sensitive, but lethal toxicity occurs generally in the range of 0.1 to 100 mg/l.

If sublethal levels of petroleum components are present in water, the most prevalent effects with marine fish species are behavioral changes, such as swimming performance and ability. Physiological effects include changes in heart rate, respiration rate, and growth. Past chronic exposure to low levels of hydrocarbons can be detected by elevated levels of enzymes responsible for metabolism, such as the mixed-function oxidase (MFO) enzyme system.

TABLE 5.5
Toxicity of Water-Soluble Aromatic Compounds from Petroleum to Marine Organisms

Organism Class	Range of LC_{50} Values (mg/l)
Fin fish	1–50
Crustaceans	0.1–10
Bivalves	1–50
Gastropods	1–100
Larvae (all species)	0.1–5
Juveniles (all species)	1–10

Benthic intertidal communities are particularly vulnerable to physical coating with oil and subsequent adverse effects. Floating oil contacts the shoreline in the intertidal zone, i.e., the area between the high and low water marks. Resident organisms such as plants and small invertebrates are relatively immobile and cannot escape. If hydrocarbons contaminate the plumage of seabirds, the trapped air providing thermal insulation and buoyancy is lost. Death can result from a number of causes. Loss of buoyancy can lead to drowning, or pneumonia can be caused by loss of thermal insulation. Starvation may occur due to an increased metabolic rate compensating for loss of body heat and also decreased feeding activity. Alternatively, toxic levels of petroleum components may be ingested as a result of excessive preening. Seabirds reproduce relatively slowly, and so losses are difficult to replace.

5.5 KEY POINTS

1. Petroleum is transported and used in large quantities and is a major contaminant in the environment. The major sources are the incomplete combustion of fuels in the atmospheric environment and sewage and urban runoff in the marine environment.
2. The chemical components of petroleum and derived products are hydrocarbons, including *n*-alkanes, cycloalkanes, and aromatic compounds, ranging in physical state from gases to various liquids. In addition, sulfur-, nitrogen-, and oxygen-containing compounds are present.
3. Petroleum spilled into water bodies is acted on by the following physical processes: dispersion, evaporation, dissolution, and emulsion formation. These processes are governed by the physical chemical properties of the compounds present, including aqueous solubility and vapor pressure.
4. Chemical transformation and degradation processes act on petroleum compounds in the environment. Photooxidation and microbial transformation are of particular importance.
5. Straight-chain compounds, such as *n*-alkanes, are more susceptible to microbial transformation by oxidation than branched-chain compounds. This is due to the oxidation process requiring the formation of a ketone on the β carbon atom in the alkanoic acid initially formed by oxidation. A ketone cannot be formed if an alkyl group is attached at this position.
6. Aromatic hydrocarbons are oxidized by biota to form *cis*- and *trans*-diols as the initial products of reaction with oxygen.
7. The aromatic components of petroleum are the most toxic compounds present. These substances are usually lethal to larvae and juveniles in the range of 0.1 to 10 mg/l and adults in the range of 0.1 to 100 mg/l.

REFERENCES

Anon., Sources, Effects and Sinks of Hydrocarbons in the Aquatic Environment, paper presented at Proceedings of the Symposium, American University, Washington, DC, August 9–11, 1976.

Jordan, R.E. and Payne, J.R., *Fate and Weathering of Petroleum Spills in the Marine Environment*, Ann Arbor Science, Ann Arbor, MI, 1980.

Payne, J.R. and Phillips, C.R., Photochemistry of petroleum in water, *Env. Sci. Technol.*, 19, 569–579, 1985.

QUESTIONS

1. A spill of a specialized petroleum product consisting of equal amounts of *n*-butane, *n*-hexane, *n*-octadecane, and benzene has occurred in a marine area subject to moderate wave action. Using information in the text, evaluate the following:
   a. Possible toxic effects in the marine system
   b. The possible composition of the product after 1 month in the marine area
2. A common constituent of many petroleum products is indan, with the structure

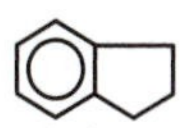

Indan

Would you expect this compound to be susceptible to photooxidation, and if so, what would the initial chemical product be?

ANSWERS

1. a. The only aromatic compound present is benzene. Table 5.4 indicates its water solubility as 22.8 mol/m^3 or 22.8 × 78 (MW) = 1780 g/m^3 or 1.78 g/l or 1780 mg/l. The LC_{50} (lethal concentration at 50%) mortality values for adults and juveniles range from 0.1 to 100 mg/l. With moderate wave action, the benzene in water concentration may initially approach its aqueous solubility. This greatly exceeds the toxicity range, so mortalities of marine organisms would be expected.

   b. The vapor pressures of *n*-butane and *n*-hexane are >0.019 atmospheres (Figure 5.3), and so would be expected to evaporate rapidly and disappear over 1 month. On the other hand, *n*-octadecane has a vapor pressure considerably lower than 0.019 atmospheres, as indicated by Figure 5.3, and would persist over 1 month. The benzene is water soluble and would be expected to have a vapor pressure in the range of *n*-hexane (i.e., >0.019 atmospheres) and, with these two properties, would be expected to disappear within 1 month. Thus, after 1 month only *n*-octadecane would remain.

2. Indan contains a benzene ring and therefore will absorb UV radiation from sunlight, which will activate the molecule. In the activated state, the methylene groups in the α position from the ring are susceptible to the addition of oxygen. Thus,

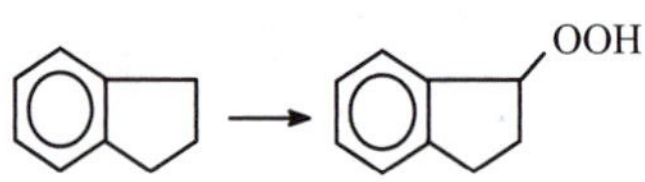

So an indan hydroperoxide would be the initial product of oxidation.

6 Polychlorinated Biphenyls (PCBs) and Dioxins

6.1 INTRODUCTION

The polychlorinated biphenyls (PCBs) and dioxins are important environmental pollutants that have a different history of environmental contamination. These substances have similar chemical structures and share a wide range of chemical, biological, and environmental properties. It is helpful to consider them together for these reasons.

The manufacture of PCBs has been carried out in many countries and the compounds marketed under trade names such as Aroclor, Chlophen, Kanechlor, and Fenclor. The applications of these substances have been quite diverse, ranging from use as plasticizers to components of printer's ink. In the 1960s and 1970s, many investigations found that PCBs occurred widely in the environment and adverse biological effects were occurring. As a result, voluntary restrictions were introduced by many industries, and many governments have acted to curb the use of these substances. The PCBs were first produced commercially in the U.S. in 1929, and approximately 600,000 tons were produced up until 1977, when manufacture in North America was halted.

Industrially, PCBs are produced by chlorination of biphenyl with molecular chlorine, using iron turnings or ferric chloride as a catalyst and giving a reaction that can be described by

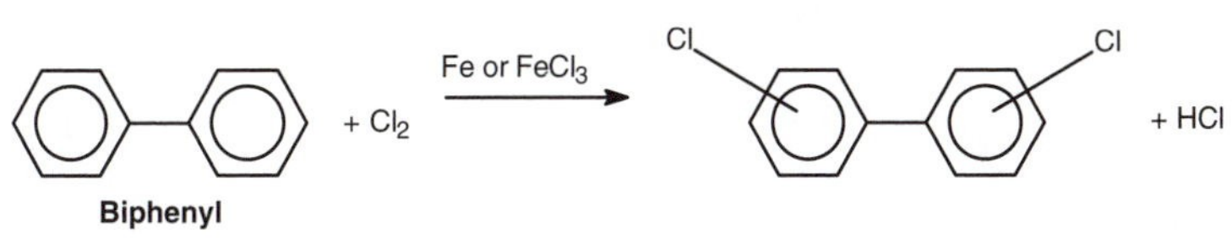

As reaction time increases, more of the 10 hydrogens on biphenyl are replaced by chlorine. A mixture of different individual PCBs is invariably produced however, and it is as mixtures that PCBs have been used commercially, and are encountered in the environment.

The first dioxin was synthesized in the laboratory in 1872, but unlike the PCBs, the dioxins or PCDDs, which stands for **polychlorinated dibenzo [1,4] dioxins**, have no known uses and have never been deliberately manufactured, except on a laboratory scale. Dioxins are generally formed as unwanted by-products in chemical processes involving chlorine (e.g., 2,4-dichlorophenoxyacetic acid (2,4-D) and 2,4,5-trichlorophenoxyacetic acid (2,4,5-T) manufacture) and combustion processes. This family of compounds has come to public attention because of the relatively high toxicity of certain members, such as **2,3,7,8-tetrachlorodibenzo [1,4] dioxin (2,3,7,8-TCCD or TCDD)**, to some test organisms. The structure of TCDD is

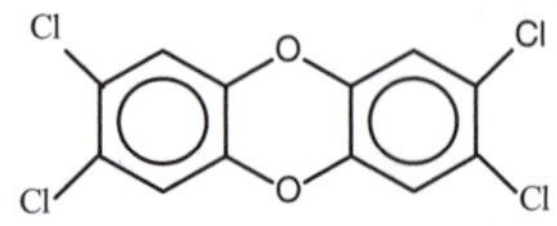

2,3,7,8-Tetrachlorodibenzo [1,4] dioxin

Like the PCBs, dioxins are usually found in the environment as mixtures. They are widely distributed environmental contaminants found, albeit in relatively low concentrations, in the atmosphere, soils, sediments, plants, and animals, including humans.

6.2 THE NAMING SYSTEM FOR PCBS AND PCDDS: NOMENCLATURE

Altogether, there are some 209 different individual PCB compounds, although only about 130 of these are found in commercial mixtures. PCBs with the same number of chlorine substituents, but different substitution patterns, are referred to as isomers, while PCBs with different numbers of chlorines are described as congeners. It is important to have a nomenclature system for the 209 PCB congeners so that they can be unambiguously identified. First, the system is somewhat complex in order to accommodate all the potential variations and permutations. A few of the more important rules are outlined below. The position of attachment of the chlorine atoms is denoted by numbering the carbon atoms of the phenyl rings as follows:

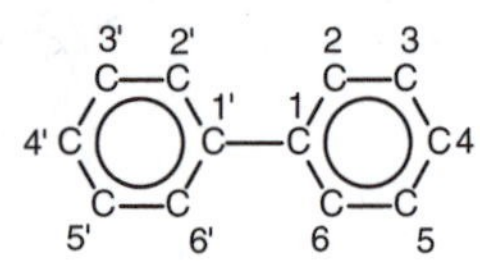

The ring with the smaller number of chlorine substituents is the one given the prime or dashed number scheme. The carbon atoms of each ring are numbered in such a way that the chlorine substituents are attached to carbon atoms with the lowest possible number. When writing the name of a particular congener, the numbers are arranged in ascending order, with dashed numbers following the equivalent nondashed numbers, e.g., **2,2′,4,4′,5,6′-hexachlorobiphenyl**. Examples of named PCB congeners are shown in Figure 6.1.

On a worldwide scale, the Monsanto Corporation was the principal manufacturer of PCBs. Their **Aroclors** were a series of mixtures, largely comprising PCB congeners, each with a four-digit code. For example, **Aroclor 1242** is a mixture of di-,

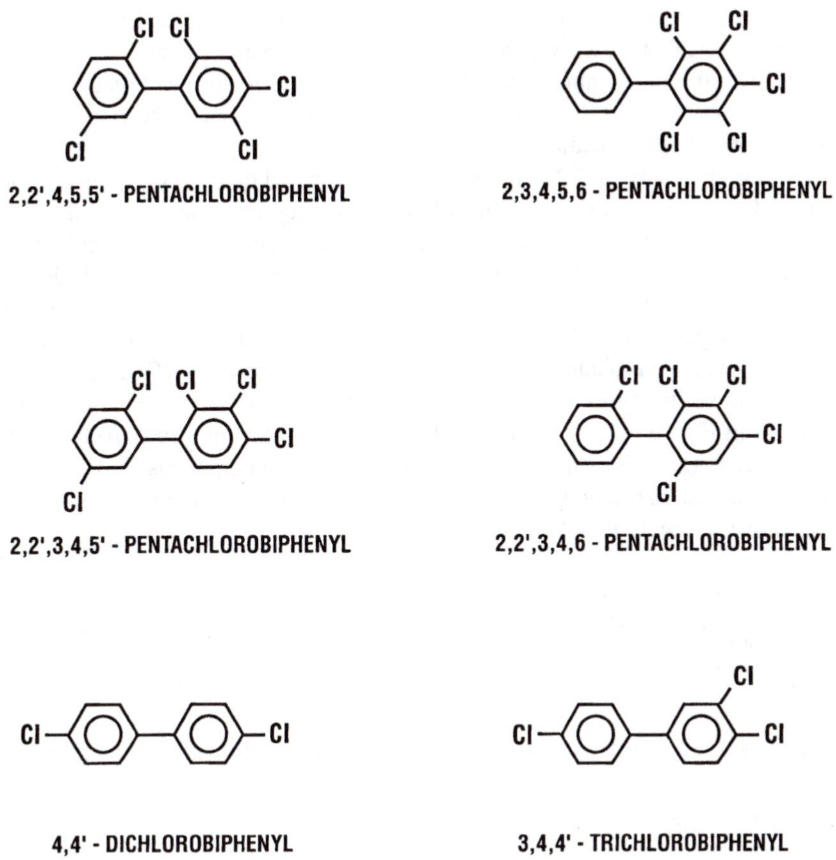

FIGURE 6.1 Molecular structures and names of some pentachlorobiphenyl isomers and other PCB congeners.

tri-, tetra-, and pentachlorinated biphenyls, with the 12 indicating PCBs as the principal component and the last two digits of the code indicating the approximate chlorine content as a mass percentage, i.e., 42% in the case of Aroclor 1242.

Most correctly, dioxins are polychlorinated derivatives of dibenzo [b,e][1,4] dioxin (PCDDs). The [1,4] dioxin portion of the name refers to the middle ring of the tricyclic structure. The molecular structure of [1,4] dioxin itself, as well as dibenzo [b,e][1,4] dioxin, is shown below :

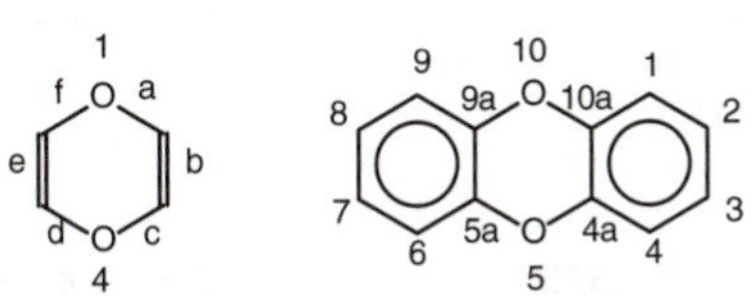

1,4 dioxin **dibenzo[b,e][1,4] dioxin**

The [b,e] denotes the sides of the [1,4] dioxin ring to which the phenyl rings are fused. As with the PCBs, the carbon atoms of the parent structure are numbered.

However, dibenzo [b,e][1,4] dioxin also contains oxygen atoms, and these are also numbered. Note that these oxygen atoms are numbered 5 and 10, but the name itself still contains the [1,4] pattern from the original dioxin molecule. Throughout this chapter, the chlorinated derivatives of dibenzo [1,4] dioxin will be referred to as dioxins, a slight simplification of their correct name.

In total, there are 75 different individual chlorinated dibenzo [1,4] dioxin compounds. These range from the monochloro isomers up to octachlorodibenzo [1,4] dioxin, where all eight hydrogens on the parent molecule have been replaced by chlorine. In naming these compounds, the pattern of chlorine substitution is denoted by listing, in ascending order, the number of each carbon to which a chlorine atom is attached, e.g., 2,3,4,6,7,8-hexachlorodibenzo [1,4] dioxin. The number assigned to each carbon atom is obtained from the scheme shown previously. Further examples of the chemical structures of polychlorinated dibenzo [1,4] dioxins and their names are found in Figure 6.2. Also found in Figure 6.2 are examples of polychlorinated dibenzofurans (PCDFs), another family of chlorinated compounds closely related to the dioxins but based on furan. These substances often co-occur with the dioxins, but in lower concentrations.

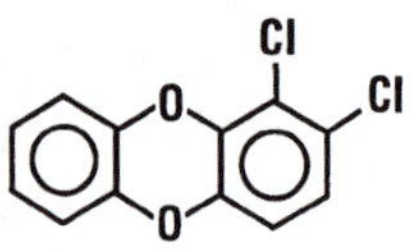

1,2 - DICHLORODIBENZO[1,4]DIOXIN

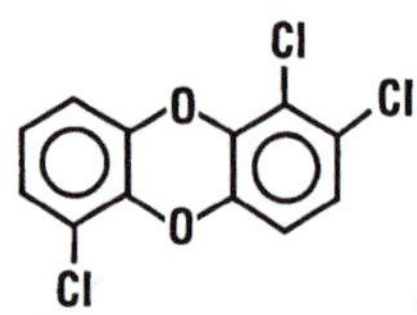

1,2,6 -TRICHLORODIBENZO[1,4]DIOXIN

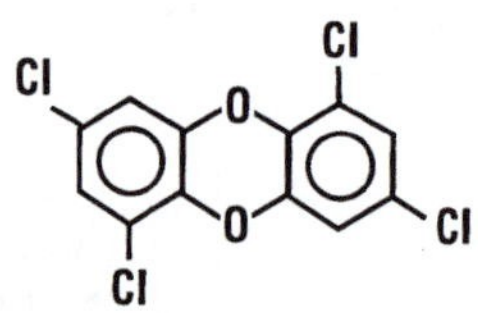

1,3,6,8- TETRACHLORODIBENZO[1,4]DIOXIN

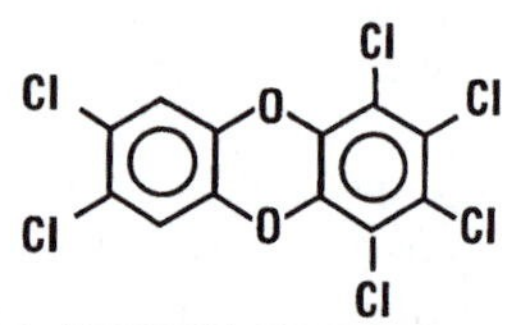

1,2,3,4,7,8 - HEXACHLORODIBENZO[1,4]DIOXIN

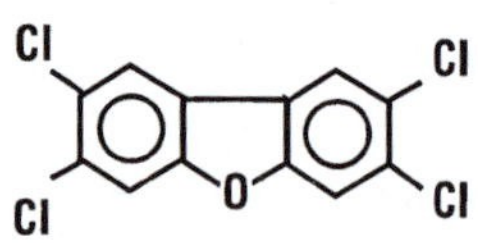

2,3,7,8 - TETRACHLORODIBENZOFURAN

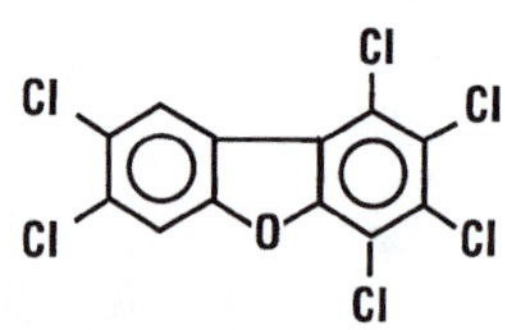

1,2,3,4,7,8 - HEXACHLORODIBENZOFURAN

FIGURE 6.2 Molecular structures and names of some dioxin and PCDF congeners.

6.3 SOURCES OF ENVIRONMENTAL CONTAMINATION

6.3.1 PCBs

The sources of the PCBs found in the environment today are related to their uses both now and in the past, since a substantial proportion of PCBs from past uses is still present in the environment. PCBs have found a wide variety of commercial uses, particularly in electrical equipment such as capacitors and transformers. In these applications, PCBs have been employed as a heat transfer fluid, transferring heat to the outer casing of the equipment, where it can be dissipated to the atmosphere. This reflects the relatively high thermal stability of PCBs. They have also been used because they are an insulating medium or dielectric, and because their nonflammability reduces the risk of explosions and fires in the event of a spark. Frequently, the fluids contained not only PCBs, but also chlorinated benzenes.

Such uses, particularly with large capacitors and transformers, are termed **controllable closed systems**. The quantities of PCBs involved are generally such that there is an incentive for recovery and ultimately disposal of the material. In addition, unless there is an accidental leakage, the PCBs are not in direct contact with the environment. Smaller capacitors, for example, those associated with lighting systems, air conditioners, pumps, and fans, may also contain PCBs. Recovery of relatively small volumes of PCB mixtures from widely dispersed units is difficult. Small PCB-containing capacitors are disposed of, often to landfills, along with the equipment in which they are incorporated, when the equipment reaches the end of its working life. This use is termed a **noncontrollable closed system** — closed in theory, but in practice there is little recovery and proper disposal. Other uses for PCBs that can be classified as noncontrollable closed systems, i.e., not contained in a closed system but in direct contact with the environment, include hydraulic fluids and lubricants.

In 1971, closed-system applications, whether controllable or noncontrollable, accounted for some 90% of PCB usage in the U.S. Open uses, with direct environmental contamination in applications such as plasticizers and fireproofing agents in paints, plastics, adhesives, inks, and copying paper, have largely ceased. Table 6.1 summarizes closed and open usage categories of PCBs. Given these uses, the principal sources of PCBs encountered in the environment include:

TABLE 6.1
Usage Categories of PCBs

Controllable Closed Systems	Noncontrollable Closed Systems	Open Uses
Large capacitors Large transformers	Small capacitors Hydraulic fluids Lubricating fluids	Plasticizer or fireproofing agent in paints, plastics, adhesives, inks, and copying paper

- Open burning or incomplete combustion of PCB-containing solid waste. Refuse incinerators can emit PCBs because of inadequate combustion conditions. Chemical incinerators have more rigorous combustion requirements, and a PCB destruction efficiency of 99.9999% for incinerators burning PCB-containing materials is mandatory in some countries.
- Vaporization of PCBs in open applications.
- Accidental spills or leakages of PCBs in closed-system applications.
- Disposal into sewage systems and subsequent dispersal of sewage sludge.

6.3.2 Dioxins

Sources of dioxins in the environment can be subdivided into two types:

1. Chemical processes involving chlorine
2. Combustion processes

Chemical processes involving chlorine include manufacture of organochlorine compounds, bleaching of pulp in pulp and paper mills, and other industrial processes. The manufacture of some organochlorines may result in the formation of **dioxins** (and **polychlorinated dibenzofurans**) as unwanted by-products. A well-known example of this occurs in the synthesis of the pesticide **2,4,5-trichlorophenoxyacetic acid (2,4,5-T)**. One method of preparing this compound involves the high-temperature alkaline hydrolysis of 1,2,4,5-tetrachlorobenzene to 2,4,5-trichlorophenoxide (a derivative of 2,4,5-trichlorophenol formed under alkaline conditions), followed by reaction of the phenoxide with chloroacetic acid, as shown below:

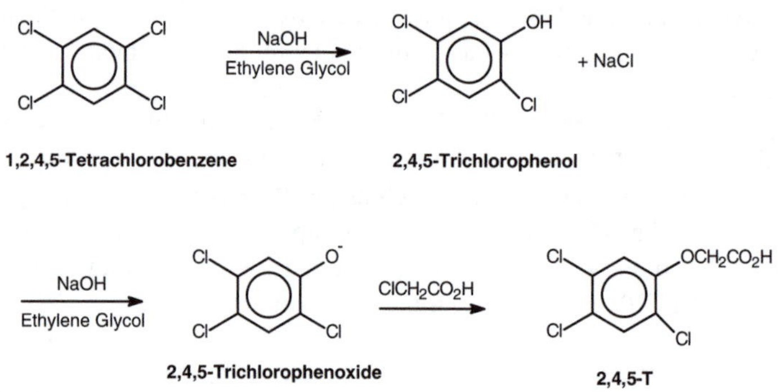

Unfortunately, a side reaction can also occur in which two phenoxide molecules react together to form 2,3,7,8-TCDD:

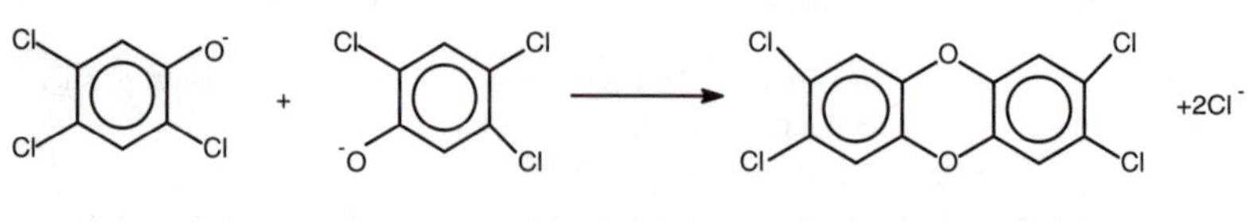

The higher the reaction temperature, the greater the amount of the dioxin formed as a by-product. Related phenoxy acids, used as pesticides, such as **MCPA (2-methyl-4-chlorophenoxyacetic acid)** and **2,4-D (2,4-dichlorophenoxyacetic acid)**, may also potentially contain trace amounts of dioxin congeners.

Chlorophenols themselves are employed as fungicides, mold inhibitors, antiseptics, disinfectants, and in leather processes, and whether prepared by the direct chlorination of phenol or the alkaline hydrolysis of chlorobenzenes, they may contain dioxins as contaminants. In general, ortho-halogenated phenols, compounds that contain this structure, or compounds that involve ortho-halogenated phenols in their preparation are highly likely to be associated with the presence of dioxins. Below is the chemical structure of an ortho-halogenated phenol:

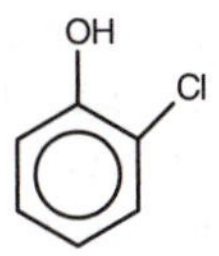

This is particularly true when reaction conditions such as high temperature, alkalinity, or the presence of free halogens are involved. PCBs tend to contain PCDFs as contaminants introduced during manufacture.

In pulp and paper mills, bleaching of pulp with chlorine may result in the formation of dioxins from naturally occurring phenolic compounds such as lignin. As well as being found in the paper products, the effluent from mills employing a chlorine bleaching process may contain dioxins. Some reports indicate that a characteristic congener pattern is observed with the dioxins formed. The pattern tends to be dominated by 2,3,7,8-TCDD, and its chlorinated dibenzofuran analog 2,3,7,8-TCDF. Use of alternative bleaching agents such as chlorine dioxide (ClO_2) and hydrogen peroxide (H_2O_2) or modification of the traditional process leads to substantial reductions in dioxin levels.

A number of industrial processes have also been identified as sources of dioxins. These generally involve high temperatures and the presence (often accidentally) of chlorine-containing compounds. An example is the melting of scrap metals contaminated with polyvinyl chloride (PVC). Another is dioxin contamination of the sludge formed when graphite is used as an anode in the electrolysis of brine to produce chlorine, a method known as the chloralkali process.

Apart from chemical processes involving chlorine, the second major source of dioxins is combustion processes. Combustion sources produce a variety of PCDD and PCDF congener profiles. Even for a single source, the profile will vary with the fuel used and the combustion temperature. Combustion sources include municipal solid-waste incinerators, coal-burning power plants, fires involving organochlorine compounds (including PCBs), or even burning of vegetation treated with phenoxyacetic acid herbicides. Vehicle exhaust emissions are another source. This is particularly true when vehicles run on leaded fuel, presumably because of the organochlorine fuel additives such as 1,2-dichloroethane.

The presence of dioxins and related compounds in combustion emissions is thought to arise in three ways:

1. They are already present in the combusted material and are not destroyed by the combustion process.
2. They may be formed from organochlorine precursors (e.g., chlorophenols, PCBs) present in the combusted material, during the combustion process.
3. They are formed from high-temperature reactions between nonchlorinated organic molecules and chloride ions.

The last process suggests that dioxins could have been present in the environment for a relatively long period, certainly since anthropogenic combustion of coal and wood commenced. It is conceivable that burning of wood contaminated with chloride ion could constitute a natural source of trace levels of dioxins. Analysis of archived soil samples from as long ago as the 1840s, as shown in Figure 6.3, tends to confirm this. Tissue analysis of 2800-year-old Chilean mummies or of Eskimos trapped in ice for more than 400 years might be expected to show an influence from this natural source, or from burning of wood or oil in cooking and heating fires. Analysis of the Eskimo tissue has in fact revealed trace (pg/g or parts per 10^{12}) levels of hepta- and octachlorodibenzo [1,4] dioxin. It is generally believed, however, that the vast majority of the PCDDs and PCDFs found in the environment today have an anthropogenic source. In the U.S., studies of sediment cores show relatively low concentrations up

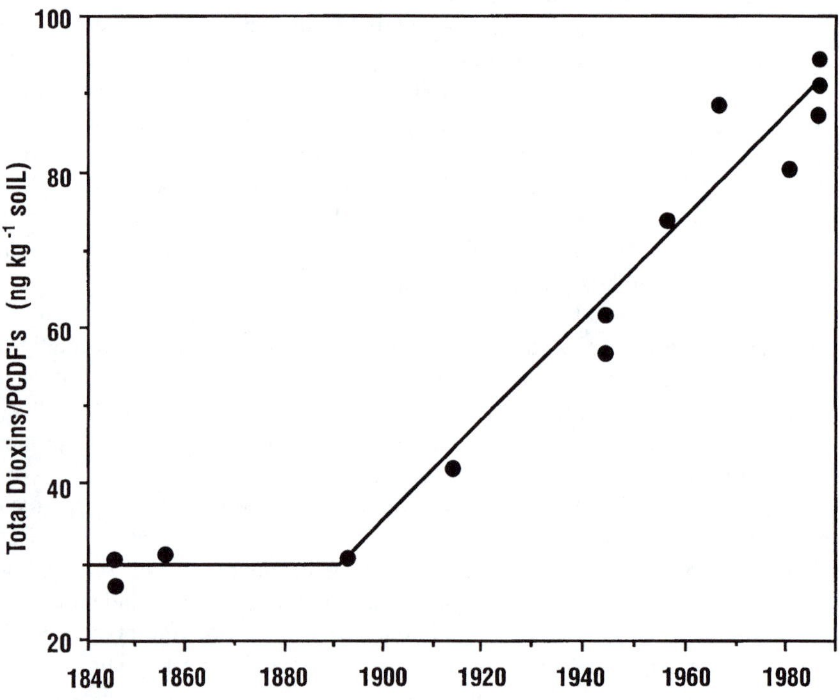

FIGURE 6.3 Trends in total dioxin and PCDF concentration in archived soil samples with time. (From Kjeller, L. et al., *Environ. Sci. Technol.*, 25, 1619, 1991. Copyright © American Chemical Society. With permission.)

until the 1940s. After this time, an increasing concentration, possibly due to combustion of material containing chlorinated aromatic compounds, is observed.

6.4 PHYSICAL-CHEMICAL PROPERTIES OF PCBS AND DIOXINS

Individually, most PCB congeners are colorless, crystalline solids at room temperature and atmospheric pressure, but some of the less chlorinated congeners are liquids under these conditions. PCBs have a number of physical and chemical characteristics that have contributed to their widespread use:

Low aqueous solubility
Nonflammability
Resistance to oxidation
Resistance to hydrolysis
Low electrical conductivity

As the degree of chlorination increases, aqueous solubility, flammability, and reactivity tend to decrease. This is shown in Table 6.2, in which some properties of biphenyl, a pentachlorobiphenyl, and decachlorobiphenyl are compared. The lack of flammability is important in terms of disposal by chemical incineration. PCBs need to be mixed with a large excess of combustible material (e.g., hydrocarbons) in order to generate the high temperatures necessary for decomposition. While PCBs are much more soluble in lipid material than water, solubility in lipid also tends to decrease with increased chlorine substitution, though not to the same degree as that observed for aqueous solubility. Despite this, K_{OW} values increase with the number of chlorine substituents, as seen in Table 6.2. Among PCBs with equal numbers of chlorines, those with substituents in one or more of the 2,2′,6 and 6′ positions possess relatively low K_{OW} values and high aqueous solubility.

Unlike the individual PCB congeners, commercial PCB mixtures have different physical appearances. Aroclors 1221, 1232, 1242, and 1248 are colorless or almost

TABLE 6.2
Some Physical-Chemical Properties of Biphenyl, 2,2′4,5,5′-Pentachlorobiphenyl, and Decachlorobiphenyl

	Biphenyl	2,2′4,5,5′-Pentachlorobiphenyl	Decachlorobiphenyl
Molecular weight	154.2	326.4	498.7
log K_{OW}	3.76	6.38	8.20
Aqueous solubility (mol m^{-1})	4.4×10^{-2}	4.7×10^{-5}	1.3×10^{-9}
Vapor pressure at 25°C (Pa)	9.4×10^{-1}	1.1×10^{-3}	6.9×10^{-9}
Melting point (K)	344	349.5	578.9

colorless mobile oils, due to mutual depression of melting points of their components. The more highly chlorinated Aroclor 1254 is a light yellow viscous oil, while Aroclor 1260 and 1262 are sticky resins. The density of these mixtures also increases with the extent of chlorine substitution but is always greater than 1 g cm^{-3}. PCB mixtures can usually be distinguished from lower-density mineral oils by the fact that PCBs settle to the bottom of water and do not float as pure hydrocarbon-based mineral oils do.

Combustion of PCB mixtures at relatively low temperatures (<700°C), in the presence of oxygen, can lead to the formation of PCDFs. For example,

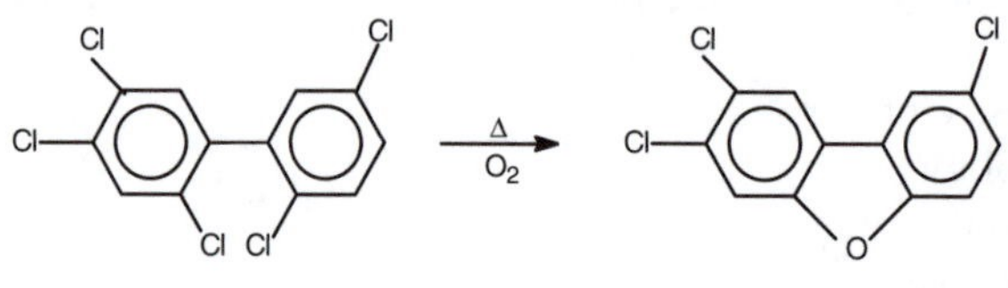

Thermal events such as fires and explosions with PCB-containing oils, for example, can result in the formation of PCDFs in any remaining oil, in combustion effluent, and in soot that is formed.

The physical-chemical properties of the 75 chlorinated dioxins are generally similar to those of the PCBs. Individually, most are colorless solids of decreasing volatility and aqueous solubility, with increasing molecular size. Using log K_{OW} values as a measure of hydrophobicity, they vary from 4.20 for the parent nonchlorinated compound dibenzo [1,4] dioxin itself up to 8.20 for octachlorodibenzo [1,4] dioxin. Characteristics of the most toxic dioxin congener, 2,3,7,8-TCDD, are shown in Table 6.3. It is interesting to note that the log K_{OW} value of 6.80 is similar to that of DDT at 6.20, which suggests maximum biological activity around this value for chlorhydrocarbons and closely related compounds.

TABLE 6.3
Some Physical-Chemical Properties of 2,3,7,8-TCDD

Molecular weight	322
log K_{OW}	6.80
Aqueous solubility (mol m^{-1})	6×10^{-8}
Solubility in methanol (mol m^{-1})	3×10^{-2}
Solubility in benzene (mol m^{-1})	1.8
Vapor pressure at 25°C (Pa)	2×10^{-7}
Henry's law constant (Pa m^3 mol^{-1})	3.34
Melting point (K)	578
Decomposition temperature (K)	>973

6.5 ENVIRONMENTAL DISTRIBUTION AND BEHAVIOR

PCBs were first recognized as widespread or ubiquitous environmental contaminants in 1966. One estimate is that about 400,000 tons of PCBs are presently in the environment, which represents about 30% of the total world production. Of this total burden, approximately 60% is to be found in the hydrosphere, principally the oceans, about 1% in the atmospheric, and the remainder in the terrestrial environment. Despite relatively low vapor pressures, atmospheric transport and dispersal are important and account for the occurrence of PCBs in areas as remote as the polar regions. Individual PCB congeners are colorless materials, so they do not absorb visible (VIS) radiation, though they do absorb slightly at the low-wavelength end of the ultraviolet (UV) solar radiation received on the Earth's surface. A UV/VIS absorption spectrum of biphenyl itself is shown in Figure 6.4. This shows a main absorption band near 200 nm and a weaker absorption around 240 nm. The wavelength for UV absorption varies according to the number and location of the chlorine substituents. Phototransformation is not a major fate for vapor phase PCBs, except perhaps for highly chlorinated congeners.

A more important fate of vapor phase PCBs is reaction with hydroxyl radicals (OH·). The principles of the kinetics of environmental transformation and degradation processes are described in Chapter 3. Since the rate of reaction is proportional to the product of the reactant concentrations, then with the reaction of PCBs with OH·, the rate of reaction may be described by:

$$\text{Rate} = -\frac{d[\text{PCB}]}{dt} = k[\text{PCB}][\text{OH}\bullet] \tag{6.1}$$

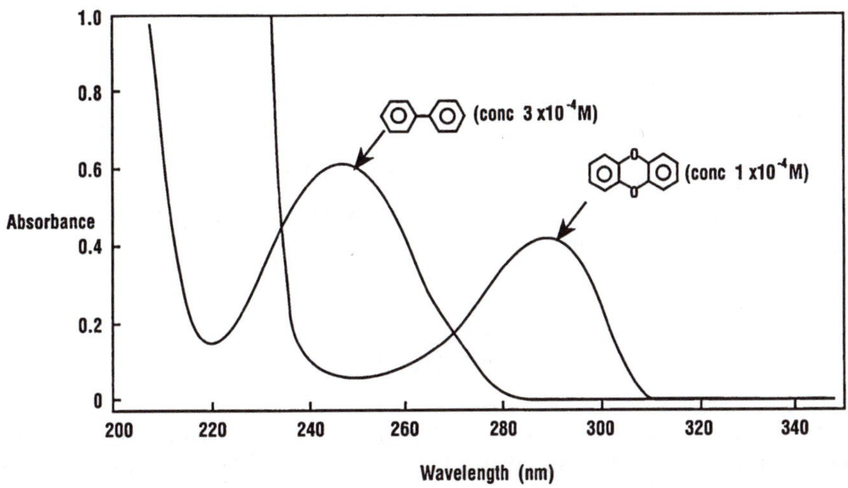

FIGURE 6.4 Ultraviolet absorption spectra of methanol solutions of biphenyl and dibenzo [1,4] dioxin.

where k is a second-order rate constant with units of concentration time (e.g., cm^3 molecule s^{-1}). The concentration of OH· in the atmosphere can be considered to be effectively constant, and so the kinetic situation can be treated as pseudo-first order, where the rate is essentially dependent only on vapor phase PCB concentrations. Equation 6.1 becomes:

$$\text{Rate} = -\frac{d[PCB]}{dt} = k'[PCB] \tag{6.2}$$

where k′ is a pseudo-first-order rate constant with units of $time^{-1}$. Integrating Equation 6.2 and rearranging, we have:

$$\ln\left(\frac{[PCB]}{[PCB]_0}\right) = -k't$$

$$\text{or} \tag{6.3}$$

$$\left(\frac{[PCB]}{[PCB]_0}\right) = e^{-k't}$$

where $[PCB]_0$ is the initial PCB concentration and [PCB] is the concentration at some later time, t. A measure of the persistence of a compound is the lifetime (T_L). A lifetime is defined as the time required for the concentration to fall to 1/e or approximately 37% of its initial value. After one lifetime, when t is T_L, then [PCB] = $[PCB]_0/e$, and by substitution into Equation 6.3,

$$\left(\frac{[PCB]}{[PCB]_0}\right) = \frac{1}{e} = e^{-1} = e^{-k'T_L}$$

Therefore, the lifetime (T_L) = 1/k′.

Although [OH·] was considered to be constant and thus incorporated into k', the magnitude of k' is k' = k[OH·] and

$$\frac{1}{k'} = \frac{1}{k[OH\bullet]}$$

Using this approach, and a OH· concentration of 5×10^5 molecules cm^{-1}, it was found that the more highly chlorinated PCB congeners are less reactive toward the hydroxyl radical OH·. Calculated tropospheric lifetimes range from 5 to 11 days, depending on the particular isomer, for dichlorobiphenyls, and up to 60 to 120 days for pentachlorobiphenyls.

Dioxins are a ubiquitous, widespread family of compounds, generally present in the environment at trace levels. They appear to be more common in industrialized areas of the world. Dioxins associated with particles from incinerator emissions and other combustion processes may reside in the atmosphere for considerable periods, which may partly account for their widespread occurrence.

Compared with PCBs, a larger fraction of dioxins found in the troposphere is associated with particles, with the fraction increasing with the extent of chlorine substitution. Octachlorodibenzo [1,4] dioxin is found in the troposphere almost entirely sorbed to particles.

For dioxins (and also PCDFs), phototransformation of vapor phase material is a much more important fate than for PCBs. The UV/VIS spectrum of the parent or unsubstituted dioxin, dibenzo [1,4] dioxin, is shown in Figure 6.4. In comparison with biphenyl, there is much more absorbance at longer wavelengths. As the number of chlorine substituents increases, the absorbances tend to shift to longer wavelengths. This means increasing overlap with the solar emission spectrum and increasing absorption of solar energy, leading to greater reactivity of the molecule.

Products of phototransformation of PCBs in the troposphere are often hydroxylated biphenyls. These probably arise by an initial homolytic cleavage of a C–Cl bond to form a radical intermediate, followed by reaction with oxygen. For example,

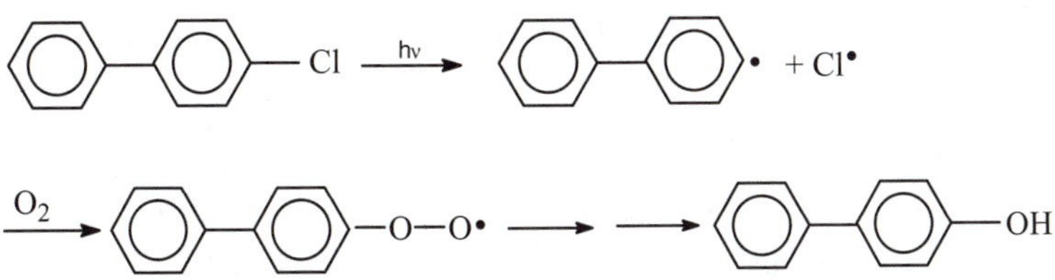

While an analagous process is possible with dioxins, a more important mechanism is probably C–O bond cleavage. As for reaction of OH· with dioxins, an estimate of the tropospheric lifetime for the parent compound is approximately 7 hours, while for 2,3,7,8-TCDD an estimated lifetime is about 3 days. The reaction of OH· with molecules such as dioxins is complex. The first step is usually addition of OH· to an aromatic ring, as shown:

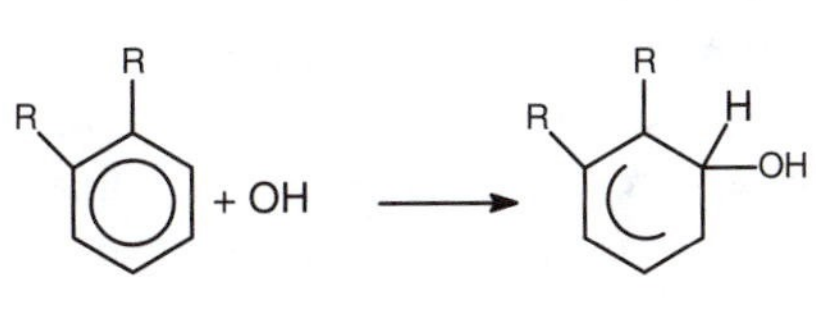

Products may include hydroxyl derivatives and molecules in which an aromatic ring has been cleaved open.

Both PCBs and dioxins become less soluble in water and have increased K_{OW} values as the number of chlorine substituents increases, as shown in Table 6.2. In water bodies, these molecules may sorb to bottom sediments or suspended particules, volatilize from water, and, in the upper regions where light penetrates, undergo phototransformation. The half-life for phototransformation of 2,3,7,8-TCDD in surface water in summer at 40°N latitude has been calculated to be about 6 days. The C–Cl bond found in PCBs and dioxins tends to be resistant to hydrolysis. On account of the relatively low aqueous solubility and high K_{OW} values, most of these substances will be found sorbed to sediment near discharges. Typically, sediment samples show octachlorodibenzo [1,4] dioxin to be the most prevalent dioxin congener.

The transport of PCBs and dioxins within rivers is by desorption into solution and subsequent resorption downstream, as well as by movement of sediment itself. If sufficient quantities of PCBs are discharged, pools or pockets of relatively pure material may be found at the sediment–water interface. As an example, in 1974 an electrical transformer accidentally fell into the Duwamish River near Seattle, WA, during loading operations, resulting in a spill of some 950 l of Aroclor 1242 (density of 1.38 to 1.39 g cm^3). Divers observed pools of this material on the bottom being moved about by current and tidal activity. Dredges, including handheld suction dredges, ultimately recovered about 90% of the spilled PCB mixture.

In the terrestrial environment, PCBs and dioxins tend to sorb strongly to the organic matter fraction of soils. Greatest sorption is observed with the larger, more highly chlorinated congeners. Mobility of these compounds within soils depends on soil characteristics such as bulk density and moisture content. It is also inversely related to the hydrophobicity of a compound, as measured by log K_{OW}. This means the larger congeners also tend to be the least mobile.

Within the soil, microbial biotransformation may occur. Bacteria degrade PCBs and probably dioxins by the action of dioxygenases. Dioxygenases are enzymes that catalyze the incorporation of two oxygen atoms into a substrate molecule. In order for this to occur, there should be at least two adjacent non-chlorine-substituted carbon atoms on an aromatic ring. The possession of two adjacent unsubstituted carbon atoms is less likely with the more highly chlorinated congeners, which are usually observed to be the most resistant toward biotransformation. A possible mechanism for the microbial biotransformation of 2,4-dichlorobiphenyl involving formation of a dihydroxy derivative and then ring cleavage is shown in Figure 6.5.

Compounds such as PCBs may also be transformed by bacteria under anaerobic conditions by a process of reductive dehalogenation. This essentially means the replacement of a chlorine substituent by a hydrogen from water. In this way, hexachlorobiphenyls are converted into pentachlorobiphenyls and so on. This process may occur in anaerobic river sediments and also during the anaerobic digestion of sewage sludge. The products are less chlorinated PCBs that are more susceptible to subsequent aerobic biotransformation than the original more highly chlorinated compounds.

Higher organisms such as mammals are also able to biotransform some of these compounds. Such organisms possess a monooxygenase system, also known as the mixed-function oxidase (MFO) system. The incorporation of one atom of molecular oxygen into the substrate PCB or dioxin results in the formation of a three-membered oxygen-containing ring, which then undergoes further reaction. For example,

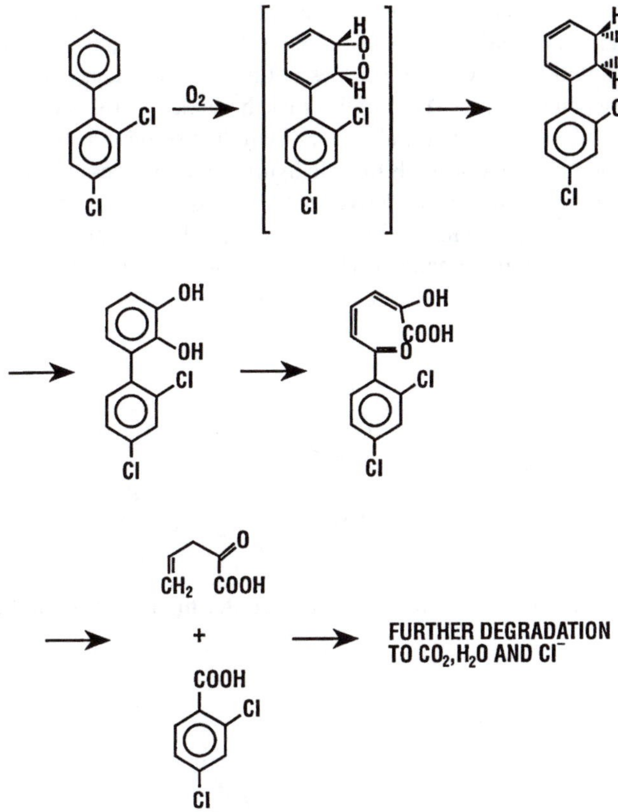

FIGURE 6.5 Possible microbial biotransformation pathway for 2,4-dichlorobiphenyl.

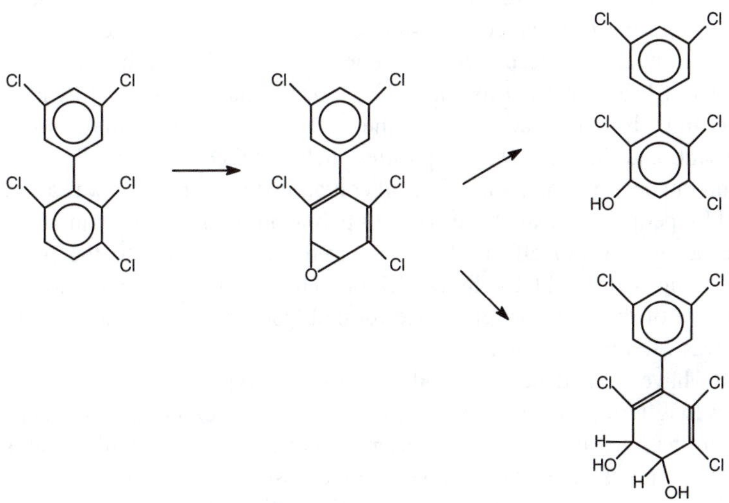

The unsubstituted parent dioxin, dibenzo [1,4] dioxin, may undergo biotransformation by an analagous process, but 2,3,7,8-TCDD is relatively resistant. It is interesting to note that this compound does not possess two adjacent carbon atoms without chlorine substituents. This implies that the toxicity of this congener is due to the unmetabolized compound, rather than a product of biotransformation.

Overall, biotransformation tends to yield hydroxyl derivatives that are more polar than the original substrate. These derivatives are readily conjugated (joined) with compounds such as glucuronic acid to form water-soluble molecules that can be readily eliminated from the organism. These processes are described in more detail in Chapter 3. The structure of glucuronic acid is

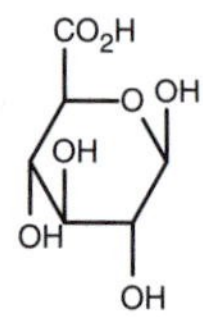

Glucuronic acid

The many polar groups on glucuronic acid confer high polarity on the conjugation product.

6.6 TOXICITY

6.6.1 PCBs

It is generally agreed that the acute toxicity of PCBs is relatively low. Acute (i.e., single-dose) oral LD_{50} (lethal dose for 50%) values for a series of Aroclor mixtures with rats are reportedly of the order of 10 g kg^{-1} of body weight. Put in context, this means that these mixtures are less toxic than DDT and other chlorohydrocarbon pesticides, and the LD_{50} values are similar to that of aspirin. One report suggests that the minimum dose necessary to produce clinical effects in humans is approximately 0.5 g. It is important to remember, though, that commercial PCB mixtures may vary in their toxicity according to the relative proportion of the various congeners present, and the amount of impurities such as PCDFs present. Events such as the Yusho incident of 1968, when rice oil contaminated with PCBs was accidentally ingested by people in Japan, have come to public attention. More than 1500 people suffered adverse health effects, but most were probably due to the presence of PCDFs, along with the PCBs, in the rice oil. The PCDFs could have arisen during manufacture of the PCBs, or else were formed from the PCBs when contaminated rice oil was used for cooking.

PCBs have a tendency to produce chronic toxic effects. Chronic toxicity describes long-term, low-level exposure effects. Chronic exposure to PCBs produces damage to the liver as well as a condition known as chloracne. Chloracne is a skin condition that has long been known to be associated with chronic exposure to chlorinated aromatic compounds, including PCBs. It is characterized by the appearance of comedones (blackheads) and cysts, largely on the face and neck, but in severe cases, also on the chest, back, and genitalia. As well as actual dermal (skin) contact, chloracne can also result from oral ingestion of these compounds. Its duration is dependent upon the degree of exposure, but is often prolonged, lasting several years. Other symptoms of chronic exposure to PCBs include nausea, vomiting, jaundice, and fatigue. Biochemical changes involved include vitamin A depletion, alteration to lipid metabolism, and hormonal changes. Available evidence suggests that if PCBs themselves are, in fact, carcinogenic, activity is weak.

Individual PCB congeners demonstrate a range of toxicities. The most toxic have chlorine substituents in some or all of the 3,3′,4,4′,5,5′ positions, as shown below:

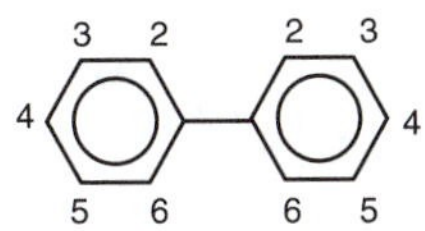

These congeners are also known as non-ortho or coplanar PCBs, as shown in Figure 6.6. Non-ortho means that the molecules lack chlorine substituents on the carbon atoms adjacent to the carbon–carbon bond joining the two phenyl rings that is the ortho position. Altogether, there are four ortho positions (2,2′,6,6′) in a PCB molecule. The second carbon atom from the carbon joining the two rings is described as being in the meta position, while the third carbon is described as being in the para position.

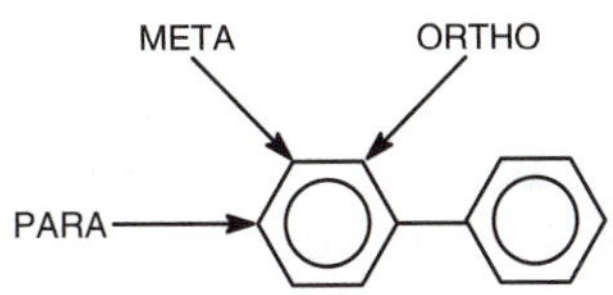

Non-ortho-chlorine-substituted PCB congeners are referred to as coplanar because the phenyl rings, each of which is essentially flat, can rotate separately about the carbon–carbon bond joining the two rings. In doing so, the whole molecule may achieve coplanarity, where the phenyl rings are in the same plane, though this is not the most stable configuration. Chlorine atoms are much larger than hydrogen atoms. Ortho-substituted PCB congeners have one or more chlorine substituents in the 2,2′,6,6′ positions. Separate complete rotation of the phenyl rings is hindered because of a steric interaction (crowding) between an ortho-chlorine substituent on one ring and an ortho-substituent (hydrogen and particularly chlorine) on the other, as shown in Figure 6.6. Coplanarity is difficult, if not impossible, to achieve for these molecules.

The non-ortho or coplanar congeners may be the most toxic because PCBs can only interact with a receptor or active site in cells when in a coplanar or flat conformation. The dioxin 2,3,7,8-TCDD is some 200 times more toxic (single oral dose) to the guinea pig than the non-ortho 3,3′,4,4′,5,5′-hexachlorobiphenyl, which

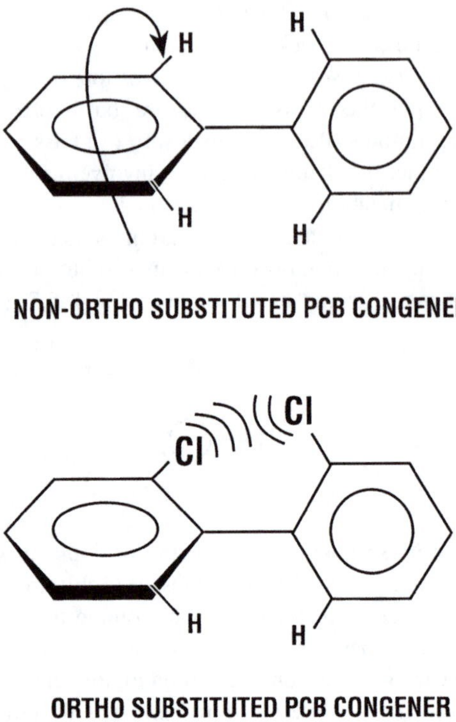

FIGURE 6.6 Separate rotation of phenyl rings in a non-ortho-substituted PCB congener, but hindered rotation in an ortho-substituted PCB congener.

is one of the more toxic PCBs. It is interesting to note from Figure 6.7, however, the close correspondence in molecular structure between the planar conformation of this PCB and 2,3,7,8-TCDD. The same receptor may be involved with both molecules, but the dioxin binds more strongly, or fits the active site better. Ortho-substituted PCBs may be less toxic because they cannot achieve coplanarity and interact with the receptor, or because they exert their toxicity by a different mechanism. These compounds are the major components of commercial mixtures. Non-ortho congeners are generally present in only small amounts.

6.6.2 Dioxins

Like the PCBs, the individual dioxin or PCDD congeners possess a range of toxicities, with some more toxic than others. The most toxic congener is 2,3,7,8-TCDD. (Among the PCDFs, 2,3,4,7,8-pentachlorodibenzofuran is the most toxic.) The compound 2,3,7,8-TCDD is among the most toxic compounds made by humans, although this is usually inadvertent. However, natural toxins produced by bacteria associated with tetanus and botulism (a kind of food poisoning) are much more toxic on the basis of minimum lethal dose (mol kg^{-1} of test organism).

The acute toxicity of 2,3,7,8-TCDD shows considerable variation between species. Guinea pigs are among the most sensitive, while hamsters are comparatively

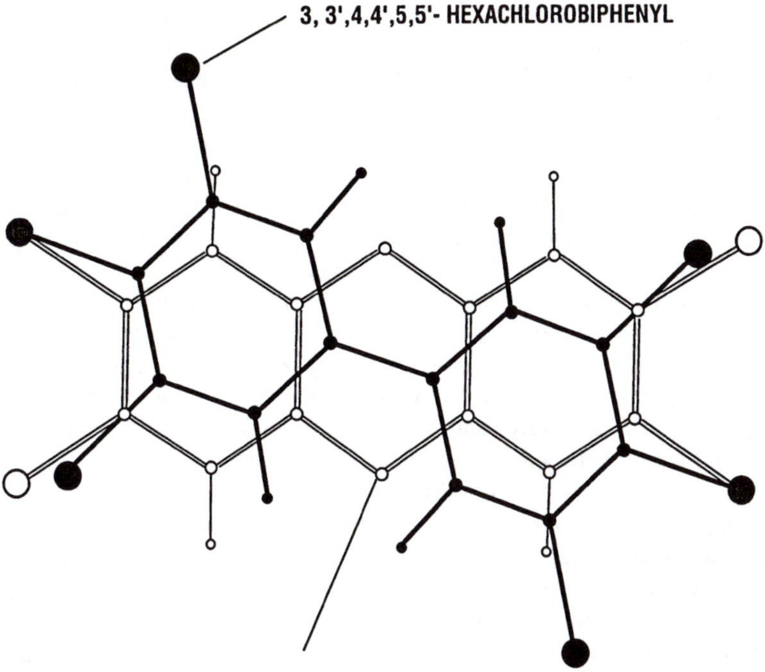

FIGURE 6.7 Correspondence in molecular dimensions as shown by an overlay of the planar conformation of 3,3′,4,4′,5,5′-hexachlorobiphenyl (l) with 2,3,7,8-TCDD (O). (From McKinney, J.D. and Singh, P., *Chem. Biol. Interact.*, 33, 277, 1981. Copyright © Elsevier. With permission.)

resistant, and mice, rats, and rabbits show intermediate susceptibility. Available evidence indicates that humans are relatively insensitive. However, the toxicity of this dioxin was demonstrated by laboratory workers involved in the first reported syntheses of both 2,3,7,8-TCDD and its brominated analog 2,3,7,8-tetrabromodibenzo [1,4] dioxin being hospitalized as a result of accidental exposure.

Another feature of 2,3,7,8-TCDD is its delayed lethality. Following administration of a single lethal dose, test animals lose weight and death does not occur for a number of days or even weeks. In some test species, the main target organ is the liver.

It is thought that 2,3,7,8-TCDD exerts its toxic effects by binding to an intracellular protein molecule that is soluble and not bound to the cell membrane. This protein is known as the Ah receptor. This name arises because, as we shall see, this receptor causes the organism to become extremely responsive toward the presence of aryl (or aromatic) hydrocarbons. It seems that 2,3,7,8-TCDD binds extremely strongly to this receptor molecule. This particular dioxin congener optimally fits a binding site on the surface of the protein. Other dioxins, PCDFs, and PCBs bind less strongly, fit less well, and are observed to be less toxic. A dioxin receptor is found in many species, including humans, most mammals, and even some fish.

Although there is some structural variation of this receptor between different species, this fact alone does not account for the observed varying toxicity.

Following binding of 2,3,7,8-TCDD to the Ah receptor, a poorly understood transformation process takes place during which the structure of the receptor is modified. The dioxin–Ah receptor complex then moves into the nucleus of a cell and binds to DNA at specific sites with high affinity. These sites are just upstream of a cytochrome P_{450} gene. Binding to the DNA induces the transcription of specific messenger RNA molecules, which direct the synthesis of new cytochrome P_{450}. As mentioned in Chapter 3, cytochrome P_{450} is a family of closely related enzymes (or isozymes) that catalyze the biotransformation of many lipophilic substances, including hormones, fatty acids, some drugs, and hydrocarbons such as polycyclic aromatic hydrocarbons (PAHs). However, the action of cytochrome P_{450} on PAHs may produce reactive intermediates that are carcinogenic or exhibit toxicity by interacting with critical target molecules in various organs. This relatively complex mechanism of action of 2,3,7,8-TCDD, summarized in Figure 6.8, may in part explain the delayed lethality observed in test organisms. Despite the ability of 2,3,7,8-TCDD to trigger the biotransformation of many other compounds, it is itself relatively resistant to biotransformation.

Recent evidence suggests that binding of the dioxin–receptor complex to DNA may bend or distort the DNA helix, making it more accessible to other proteins associated with induction of other enzymes. It is known that 2,3,7,8-TCDD induces the formation of other enzymes apart from cytochrome P_{450}, although it can also suppress the formation of some enzymes.

Currently, there is considerable debate as to whether exposure of humans to dioxins such as 2,3,7,8-TCDD results in cancer or birth defects. Epidemiological studies finding increased occurrences have been vigorously contested. It is often difficult, for example, to rule out the possibility that observed effects are due to coincidental exposure to other chemicals. There is little doubt, however, that chronic exposure of humans can result in effects such as some liver damage, nerve damage, and chloracne. In addition, since dioxins are hydrophobic compounds, they may accumulate in adipose or fatty tissue of humans and in breast milk of nursing mothers.

Octachlorodibenzo [1,4] dioxin (structure below) is usually the congener found in greatest concentration in breast milk. Typically, more than 90% of the dioxin burden accumulated by an adult human is from food.

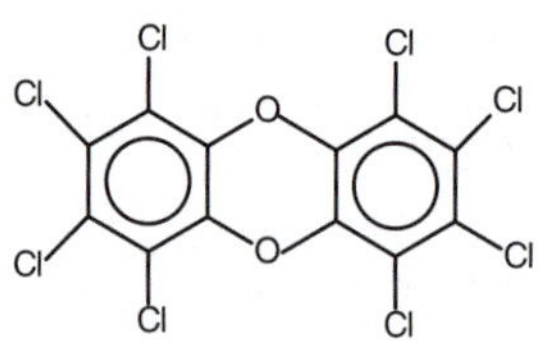

Octachlorodibenzo[1,4] dioxin

Dioxins are invariably encountered in the environment as mixtures of congeners, and the hazard potential of these mixtures is difficult to assess. It is also difficult to compare the hazard potentials of two different mixtures. One approach to this problem is the concept of International Toxic Equivalence Factors (I-TEFs), in which

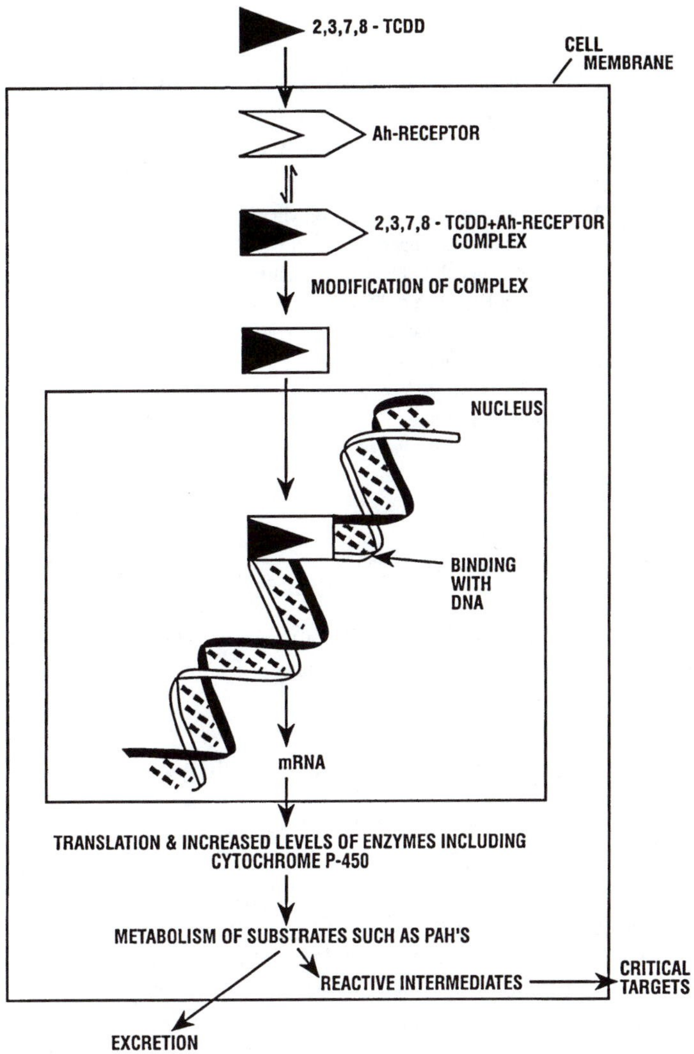

FIGURE 6.8 Possible mechanism of toxic action of 2,3,7,8-TCDD.

a toxic measure of a given congener (e.g., EC_{50} with a test organism) is compared with the same toxic measure for 2,3,7,8-TCDD:

$$\text{I} - \text{TEF} = \frac{EC_{50}(2,3,7,8 - \text{TCDD})}{EC_{50}(\text{congener of interest})}$$

By definition, therefore, the I-TEF value for 2,3,7,8-TCDD is unity. The other congeners are less toxic, and accordingly, their I-TEF values are less than unity. The PCDF family and PCBs can also be considered in this way. Table 6.4 contains I-TEF

TABLE 6.4
I-TEF Values for Some Dioxin, PCDF, and PCB Congeners

Congener	I-TEF Value
2,3,7,8-Tetrachlorodibenzo [1,4] dioxin	1
2,3,4,7,8-Pentachlorodibenzo [1,4] dioxin	0.5
1,2,3,4,7,8-Hexachlorodibenzo [1,4] dioxin	0.1
1,2,3,4,6,7,8-Heptachlorodibenzo [1,4] dioxin	0.01
Octachlorodibenzo [1,4] dioxin	0.001
2,3,7,8-Tetrachlorodibenzofuran	0.1
2,3,4,7,8-Pentachlorodibenzofuran	0.5
Octachlorodibenzofuran	0.001
3,3′,4,4′-Tetrachlorobiphenyl	0.01
2,3,3′,4,4′-Pentachlorobiphenyl	0.001
3,3′,4,4′,5-Pentachlorobiphenyl	0.1
3,3′,4,4′,5,5′-Hexachlorobiphenyl	0.05
2,3,3′,4,4′,5-Hexachlorobiphenyl	0.001

values for some dioxin, PCDF, and PCB congeners. The I-TEF values are then multiplied by the concentration of the particular compound to give a toxic equivalent (TEQ) of 2,3,7,8-TCDD:

$$\text{TEQ} = \text{I-TEF} \times \text{Concentration}$$

For example, the I-TEF value of 1,2,3,4,7,8-hexachlorodibenzo [1,4] dioxin is 0.1. This means it is estimated to be 1/10 as toxic as 2,3,7,8-TCDD. Suppose it was present in adipose tissue at a concentration of 10 pg g^{-1}; this concentration is equivalent to 1 pg g^{-1} of 2,3,7,8-TCDD. This procedure can then be repeated for other mixture components, until the whole mixture has been converted into an equivalent concentration of 2,3,7,8-TCDD in terms of toxicity.

When converting observed concentrations into toxic equivalents, it should be remembered that a number of assumptions are involved. These include a common toxic mechanism of action for all mixture components, as well as additive interactions between components. Toxic equivalent data should be regarded as approximate only because assumptions involved in calculations may not always be valid.

6.7 KEY POINTS

1. The PCBs and dioxins have similar chemical structures, physical-chemical properties, and environmental behavior.
2. The PCBs are industrial chemicals used in a wide range of applications from electrical equipment to lubricants. Current uses are restricted, but

past usage has caused contamination at the present time due to the persistence of the compounds.

3. The dioxins are not deliberately produced as environmental contaminants, but are inadvertently produced by combustion and in some syntheses of commercial chemicals.
4. The physical-chemical properties of the PCBs include low aqueous solubility [4.4×10^{-2} (biphenyl) to 1.3×10^{-9} (decachlorobiphenyl) mol m^{-3}] and declining solubility with increasing molecular weight and high log K_{OW} values [3.76 (biphenyl) to 8.20 (decachlorobiphenyl)].
5. The most important dioxin from an environmental perspective is 2,3,7,8-tetrachlorodibenzo [1,4] dioxin (2,3,7,8-TCDD or dioxin). It has low aqueous solubility (6×10^{-8} mol m^{-3}) and a high log K_{OW} value (6.80).
6. An important process for the transformation and degradation of PCBs is reaction with the hydroxyl radical in the atmosphere. With increasing chlorination, the PCBs become increasingly resistant to this process. The tropospheric lifetimes of these compounds range up to many days.
7. The dioxins have a UV/VIS absorption spectrum that results in significant absorption from solar radiation compared to the PCBs. As a result, shorter lifetimes in the atmosphere occur. For example, 2,3,7,8-TCDD has a tropospheric lifetime of about 3 days.
8. The acute toxicity of the PCBs is relatively low, with LD_{50} values about 10 g kg^{-1} of body weight of rats. However, a range of long-term chronic effects occur.
9. The dioxin 2,3,7,8-TCDD is a highly toxic compound, with the other congeners exhibiting lower toxicities.
10. The dioxins always occur as mixtures and toxicity can be estimated in 2,3,7,8-TCDD equivalents described as International Toxic Equivalence Factors (I-TEFs).

REFERENCES

Connell, D.W. and Miller, G.J., *Chemistry and Ecotoxicology of Pollution*, John Wiley & Sons, New York, 1984.

Esposito, N.P., Teirnan, T.O., and Dryden, F.E., *Dioxins*, Industrial Environmental Research Laboratory, Office of Research and Development, U.S. Environmental Protection Agency, Cincinnati, OH, 1980.

Niimi, A.J., PCBs, PCDDs and PCDFs, in *Handbook of Ecotoxicology*, Vol. 2, Carlow, P., Ed., Blackwell Scientific, Oxford, 1994.

QUESTIONS

1. Using the rules outlined in the text, devise names for the following compounds:

A.

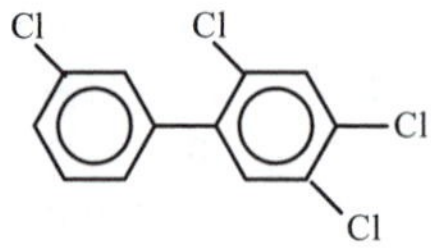

B.

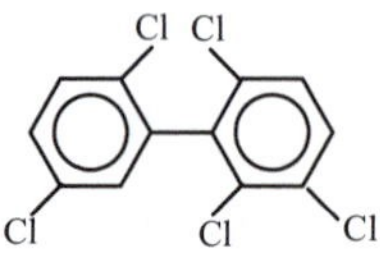

C.

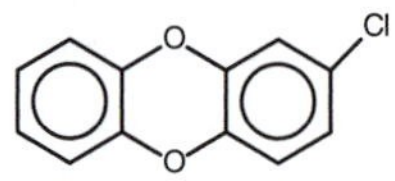

2. Using the data in Table 6.2, estimate the approximate log K_{OW} value of a hexachlorobiphenyl and a trichlorobiphenyl.
3. Which dioxin congener could potentially be formed if 2,4-D (2,4-dichlorophenoxyacetic acid) is prepared by alkaline hydrolysis of 1,2,4-trichlorobenzene, followed by reaction with chloroacetic acid?
4. How many half-lives does a lifetime correspond to?
5. You are evaluating the toxicity of two PCB isomers with the following structures:

A.

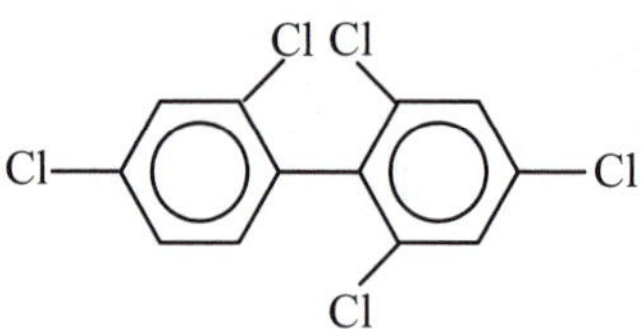

B.

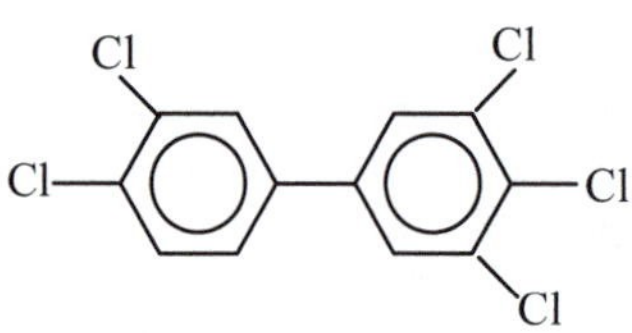

Which of these would you expect to be the most toxic and why would you expect this?

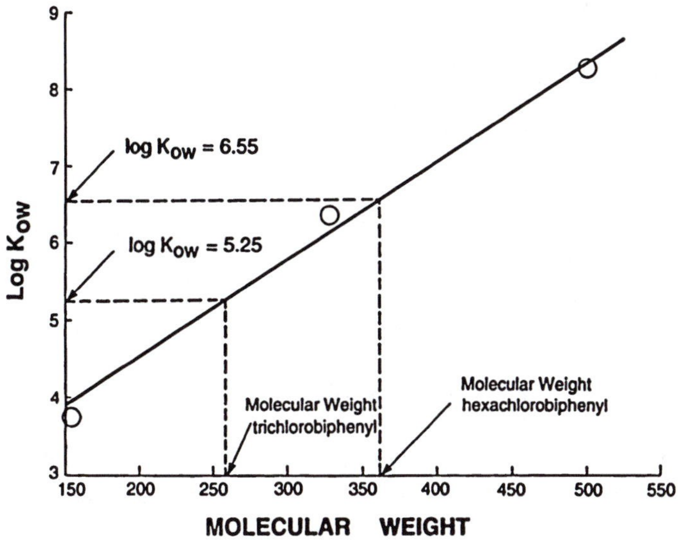

FIGURE 6.9 Plot of molecular weight against log K_{OW} for the PCBs in Table 6.2 with interpolations to estimate the log K_{OW} values of a hexa- and trichlorobiphenyl (answer to question 2).

ANSWERS

1. a. 2,3′,4,5-tetrachlorobiphenyl
   b. 2,2′,5,5′,6-pentachlorobiphenyl
   c. 2-chlorodibenzo [1,4] dioxin
2. The data can be plotted and estimations made as shown in Figure 6.9, giving:
   - Hexachlorobiphenyl: log K_{OW} = 6.55 (experimental values for isomers 6.2 to 7.2)
   - Trichlorobiphenyl: log K_{OW} = 5.25 (experimental values for isomers 5.16 to 5.58)
3. The following reaction sequence for the formation of a dioxin congener during synthesis of 2,4-D can be formulated as:

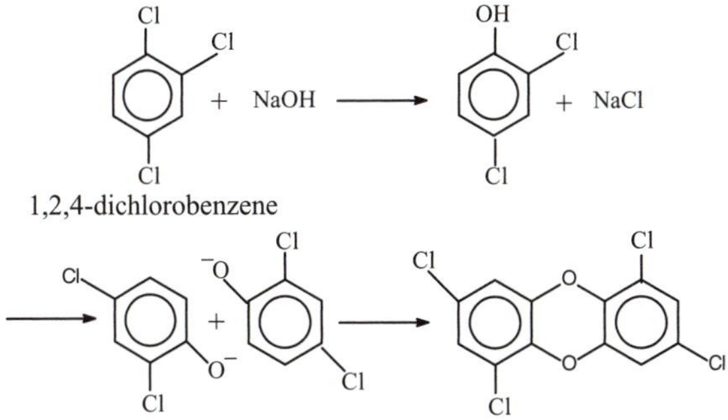

So, 1,3,6,8-tetrachlorobibenzo [1,4] dioxin could be formed in this process.

4. Half-life ($t_{1/2}$) = 0.693/k′ (see Chapter 3)
   Lifetime (TL) = 1/k′
   Number of half-lives = TL/$t_{1/2}$ = 1.44
5. Compound A would be expected to be the most toxic for the following reasons:
   - The molecular weights are the same and this factor would not be expected to influence the comparative toxicity.
   - Compounds A and B both have the same number of chlorine atoms, and so this would not influence the comparative toxicity.
   - Compound A has three chlorines in the 2,2′,6,6′ positions; compound B has none in these positions, and compound B would be expected to be coplanar.
   - Coplanar PCBs are usually the most toxic, so compound B would be expected to be the more toxic.

7 Synthetic Polymers: Plastics, Elastomers, and Synthetic Fibers

7.1 INTRODUCTION

In our society synthetic polymers are among the major chemical products used. Most of these substances were unknown 50 years ago, but now are used in almost every aspect of our daily life. In the home, plumbing, textiles, paint, floor coverings, and many other items are made from **synthetic polymers**, some of which are described in everyday language as **plastics** (Nicholsen, 1991). A summary of some plastics and their uses is shown in Table 7.1.

One of the properties of synthetic polymers that enhances their usefulness is their resistance to biotic and abiotic processes of transformation and degradation in usage situations. This means that plastic plumbing, paint, floor coverings, and so on, tend to last longer than other products. Most of these products then become solid wastes and are discarded in solid-waste disposal facilities, or some are disposed of directly into the environment and become litter. Their resistance to transform and degradation processes can be an environmental management problem. So their chemistry and properties of plastics and other synthetic polymers are aspects of environmental chemistry.

Synthetic polymers, principally in the form of plastics, are being increasingly used in all areas of life. The volume of plastics consumed each year now exceeds the volume of steel. Most modern automobiles now contain over 100 kg of plastics.

TABLE 7.1
Some Common Plastics and Their Uses

Plastic	Uses
Polyethylene	Packaging, electrical insulation
Polyvinyl chloride (PVC)	Credit cards, floor coverings, rainwear
Polyvinyl acetate (PVA)	Latex paints
Polybutadiene (BUNA) rubbers	Car tires, hoses
Poly(methyl methacrylate) (Plexiglas, Lucite)	Transport equipment
Polyamides (nylon)	Clothing

The widespread use of plastics occurs because they are lightweight yet do not break like glass, can be made virtually inert to chemical and microbiological breakdown, and are versatile and cost effective. Packaging is the largest and fastest-growing single market for plastics. Packaging material, however, typically has a lifetime of less than a year before it appears as discarded waste.

7.2 THE NATURE OF SYNTHETIC POLYMERS

The term *polymer* is derived from Greek words meaning "many parts." Polymers are, in fact, macromolecules built up from smaller molecules linked together covalently. The small molecules that are the basic building blocks for these macromolecules are known as **monomers**. The process of combining monomers together to form a polymer is known as **polymerization**. The number of monomers in the polymer, or the degree of polymerization, may be in the hundreds or thousands. Monomers may be linked together such that they form a linear, branched, or cross-linked polymer structure, as shown below:

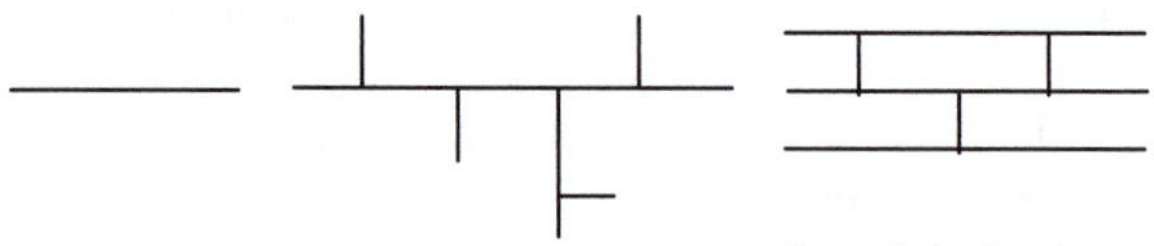

Linear polymer **Branched polymer** **Cross linked polymer**

Although there is no clear threshold, polymers typically possess molecular masses of 1000 or more. A striking difference between polymers and smaller, relatively simple molecules is the inability to assign an exact molecular mass to a given polymer. When synthesizing linear polymers, for example, molecules of chain lengths, and hence molecular weights or molar masses, are generally formed. Such molecules are written as $[CRU]_{n'}$, where CRU is the **constant repeating unit** in the chain.

It is important to recognize that there are many naturally occurring polymers. Cellulose, for example, is the main structural component of the cell walls of plants. It is a linear polymer of glucose. One of the most important and familiar commercial natural polymers is rubber, obtained as an aqueous suspension called latex, largely from the tree *Hevea brasiliensis*. Rubber is a polymer of isoprene, as shown below:

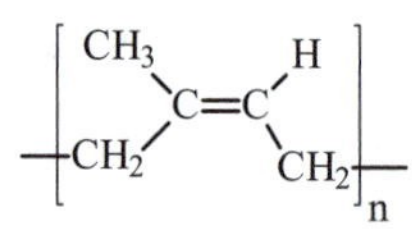

Rubber or *cis*-1,4-poly(isoprene)

Other natural polymers include polynucleotides such as DNA and RNA and proteins that are essentially linear polyamides formed from α-amino acids.

The first synthetic polymers were produced over 100 years ago, but initially were often modifications of natural polymers such as cellulose. As understanding of the nature of polymers increased, so did the development of synthetic polymers. The vast majority of polymers in commercial use are organic compounds.

They may comprise, in addition to carbon and hydrogen, elements such as oxygen, nitrogen, sulfur, and the halogens. A notable exception is the silicone polymers, linear macromolecules with a backbone consisting of alternating silicon and oxygen atoms.

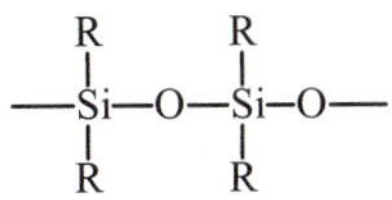

The physical and chemical properties of a polymer (e.g., melting point, chemical reactivity, flexibility, resistance to solvents) depend, among other factors, on its:

- Molecular structure
- Molecular weight
- Component monomers

Monomers can be combined in different ways so that polymers of different properties can be produced from a single monomer.

7.3 CLASSIFICATION AND DEFINITIONS

Polymers possess a diverse array of chemical structures, properties, and behaviors. It is useful to classify them into groups based on some of these characteristics. The classification scheme for synthetic polymers shown in Figure 7.1 is based on physical properties and behavior. They may be divided into three major classes: elastomers, fibers, and plastics. It should be remembered, though, that some polymers may belong to more than one class. For example, a polyester (a polymer where the monomer units are joined by ester functional groups) may form a useful plastic, but if melted, drawn, and stretched, it may also make a good fiber.

Elastomers may be defined as materials that exhibit rubbery or elastic behavior. Following application of stretching or bending forces, these materials can regain their original shape upon removal of the distorting forces (provided they have not been distorted beyond some elastic limit). An example of a synthetic polymer that is an elastomer is neoprene, which is a polymer of 2-chloro-1,3-butadiene, as shown below:

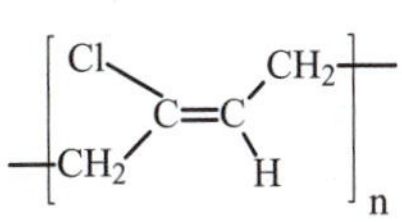

Neoprene

Neoprene tends to be more resistant toward attack by organic solvents than natural rubber. It has a similar chemical structure except that there is a *trans*-arrangement

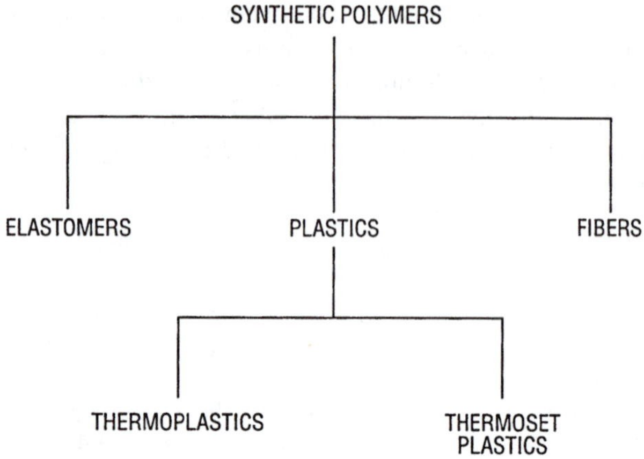

FIGURE 7.1 A classification scheme for the synthetic polymers.

around the carbon–carbon double bond compared to the *cis*-arrangement in natural rubber.

A **fiber** is a polymer with a high length-to-diameter ratio. Most polymers capable of being melted or dissolved can be drawn into filaments or fibers. However, to make a commercially useful fiber, the material should possess high tensile strength (i.e., resistance to breaking under tension), pliability, and abrasion resistance. **Nylons**, which are **linear polyamides**, make excellent fibers, as do **linear polyesters** such as **terylene** (also known as dacron or poly(ethylene terephalate)).

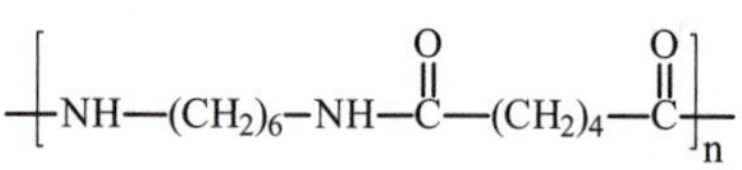

Nylon -6,6

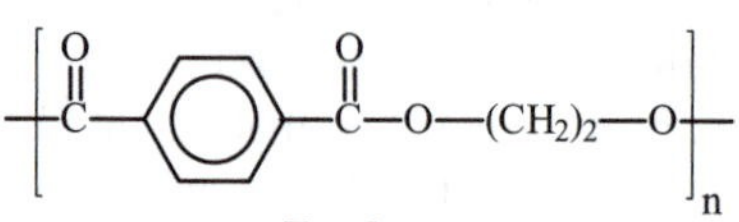

Terylene

The word *plastic* comes from a Greek word meaning to mold. In the context of polymers, plastics are defined as polymeric materials capable of changing their shape on application of a force, but retaining this shape on removal of the force. Plastics may be subdivided into those that are thermoplastic and those that are thermoset.

Thermoplastics are polymers that will soften and eventually melt when heated. They can then be molded and shaped, and on cooling will harden in this shape.

Important thermoplastics include polyethylene, polypropylene, Perspex or poly(methyl methacrylate), and polystyrene. Although classed as a thermoplastic, Teflon or polytetrafluoroethene demonstrates unusual melting behavior. It does not flow at its melting point (372°C), but forms a very viscous, translucent material that fractures rather than flows when a force is applied.

The chemical structures of some specific thermoplastic polymers are shown below:

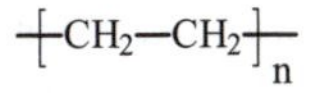

Polyethylene

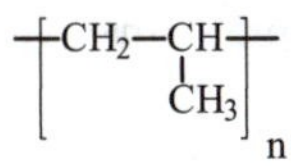

Polypropylene

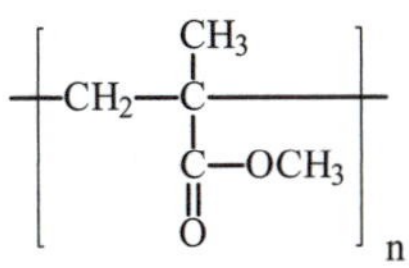

Poly(methyl methacrylate)

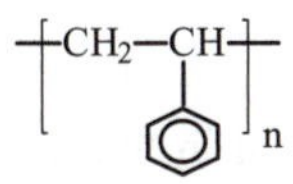

Polystyrene

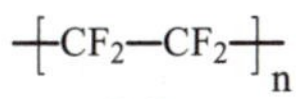

Teflon

Thermoplastic polymers are generally linear or only lightly branched in structure. Most of the plastics produced commercially in many countries are thermoplastics.

In contrast to thermoplastics, **thermosets** usually have a cross-linked structure. Because of this structure, thermosets do not melt and so must be shaped during the cross-linking process. The cross-linking may be initiated by heating, UV light, or a chemical catalyst. Thermosetting polymers are used in paints and surface coatings where oxidation by molecular oxygen (O_2) during drying forms a tough, resistant, cross-linked film. Because they do not melt, thermosets are often employed in high-temperature applications, e.g., handles for cooking utensils, automotive transmission, and brake components. With extremely high temperatures, however, thermosets decompose irreversibly, blistering as a result of gas evolution and eventually charring. Thermosetting plastics are generally rigid, and bending usually results in a sharp fracture of the material. In contrast, thermoplastics are usually less rigid, but if bent too far will crack and then separate into fragments.

7.4 SYNTHESIS OF SYNTHETIC POLYMERS

7.4.1 Polymerization Processes

Polymerization reactions were historically classified into two types:

1. Condensation polymerization, where the monomers are linked together with the simultaneous elimination of a small molecule, such as water at each step. Below is an example the synthesis of terylene:

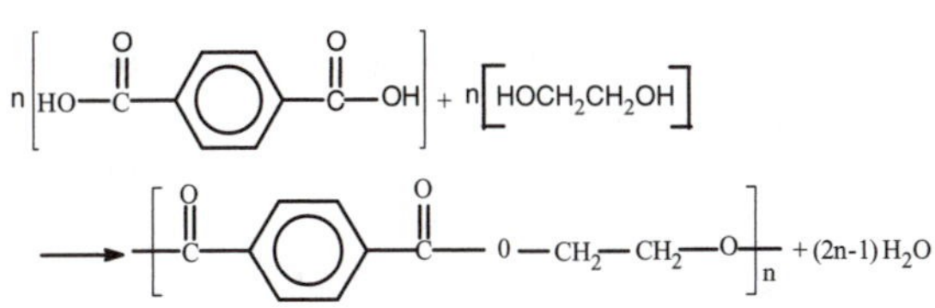

Polymers prepared in this way were called **condensation polymers**.

2. Addition polymerization, where the monomers are literally added together, so that the empirical formulae of monomer and polymer are the same. An example is the synthesis of poly(ethylene):

$$n\,CH_2{=}CH_2 \longrightarrow -\!\!\left[CH_2-CH_2\right]_n\!\!-$$

Polymers prepared in this way were called **addition polymers**.

Unfortunately, this classification scheme can lead to confusion. For instance, some polyesters can be formed by both addition and condensation reactions, and therefore, it is difficult to know how to classify the polymer.

A more modern and useful classification scheme is based on how the chain actually grows. Thus, polymers may be grouped according to whether the polymerization occurs in a stepwise manner (**step growth**) or from a growing chain (**chain growth**). This scheme eliminates the problems and anomalies identified above. Let us now examine in more detail the step growth and chain growth processes.

Step growth polymerization usually occurs with monomers containing functional groups such as hydroxyl (–OH), amino ($-NH_2$), and carboxylic acid (–C(=O)–OH). Polymers such as polyesters and polyamides are produced by a series of condensation reactions, so step growth polymers are essentially the same as the condensation polymers of the early classification scheme.

On the other hand, chain growth polymerization begins with the generation of reactive centers on a relatively small number of monomer molecules. The reactive center may be a radical, a carbocation (a cation where carbon carries the positive charge), or a carbanion (an anion where carbon carries the negative charge). This process constitutes initiation, the first of the three phases of chain growth polymerization:

1. Initiation
2. Propagation
3. Termination

Chain growth polymers are essentially equivalent to the old addition polymer classification. Chain growth polymerization tends to produce thermoplastics only. The monomers involved are typically unsaturated, possessing a carbon–carbon multiple bond, e.g., $CH_2{=}CR_1R_2$.

The most important type of chain growth polymerization, in terms of both versatility and the amount of polymers produced commercially on an annual basis, is the radical-initiated variety. Here, a small trace of initiator substance that readily fragments into radicals is used. The most common initiators are benzoyl peroxide and azobisisobutyronitrile (AIBN), which fragment when heated to around 100°C, or when irradiated with near-UV light, as shown below:

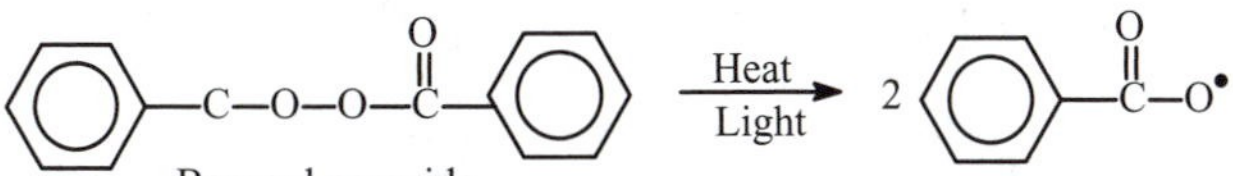

Benzoyl peroxide

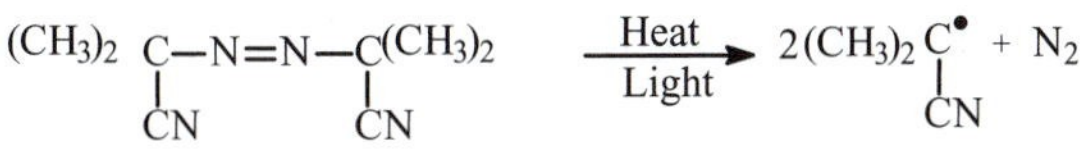

AIBN

Once produced, the radical reacts rapidly with a monomer molecule, forming a new, larger species that is also a radical, and is called the chain carrier.

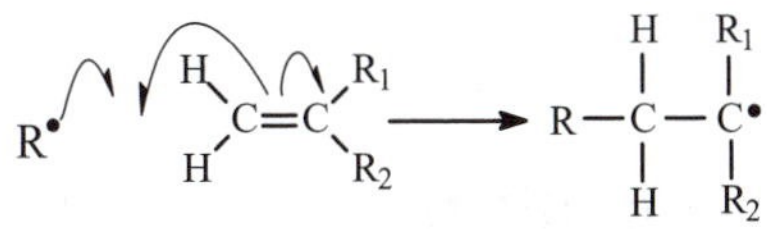

In the process of propagation, the radical center at the end of the growing chain reacts with another monomer molecule, as shown below, to continue chain propagation.

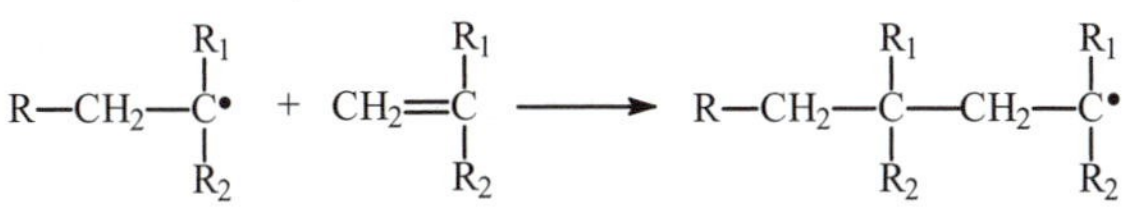

Reaction of the chain carrier with a monomer molecule can occur repeatedly, forming a polymer chain of increasing length. Each individual reaction depends on the previous one. Chain formation occurs very quickly, since the average lifetime of a growing chain is relatively short. Several thousand monomers can be linked together in just a few seconds.

With chain growth polymerization, high-molecular-weight polymer chains are reached with relatively low percentages of monomer reaction. This is because the reaction mixture contains only the growing polymer chains and monomer molecules, and the concentration of growing chains is maintained at a low level. This is in contrast to step growth polymerization, where in addition to monomer and large polymer molecules, many species of intermediate size are also present.

Propagation or chain growth does not continue until all the monomers present have reacted. Termination of growth occurs when the reactive radical centers on two different chains interact together, or when a radical center reacts with a solvent molecule, an initiator radical, or impurities in the reaction mixture, e.g., oxygen. Polymers typically produced by a chain growth polymerization method include polyethylene, polystyrene, and polyvinyl chloride (PVC). Different conditions lead to formation of polymers of the same monomer with different properties due to the different characteristics of the polymer chain (Figure 7.2).

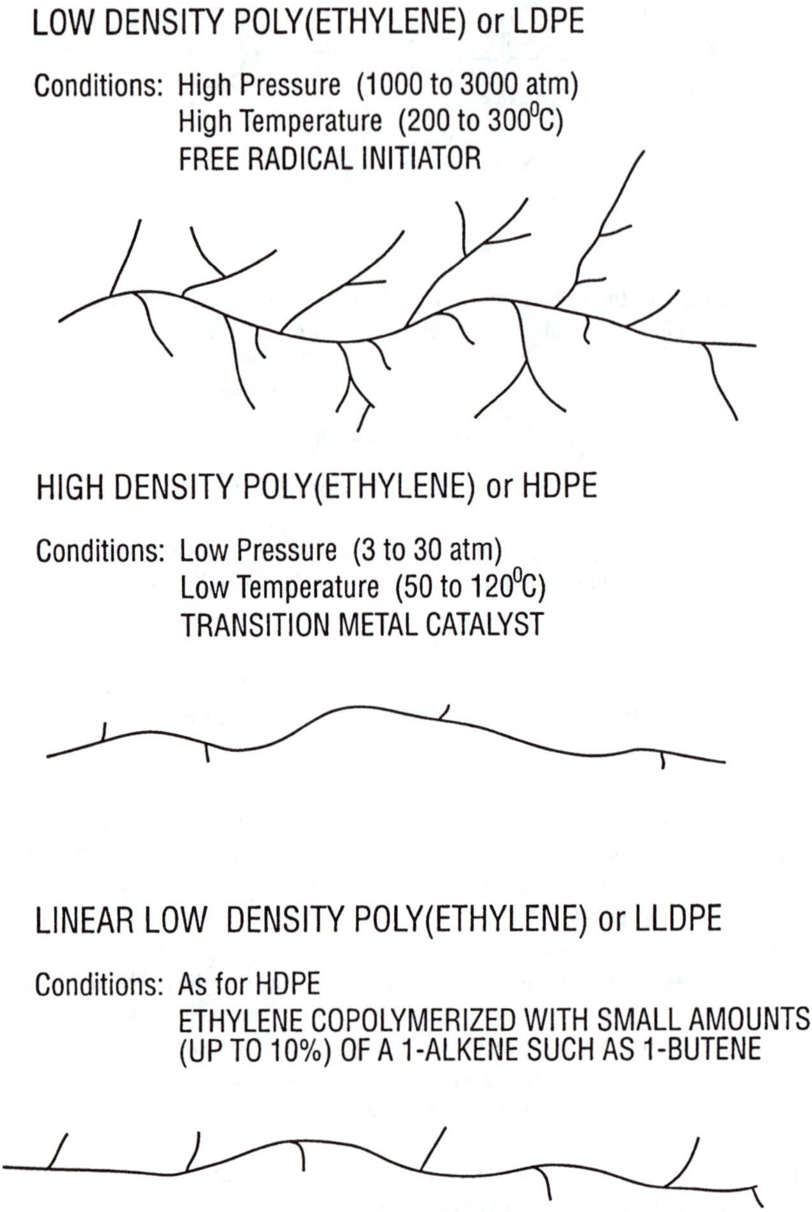

FIGURE 7.2 The various forms of polyethylene and their modes of formation.

7.4.2 Cross-Linking

Bridges comprising covalent chemical bonds that occur between separate polymer chains are known as cross-links. The presence and frequency of occurrence of cross-links can have a significant influence on the physical and chemical properties of materials in which they exist. Polymers not cross-linked are generally thermoplastic, which means that they will melt and flow at sufficiently high temperatures. In contrast, cross-linked polymers are almost invariably thermosets. They cannot melt because the cross-links restrict the motion and movement of individual chains. Cross-linking tends to impart rigidity to polymers. However, extensively cross-linked materials are usually very brittle. The dense three-dimensional network of covalent bonds allows little flexibility and hence no way for the material to take up stress.

Probably the best-known example of cross-linking is the vulcanization of natural rubber. In its raw state, natural rubber is sticky when warm and prone to oxidative degradation. By heating with a few percent of elemental sulfur in the presence of catalysts, rubber molecules are linked by short chains of sulfur atoms. The vulcanized rubber produced is a useful elastomer, harder and more resistant to oxidation and attack by solvents than its natural counterpart. With about 30% sulfur and heating to 150°C, ebonite is formed. This hard and rigid material has been used for making the casings of automobile batteries. Another commonly encountered example of cross-linking is the so-called drying of some paints, which involves cross-linking by molecular oxygen from the atmosphere, resulting in a tough resistant film.

Cross-linking also affects the solubility of polymers. A linear polymer that is not cross-linked will usually dissolve in an appropriate solvent (although it may take some time). Cross-linked polymers will not dissolve. They may, however, swell as solvent molecules diffuse into the network. Polystyrene, for instance, is soluble in many solvents, such as toluene, benzene, and carbon tetrachloride. Cross-linking by copolymerization of styrene with only a small amount of divinyl benzene produces a material that does not dissolve in these solvents, but swells by taking them up.

7.5 ENVIRONMENTAL CHARACTERISTICS

7.5.1 COMBUSTION

Polymers are organic materials, and like most organic materials, almost all of them will undergo combustion if the temperature is high enough. Combustion represents a form of degradation for polymers. When they burn, polymers, even those comprising mainly carbon and hydrogen, give off significant amounts of smoke, and in some cases toxic gases (e.g., HCl, HCN, CO, HNO_3), as well as leaving a solid, carbonaceous residue.

Combustion of polymers occurs in two stages. In the first, a source of heat (e.g., a flame) raises the temperature of a polymer sample to such a level that bonds begin to break. This generates small, volatile, low-molecular-weight products. Thermoplastics melt, which assists these products in migrating through the polymer and out into the surrounding air where they burn, which is the second stage of the process. This burning produces heat, which is fed back into the polymer. For many thermoplastics, this process can become self-sustaining and will continue until production of volatile materials from the polymer ceases. Thermosets tend to be relatively nonflammable. The presence of many cross-links between chains retards movement of the small volatile molecules. Heating may in fact cause more cross-linking to occur, which will hinder movement even further. The result is that the combustion zone surrounding the polymer is starved of organic fuel.

The small volatile molecules produced in the decomposition of thermoplastics are often the result of either depolymerization or elimination reactions. Depolymerization is essentially the reverse of polymerization, so that monomers are formed from the breakdown of polymer molecules. Combustion of poly(methyl methacrylate) or Perspex, leads to the formation of methyl methacrylate in high yield, as shown below:

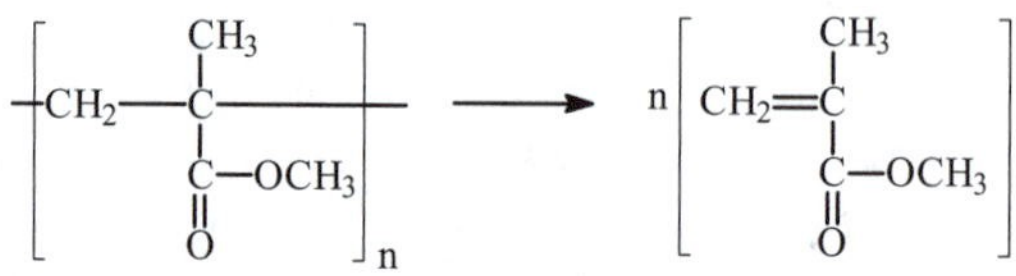

Other polymers to undergo depolymerization include polystyrene and poly(oxymethylene):

Poly(oxmethylene) **Formaldehyde**

Polyvinyl chloride provides a good example of an elimination process. When PVC is heated, the polymer begins to discolor and evolve HCl. A carbon–carbon double bond is left in the chain when the HCl is eliminated. The HCl molecule can then catalyze loss of other HCl molecules. This produces conjugated sequences or stretches of alternating carbon–carbon double and single bonds, accounting for some of the color of thermally degraded PVC.

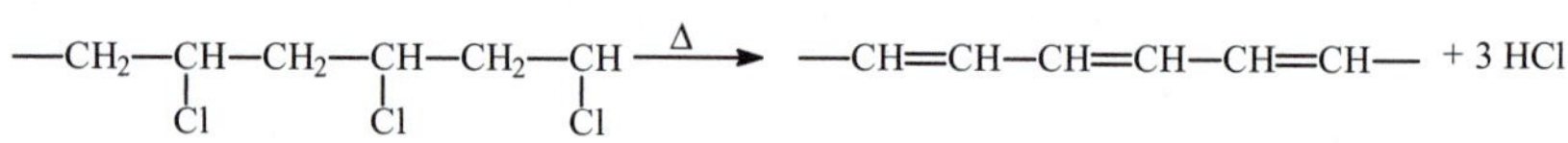

Poly(vinyl chloride)

Polymers such as polyethylene and polypropylene degrade by both depolymerization and elimination, as well as more complex mechanisms.

7.5.2 Biodegradation

Biodegradation of synthetic polymers involves breakdown, by microorganisms or higher organisms, using enzyme-catalyzed pathways. Most of the current applications of synthetic polymers, however, are based, at least partly, on their relative resistance to biodegradation, compared with natural macromolecular or polymeric substances such as starch and proteins. It would clearly be undesirable, for example, if plastic coating on telephone lines buried underground were to degrade, allowing corrosion of the metal fibers. Most of the synthetic polymers available today are bioresistant.

Biological systems tend to degrade natural polymers, e.g., starch by hydrolysis, then oxidation. It is not surprising then that most known biodegradable synthetic polymers have hydrolyzable functional groups in the main chain (e.g., ester, amide, urethane groups).

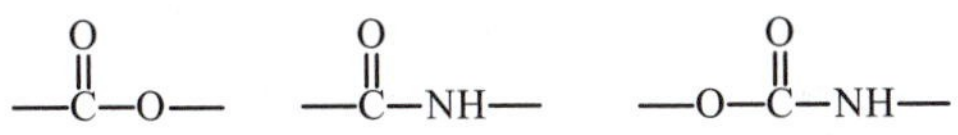

Ester **Amide** **Urethane**

Aliphatic (i.e., nonaromatic) polyesters are among the most biodegradable. This is attributable not only to the possession of ester groups, but also the flexibility of the main chain. This flexibility facilitates the polymer chain binding to the active site of enzymes. The degradable and absorbable sutures employed in surgical procedures today are often made of poly(glycolic acid), while another aliphatic, polyester polycaprolactone, is used as a matrix for the controlled release of drugs in the body.

Glycolic acid **Poly(glycolic acid)**

Caprolactone **Polycaprolactone**

Thermoplastic materials made from polycaprolactone have been shown to almost completely degrade on burial in soil for 12 months. Here, microorganisms are presumably using the polymer as a growth substance, and converting it into microbial biomass, CO_2 and H_2O. Polyesters that contain aromatic rings in the main chain between the ester groups, e.g., poly(ethylene terephthalate) (PET), are not as degradable as their aliphatic counterparts. This is because the rigid aromatic ring restricts the flexibility of the main chain.

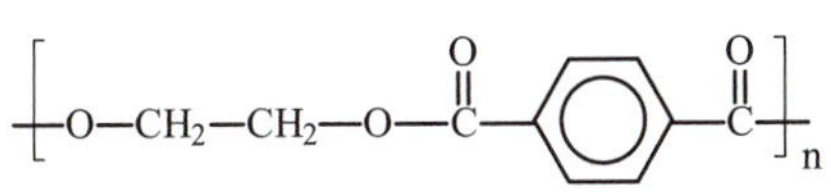

PET or poly(ethylene terephthalate)

Lack of flexibility is also the reason why polyamides are less biodegradable than polyesters with analogous structures. It is well known that the internal rotation within an amide functional group is restricted compared to an ester group, because of the considerable double-bond character in the carbon–nitrogen bond.

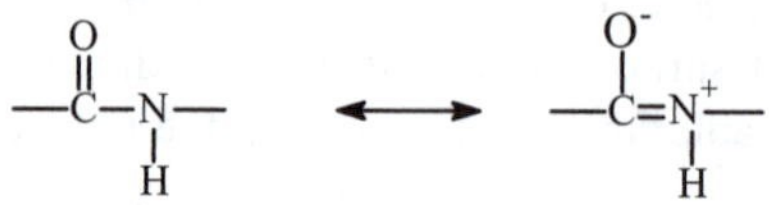

In many countries, the most widely used plastics are polyethylene, polypropylene, polyvinyl chloride, and polystyrene. The polymers involved contain only carbon atoms in their main chain, with no hydrolyzable functional groups.

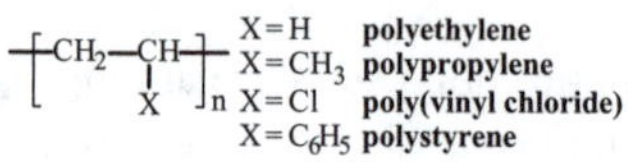

These polymers are very bioresistant. Straight-chain alkanes serve as a guide for the polymers. These compounds are degraded by microorganisms such as fungi up until the alkane molecular weight reaches 500 (n-C_{35}). For larger alkanes, biodegradation becomes effectively zero. This is shown in Table 7.2, which also shows that chain branching inhibits biodegradation (a phenomenon also observed with surfactants). Given that polymers have a distribution of molecular weights, it would be expected that materials with low molecular weights (and hence a higher proportion of chain 535 carbons) and less branching would be more susceptible to biodegradation. It must be remembered, though, that chains greater than 35 carbons in length would be inert, and this is usually the majority of a polymer sample for materials such as polyethylene and polyvinyl chloride. Cross-linking restricts access of

TABLE 7.2
Influence of Chain Length and Extent of Branching of Alkanes on Biodegradation by Fungi Mixture

Compound	Molecular Weight	Relative Biodegradability (0 = poor, 4 = good)
Hexadecane ($C_{16}H_{34}$)	226	4
Octadecane ($C_{18}H_{38}$)	255	4
Tetracosane ($C_{24}H_{50}$)	339	4
Octacosane ($C_{28}H_{58}$)	395	4
Dotriacontane ($C_{32}H_{66}$)	451	4
Hexatriacontane ($C_{36}H_{74}$)	507	0
Tetracontane ($C_{40}H_{82}$)	563	0
2,6,11-Trimethyldodecane ($C_{15}H_{32}$)[a]	212	0
2,6,11,15-Tetramethylhexadecane ($C_{20}H_{42}$)[a]	283	0

[a] Branched chain.

Source: From Potts, J.E. et al., The biodegradability of synthetic polymers, in *Polymers and Ecological Problems*, Guillet, J., Ed., Plenum Press, New York, 1973, 156–162. Copyright © Plenum Press. With permission.

enzymes to the polymer, and so most thermosets are not biodegradable. Most biodegradable plastics are technically unsuitable for uses such as packaging, and thus cannot be readily substituted for the less biodegradable substances.

It is interesting to note that natural rubber has an all carbon backbone with branching off the main chain, yet is considered biodegradable. This may be due to the possession of the carbon–carbon double bond that could be the site of enzyme attack.

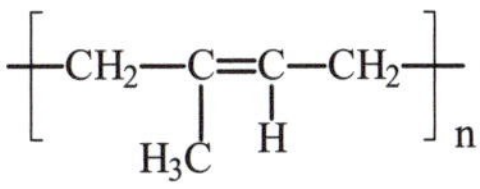

Rubber

There are a number of strategies that can be used to make synthetic polymers more biodegradable. One is to modify natural polymers, e.g., conversion of cellulose into cellulose nitrate and cellulose acetate. Cellulose nitrate has been used for movie film, and old films using this material are often in a degraded state. Subjecting polymers to elevated temperatures or making them photodegradable can sometimes cause breaking of the main chain into fragments small enough for biodegradation to occur.

Biodegradable additives may also be included. Often, these additives serve as heat stabilizers or plasticizers, but also happen to be biodegradable. Occasionally, material such as starch is deliberately added to the synthetic polymer. It should be remembered that degradation of these additives causes the plastic to lose structural integrity and become brittle. The plastic eventually forms small pieces or even a powder, but the polymer itself does not degrade.

7.5.3 Photodegradation

To undergo photodegradation in the environment, a polymer must contain a chromophore — a group capable of absorbing solar radiation. The lower wavelength limit of solar radiation reaching the surface of the earth is 290 nm. Radiation responsible for initiating photodegradation is in the 290- to 450-nm range, but particularly between 290 and 320 nm, since this is of higher energy. Even if a polymer is potentially susceptible to photodegradation, this process cannot occur unless the material is actually exposed to sunlight. Therefore, plastics that are buried in a landfill, for example, cannot undergo photodegradation.

An appropriate chromophore is the carbonyl structure >C = O. When this group occurs in the middle of a carbon chain, it is a ketone functional group and absorbance of radiation extends up to about 330 to 360 nm. A photodegradable polymer containing ketone groups can be made by copolymerizing ethylene (ethene) and carbon monoxide, as shown below.

$$CH_2{=}CH_2 + CO \rightarrow -CH_2-\overset{\overset{\displaystyle O}{\|}}{C}-(CH_2)m-\overset{\overset{\displaystyle O}{\|}}{C}-(CH_2)n-\overset{\overset{\displaystyle O}{\|}}{C}-(CH_2)-$$

Upon absorption of solar radiation, the polymer chain is broken by either a Norrish type 1 or type 2 reaction.

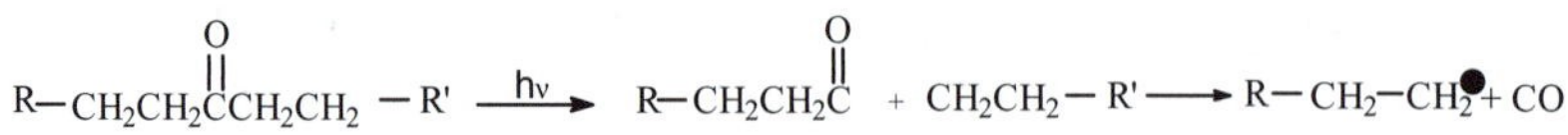

NORRISH TYPE 1

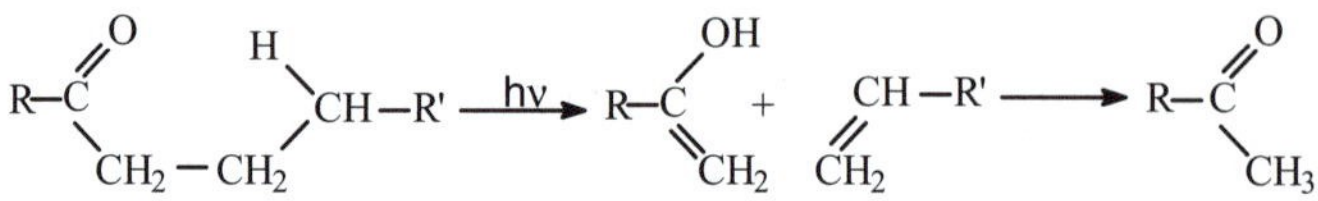

NORRISH TYPE 2

The type 1 reaction produces radicals that can undergo further reaction, and also generates carbon monoxide. The type 2 reaction yields a methyl ketone and an alkene as products. Interestingly, because of the particular range of wavelengths it absorbs, this copolymer will not photodegrade to any great extent indoors, but only outside in sunlight. Ordinary window glass absorbs most of the radiation less of than about 330 nm. No radiation less than this is produced by incandescent light bulbs, and little is produced by fluorescent lights. Therefore, a copolymer of this type should be usable behind windows under artificial light, but is photodegradable outdoors in the environment.

The ester functional group contains a carbonyl structure, but for aliphatic polyesters, absorbance is only <250 nm, which is at a lower wavelength than for ketones. Aliphatic polyesters are therefore generally not susceptible to photodegradation. For aromatic polyesters, however, where the aromatic ring is in conjugation with the carbonyl, absorption up to and perhaps greater than 300 nm can occur. Thus, PET, which is used for soft drink bottles, is potentially susceptible to photodegradation in the absence of any UV stabilizers.

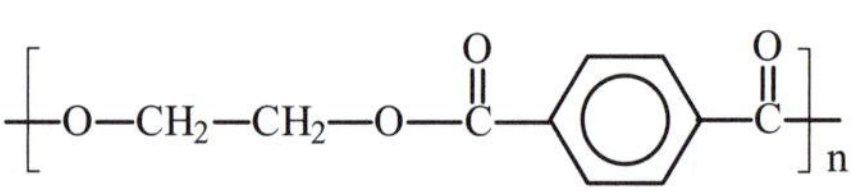

Poly (ethylene terephthalate) or (PET)

Unless stabilizers are incorporated in them, the four major thermoplastics (viz., polyethylene, polypropylene, polyvinyl chloride, and polystyrene) degrade when exposed to the UV component of sunlight. This process is quite slow, though, taking years in many cases, and so these materials can hardly be described as photodegradable. However, no photodegradation at all would be predicted by looking at the structure of these polymers, except perhaps for polystyrene. There is simply no suitable chromophore. The fact that commercial material does undergo photodegradation is due to branching and photosensitive impurities invariably introduced during polymerization and processing. Polypropylene gives a good example of why branching is important. Polypropylene has tertiary hydrogens that on removal, result in the formation of a relatively stable tertiary radical, as shown:

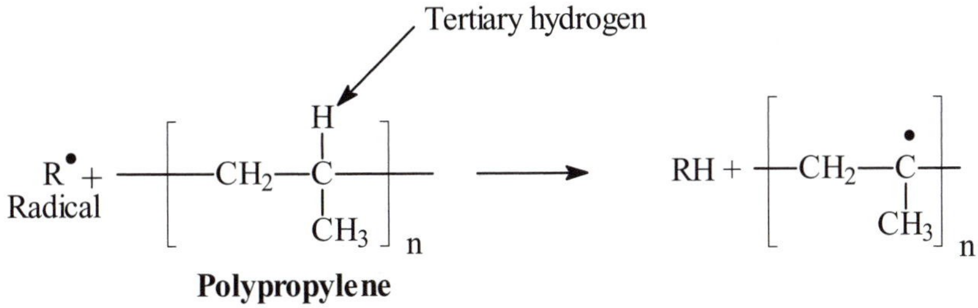

Polypropylene is therefore more susceptible to photodegradation than polyethylene. Polyvinyl chloride degrades in sunlight by a mechanism much the same as that outlined earlier for thermal degradation. This occurs on the surface of the material only. The conjugated polyene structure that is formed on loss of HCl is a good absorber of UV radiation and acts as a protective layer to the bulk polyvinyl chloride below, which is thus not directly exposed to sunlight.

In order to increase the photodegradability of polymers apart from introducing ketone groups into the main chain, introduction into branches off the main chain is sufficient. Thus, copolymers of styrene, methyl methacrylate, and from 0.3 to 10% by mass of methyl vinyl ketone (3-buten-2-one, to be more correct) produce a polymer that degrades easily in sunlight.

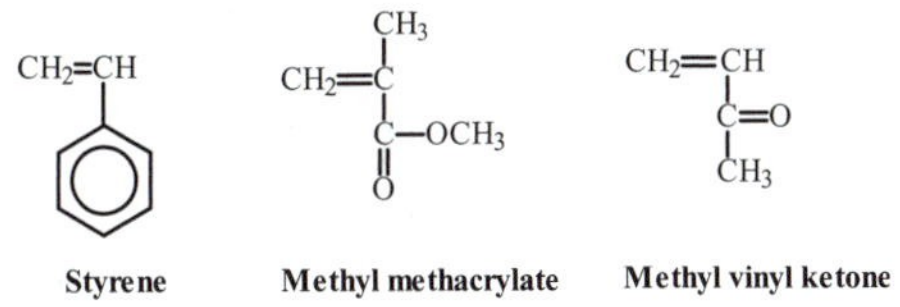

Presumably, Norrish type reactions in the side chains can lead to the formation of radicals that can then attack the main chain of other polymers. Various photosensitive additives can also be deliberately introduced to enhance photodegradation. Photosensitive additives are molecules, e.g., aromatic ketones or transition metal derivatives, that on absorption of sunlight form excited species that abstract hydrogen atoms from the main chain to form radicals and hence initiate polymer degradation. Note that photodegradation of polymers results in a weakening of the structure and a physical disintegration to small fragments. Chemically, the formation of smaller molecules means that the potential for biodegradation is increased.

7.5.4 Recycling

It would be desirable to recycle as much of the synthetic polymer waste as possible. Presently, some 30% of municipal solid wastes being placed in landfills are inorganics, e.g., glass, aluminium, or plastics that are not readily biodegradable.

TABLE 7.3
Typical Products Made from Reprocessed Polymers and Plastics

Polymer/Plastic	Reprocessed Products
Poly(ethylene terephthalate)	Carpet fiber, insulation, furniture stuffing
Polyethylene	Pipes, plant holders (pots), garbage bags
Polyurethane foam	Carpet underlay, industrial paddings
Polyvinyl chloride	Garden hose, shoe soles
Mixed plastic waste	Replacement for timber, other structural materials

Comparatively little of the energy required to produce a plastic is associated with processing, but the polymer itself represents a considerable investment of chemical energy. Some of this energy can be recovered by using the plastic as a fuel for incineration. This does not conserve as much energy as recycling, though.

In theory, thermoplastic materials can be heated, softened, and remolded repeatedly. In practice, while they are easily reprocessed, infinite recycling is impossible, as degradation gradually occurs and contaminants are introduced. Ultimately, the recycled product acquires properties that render it unsuitable for its designed use. Thermosets cannot be heated and reformed into new products. However, they can be ground to a very fine powder, mixed with new materials, and remolded. A high proportion of industrial and commercial plastic waste, perhaps greater than half, is recycled. This is because it is relatively clean, homogenous, and in sufficient volumes to make it relatively economical.

Very little household plastic waste is recycled for exactly the opposite reasons. Plastics account for about 5% by mass and 10 to 15% by volume of household garbage disposed of each year. In the U.S. in 1989, the overall recovery and reprocessing rate of plastics consumed by households was only 1 or 2%. The main plastics recycled were polyethylene (14% of quantity in household waste stream recovered and reprocessed) and PET (10%). Generally, however, it is not economical at present to collect, transport, and segregate mixed plastic waste. Typical products made from reprocessed plastics are shown in Table 7.3. Recovering and reprocessing mixed plastic waste avoids the cost of segregating plastics, but the resulting product has variable properties, depending on the input material. Mixing incompatible polymers results in a lack of adhesion between the various phases, so that stresses cannot be transmitted. The products typically lack strength. Mixed reprocessed waste has not found significant applications and markets as of yet.

7.6 KEY POINTS

1. Synthetic polymers can be classified as plastics, elastomers, and synthetic fibers and are used in many applications, including plumbing, textiles, paint, floor covering, and clothing.

2. Packaging is the largest and fastest-growing single market for synthetic polymers, with a lifetime of less than a year before being discarded as waste.
3. Synthetic polymers are generally resistant to environment transformation and degradation and may present a problem for the disposal of solid wastes and as litter.
4. Polymers consist of macromolecules built up from molecular units described as monomers. Synthetic polymers have molecular masses of 1000 or more and can have a linear, branched, or cross-linked structure.
5. There are many natural polymers, including cellulose, DNA, RNA, proteins, and rubber.
6. The physical and chemical properties of a synthetic polymer depend principally on
   - Molecular structure
   - Molecular weight
   - Component monomers
7. Some common synthetic polymers can be represented by the chemical structures below:

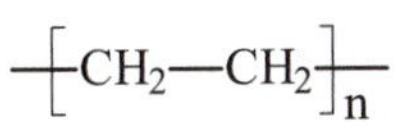

Polyethylene

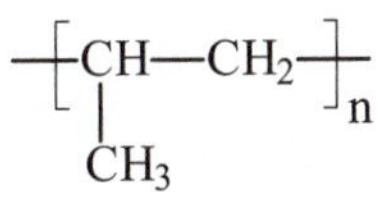

Polypropylene

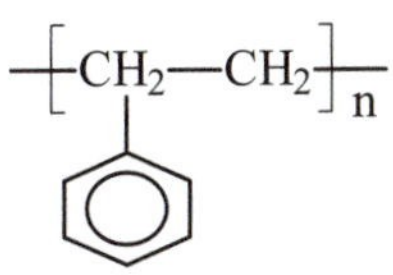

Polystyrene

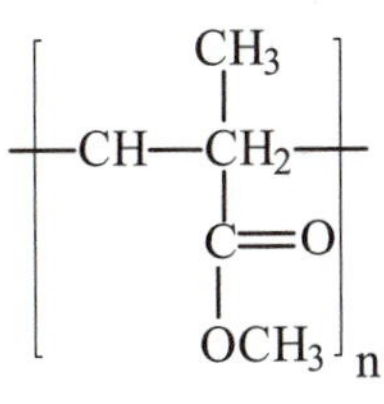

Poly(methyl methacrylate)
(Plexiglas, Lucite)

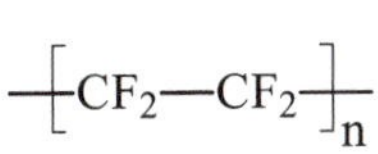

Teflon

8. Polymers can be grouped according to whether the polymerization occurs in a stepwise manner (step growth) or from a growing chain (chain growth).
9. Combustion of polymers occurs in two stages. First, a source of heat leads to the breakage of bonds and formation of small molecules, which burn

in the second stage of the process. Depolymerization reactions can occur in which molecules of the monomer are produced and then burnt. Also, elimination reactions occur, with some polymers leading to the formation of small molecules, such as HCl.

10. The commonly used synthetic polymers are generally resistant to biodegradation. Biodegradation capacity can be improved by:
    - Mixing with natural biodegradable substances, such as starch or proteins
    - Incorporating hydrolyzable functional groups, such as esters, amides, and urethanes
    - Ensuring that the main chain is flexible and thus able to facilitate binding to active sites of enzymes
11. To undergo photodegradation, the polymer should contain a chromophore capable of absorbing solar radiation, particularly in the range of 290 to 320 nm. This can be done by incorporating carbon monoxide into a polymer to yield a ketone group in the molecule. Aromatic rings conjugated with carbonyl groups, such as in poly(ethylene terephthalate) (PET), render the polymer potentially susceptible to photodegradation.

REFERENCES

Cowie, J.M.G., *Polymers: Chemistry and Physics of Modern Materials*, Intext Education, New York, 1973.

Nicholsen, J.W., *The Chemistry of Polymers*, Royal Society of Chemistry, Cambridge, England, 1991.

QUESTIONS

1. Write the chemical structure for the polymer and the additional small molecule formed from the following reactants:

a. $H_2N\text{—}[CH_2]_6\text{—}NH_2 \quad + \quad HOOC\text{—}[CH_2]_4\text{—}COOH$

b.

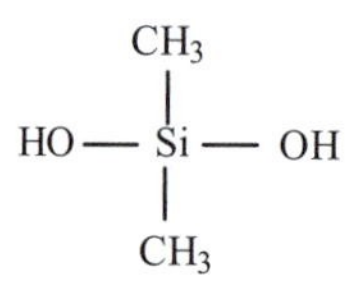

c.

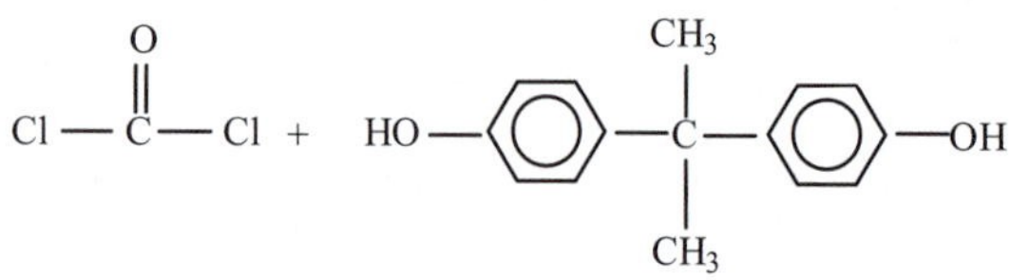

2. A polymer has the following chemical structure:

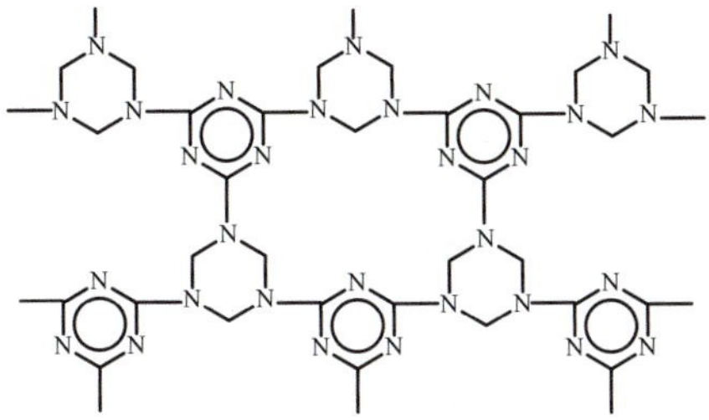

Based on this structure, briefly comment on the following properties:
a. Physical properties, such as melting point, etc.
b. Solubility in solvents
c. Combustibility
d. Biodegradability in the environment

3. A polymer has the following structure:

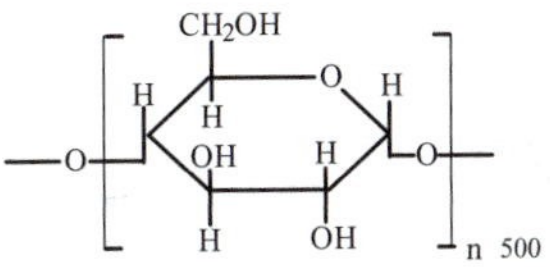

Glucose

Based on this structure, briefly comment on the following properties:
a. Physical properties
b. Solubility in solvents
c. Combustibility
d. Biodegradability

ANSWERS

1.

a. 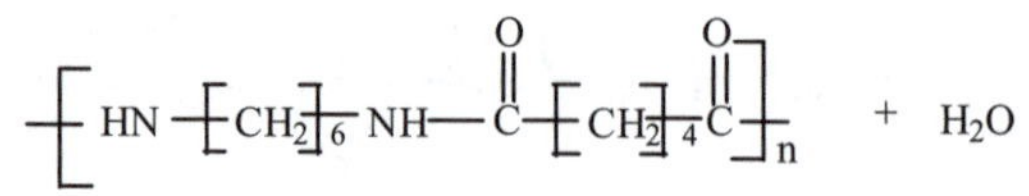

Nylon

b. 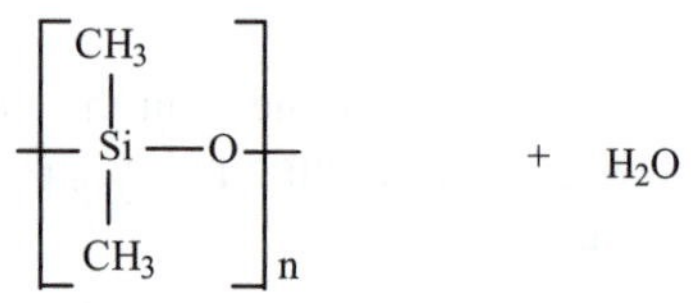

Silicone polymer or polysiloxane

c. 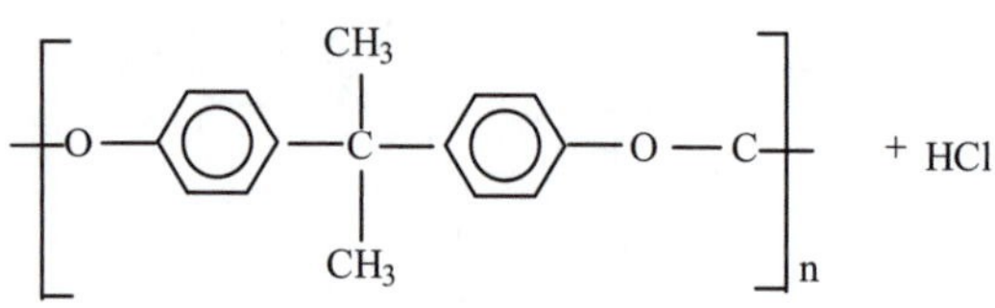

Polycarbonate

2. a. Due to many cross-links, this polymer is not likely to possess a melting point and be brittle. It will tend to decompose rather than melt.
   b. Due to the extensive cross-linking, this polymer is unlikely to be soluble in any solvent.
   c. This polymer will be resistant to combustion and the formation of small molecules by heating.
   d. This polymer lacks any functional groups that are susceptible to hydrolysis or oxidation, is inflexible, and thus is likely to be resistant to biodegradation.
3. a. This polymer has limited cross-linking and consists essentially of long chains. It would be expected to be flexible.
   b. This polymer has many polar groups (–OH) and would be expected to exhibit some solubility in polar solvents, such as water, depending on chain length.
   c. This polymer would be expected to decompose on heating to yield small combustible molecules and thus burn fairly readily.
   d. This polymer contains many groups susceptible to chemical attack, such as hydroxyl and ether groups, and thus would be expected to biodegrade.

8 Pesticides

8.1 INTRODUCTION

Pesticides are chemicals used to remove pests such as insects and weeds. Of all the environmental contaminants, pesticides have probably been the most widely criticized due to their direct use in natural systems. The nature of pesticide usage often requires broad distribution over large areas of crops. This wide treatment of the crop environment often results in treatment of adjacent areas as well. This in itself creates concern, since people and natural organisms are exposed to these chemicals. The use of pesticides in agriculture and other areas is not recent. Prior to the 1940s, insecticides such as lime, sulfur, nicotine, pyrethrum, kerosene, and rotenone were extensively used. These pesticides suffer from a number of deficiencies. In particular, they lack potency with a wide range of insects. In addition, they lack persistence in the environment, so that repeated usage is frequently required. Finally, as a group, these substances are costly to produce and use.

A major development in insect control came in 1939 when the Swiss chemist Paul Müller, working for the Geigy Company, patented DDT as an insecticide. DDT was not a new substance, but its insecticide activity was previously unknown. It was first prepared by Ziedler in Germany in 1874. It found extensive use during World War II due to its relatively long persistence, cheap cost, and potency to a wide range of insect species. It achieved such a high level of success in helping control food pests and pests bearing human diseases that Müller was awarded the Nobel Prize in 1948 for its development.

Usage of DDT and related chlorohydrocarbon insecticides rapidly accelerated during the 1940s and subsequent decades, and the organophosphate pesticides became widely used as well. Little thought was given to ecological implications, although there were a few reports of possible consequences due to the occasional "kills" of fish and other aquatic organisms associated with the use of these substances. In 1962, the book *Silent Spring* was published by Rachel Carson. This book raised many possible ecological problems that could be associated with the usage of DDT. It has had a major influence in that it initiated a large research program in the U.S. and similar programs in other countries throughout the world. As a result, there have been many scientific and governmental inquiries into the usage of DDT and other chlorohydrocarbons. It is now clear that the use of DDT has caused a range of problems, including direct lethal effects and sublethal effects, such as eggshell thinning in certain species of birds (Connell & Miller, 1984).

There are now severe restrictions on the usage of DDT and other chlorohydrocarbon compounds as pesticides due to their adverse environmental effects. Thus, usage of these chemicals has declined, and many have totally disappeared in a range of countries throughout the world. However, concern about public health and food production has lead to the manufacture and continued usage of the chlorohydrocarbon

TABLE 8.1
Some Chemical Classes in the Various Groups of Pesticides

Herbicides
Carbamates, phenoxyacetic acids, triazines, phenylureas
Insecticides
Organophosphates, carbamates, organochlorines, pyrethrins, pyrethroids
Fungicides
Dithiocarbamates, copper, mercurials

pesticides in many tropical countries. The low cost and effectiveness of these substances in the tropical environment are major factors in this continued usage.

The sale and use of a chemical in agriculture, and elsewhere, now usually require registration with a governmental agency. Many steps are currently necessary to introduce a chemical into the market as a pesticide. These start with synthesis and screening, proceed through trials and evaluation, and finally end with registration. Aside from the costs involved in selection of a compound for effectiveness and economical manufacture, there is the testing for environmental effects now required as a major component of registration. Thus, it would be expected to take about 5 years, with costs up to many tens of million dollars, to place a new pesticide on the market (Green et al., 1998).

In a bid to find alternatives without adverse environmental effects, there has been extensive testing of natural pesticides that are known to occur in plants. Many potent pesticides have been isolated, but of these, only **pyrethrum** has proven to be a commercial success, finding wide acceptance in the community. Pyrethrum is a natural pesticide prepared from the dried powdered flowers of the chrysanthemum (*Chrysanthemum cinerariaefolium*). Synthetic compounds related to these natural compounds are also prepared. In addition, a variety of other techniques have been used, such as biological control, and have helped reduce chemical usage.

Pesticides can be conveniently divided into classes depending on which particular pest they are directed toward. Thus, the main groups are the **herbicides**, **insecticides**, and **fungicides**. The various chemical classes within these groups are shown in Table 8.1. Pesticides share only one common property and that is their toxicity to organisms regarded by humans as pests. Chemically they are a very diverse group ranging from metal compounds to a range of diverse organic chemicals. The major groups of pesticides are considered below.

8.2 THE CHLORINATED HYDROCARBON PESTICIDES

This group of substances is referred to as the **chlorinated hydrocarbons**, the **chlorohydrocarbons**, or the **organochlorines**. Some caution should be adopted with the term *organochlorines* since this includes such pesticides as 2,4-dichlorophenoxyacetic acid (2,4-D) and **2,4,5-trichlorophenoxyacetic acid** (2,4,5-T), which have quite different properties. DDT is the most prominent member of the chlorohydrocarbon group, and its history and usage has had a major impact on the use of pesticides in particular and considerations of hazardous chemicals in the environment in general. After its introduction during World War II, its usage gradually expanded and extended into various parts of agriculture. It has been manufactured in many countries throughout the world in large quantities, with about 100,000 tonnes per year produced in the U.S. in the late 1950s. This declined to about 20,000 tonnes in 1971 due to concerns regarding its impact on the natural environment. The very properties that stimulated its usage had become a liability. Cheap production in large quantities and persistence in the environment reduced the need for retreatment but resulted in many problems in environmental management. With the early success of DDT, a range of related chlorohydrocarbons were produced for particular applications. Most of these substances now face similar restrictions as DDT.

8.2.1 Chemical Structure and Synthesis

The properties of the hydrocarbons were considered in Chapter 5. In general, these substances can be divided into three broad classes: **alkanes**, **alkenes**, and **aromatic hydrocarbons.** Members of all these groups can form chlorinated derivatives. The C–Cl bond can be formed by substitution reactions of the various hydrocarbons with chlorine. For example, methane can undergo substitution reactions with chlorine to form carbon tetrachloride (tetrochloromethane) and chloroform (trichloromethane). In addition, benzene can form chlorobenzenes by similar substitution reactions. The chlorohydrocarbons can also be formed by addition reactions with alkenes.

DDT was first prepared in 1874 by condensation of chloral (trichloroacetaldehyde) with monochlorobenzene, and this has remained the basis for commercial synthesis since that time. An outline of the reaction is shown in Figure 8.1. It can be seen that two isomers are produced with the more effective 4,4′ isomer produced in the greater quantities. Usually, commercial DDT contains significant amounts of the 2,4′ isomer as well.

Commercial chlorohydrocarbon insecticides, apart from DDT, include heptachlor, lindane, dieldrin, chlordane, aldrin, endrin, and mirex. Some chemical structures of these substances are shown in Figure 8.2. Within this group there is a subgroup described as the **cyclodiene group**, which has properties that enable a distinction to be made from other members. Heptachlor, dieldrin, chlordane, and aldrin are usually included in this group. Members are produced by the Diels–Alder reaction of hexachlorocyclopentadiene with cyclopentadiene, as shown in Figure 8.3, or with other substances containing at least one carbon–carbon double bond. Another compound often included in this group is **endosulfan**. Strictly, endosulfan is not a chlorohydrocarbon and not a member of the cyclodiene group because, while it contains a part of the molecule that is chlorohydrocarbon in nature, its properties are strongly influenced by the presence of a sulfur-containing group. The structure of endosulfan is shown below:

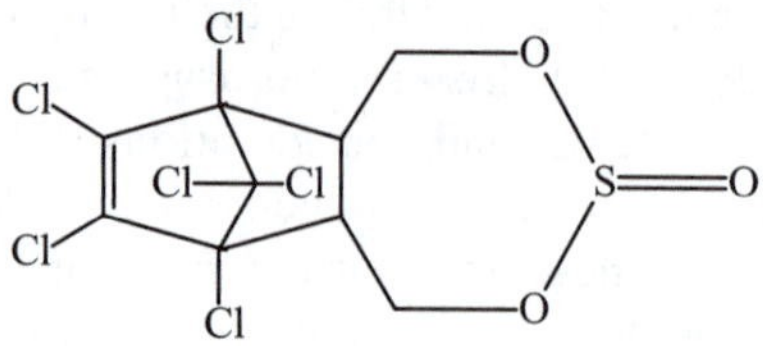

Endosulfan

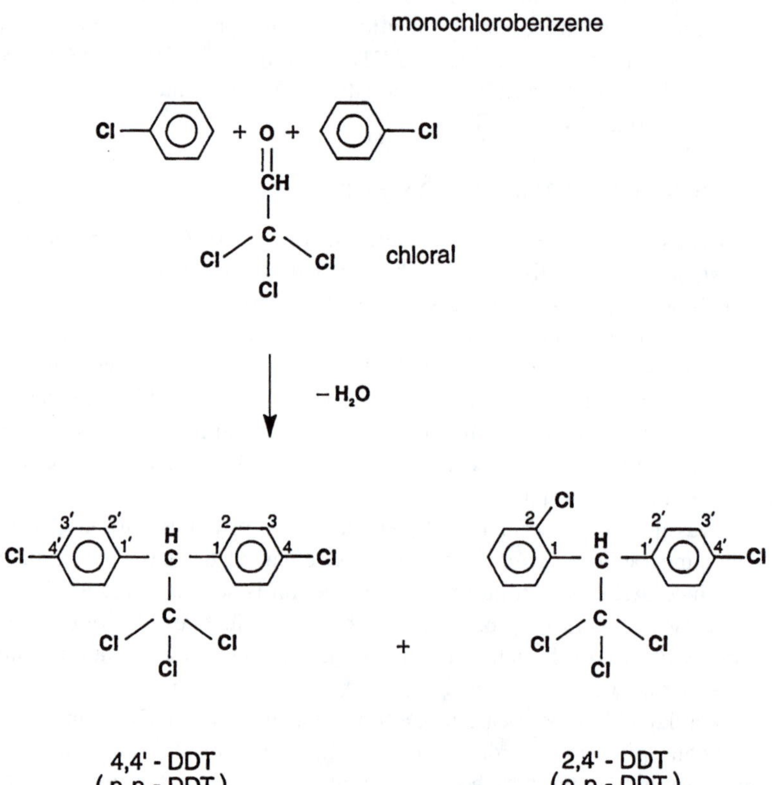

FIGURE 8.1 Synthetic reaction process for the production of DDT.

8.2.2 Physical-Chemical Properties

There is a limited range of bond types present in this group. These are the C - - - C (aromatic), C=C, C–H, and C–Cl, with lesser numbers of C–C. The symmetrical bonds, C - - - C (aromatic), C–C, and C=C, have dipole moments close to zero. Only C–H and C–Cl bonds have dipole moments, 0.4 and 0.5 Debyes respectively, though these are relatively low. This means the compounds in this group tend to have low polarity and dipole moments. These properties result in compounds in this group being **fat soluble** or **lipophilic** and having a low solubility in water.

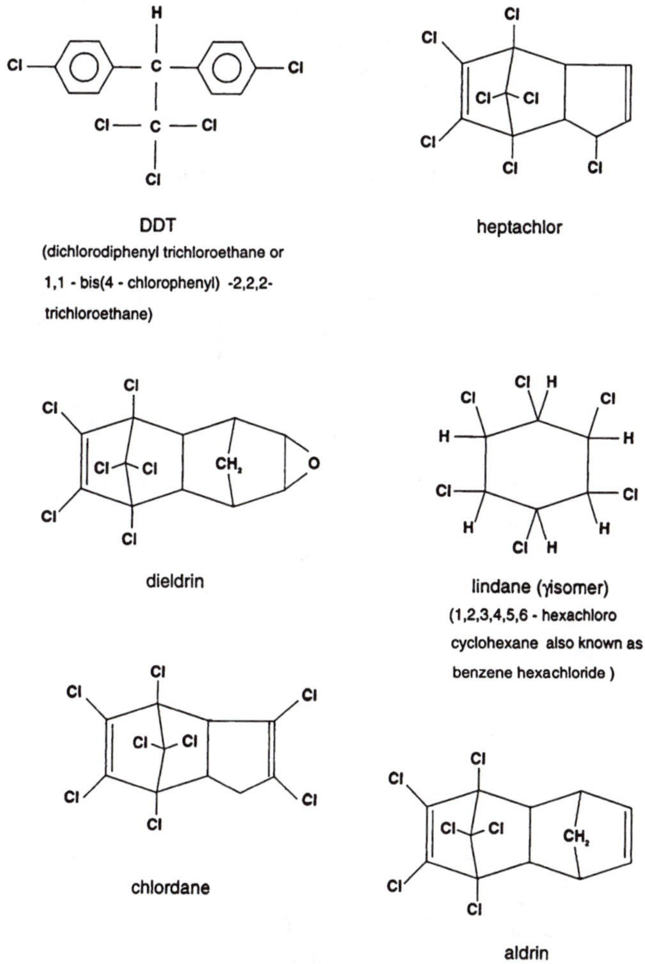

FIGURE 8.2 Examples of the chemical structures of some insecticides included in the chlorohydrocarbon group.

This is illustrated by some typical examples shown in Table 8.2. While the solubility in lipid of all of these compounds lies in the order of hundreds to thousands of grams per liter, the aqueous solubility ranges only from a few µg/l to several hundred µg/l. The lipophilicity of these compounds is indicated by the octanol–water partition coefficient (K_{OW}), which lies between 470 (log K_{OW} 2.67) and 2,300,000 (log K_{OW} 6.36).

8.2.3 Environmental Properties

The limited range of bond types present in the chlorohydrocarbon pesticides are generally relatively resistant to attack by abiotic or biotic agents in the environment. As a result, environmental degradation proceeds at a relatively slow rate. Most

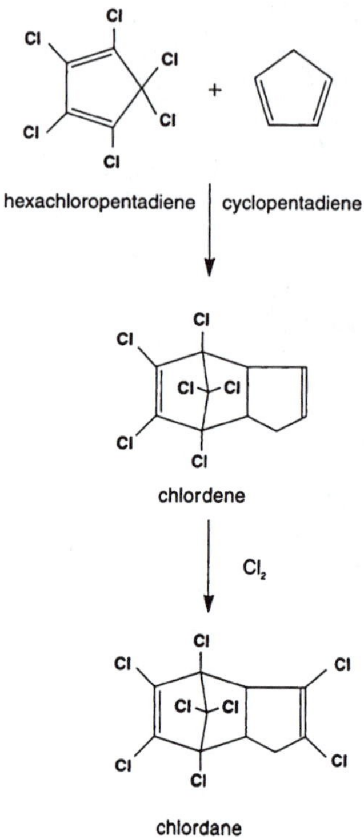

FIGURE 8.3 Synthesis of chlordane.

TABLE 8.2
Physicochemical Properties of Some Typical Chlorohydrocarbon Pesticides

Compounds	Solubility in Lipid (g/l)	Aqueous Solubility (μg/l)	Log K_{OW}
4,4′-DDT	330	3.36	6.36
Heptachlor	1000	50	4.11
Dieldrin	3700	200	3.88
Lindane	800	130	2.67

compounds in this group persist for long periods in soil and often exhibit half-lives of many years, as illustrated by the data in Table 8.3. The half-lives can range from approximately 0.04 to 15.6 years for the compounds considered. The variability is due to the range of different conditions that can occur in soil. For example, soil moisture and temperature can vary considerably, affecting the microbial population and its growth rate, resulting in a considerable impact on the degradation rate.

TABLE 8.3
Toxicity and Persistence of Various Chlorohydrocarbon Pesticides

Compound	LD_{50} (mg/kg body rats)	LC_{50} (estuarine fish µg/l; 96 h)	EC_{50}[a] (Daphnia, µg/l; 48 h)	Half-Life Range (soil, years)
4,4′-DDT	115	0.4–89	0.36	2.0–15.6
Dieldrin	50	0.9–34	250	0.5–3.0
Lindane	125	9–66	460	0.04–0.7
Aldrin	50	5–100	28	0.06–1.6

[a] Concentration that induces immobilization rather than lethality.

There are a range of possible environmental degradation patterns for the chlorohydrocarbon pesticides. A pathway for the biodegradation of DDT is shown in Figure 8.4. All pathways with all compounds would be expected to involve hydrolysis and oxidation at various stages. The ultimate products of degradation would be expected to be carbon dioxide, water, and other substances.

8.2.4 Bioconcentration Properties

Bioconcentration in aquatic organisms occurs as a result of partitioning between the organism lipid and the surrounding water. If a substance is **lipophilic**, then equilibrium occurs with a relatively high concentration in the biota lipid as compared to water. The **bioconcentration factor (K_B)** or the organism–water partition coefficient is defined as follows:

$$K_B = \frac{C_{organism}}{C_{water}}$$

The bioconcentration factor is a characteristic of a particular chemical and lipid content if its on a whole-weight basis, just as its solubility in water, vapor pressure, and melting point are also characteristics. To exhibit high bioconcentration, a compound must also persist in the organism rather than be biodegraded and removed, resulting in lower concentrations in the organism. The chlorohydrocarbons have the properties of high lipophilicity, indicated by log K_{OW} values usually between 2 and 6, as well as persistence, and thus would be expected to exhibit a strong bioconcentration capacity. For example, the K_B value of 4,4′-DDT is 52,500 (log K_B 4.72); heptachlor, 12,900 (log K_B 4.11); dieldrin, 7,600 (log K_B 3.88); and lindane, 470 (log K_B 2.67). The K_{OW} value and lipophilicity of compounds are described in Chapter 2.

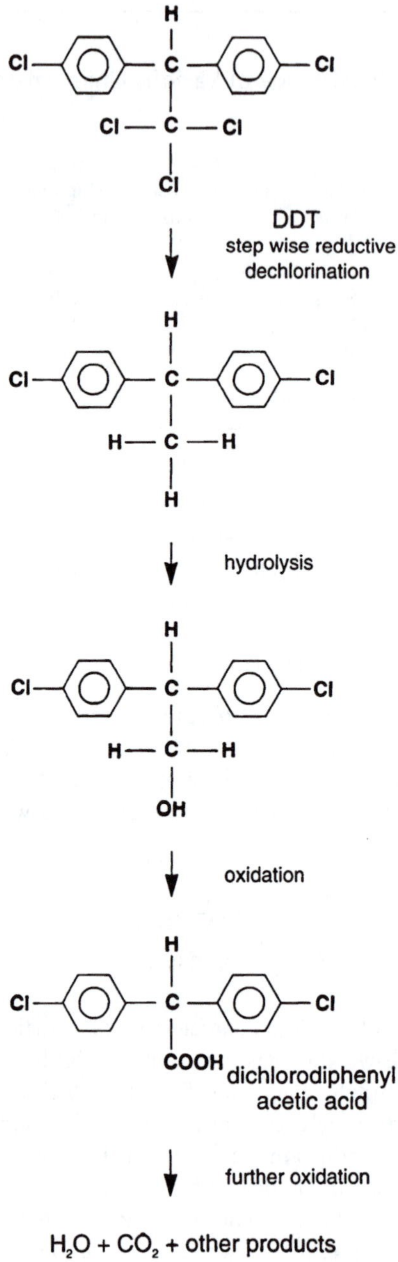

FIGURE 8.4 A pathway in the environmental degradation of DDT.

8.2.5 Toxicity

The chlorohydrocarbons as a group tend to have a wide range of activity with different insects and related organisms. Also, the members of the group are usually powerful fish toxicants, despite relatively low aqueous solubility, as illustrated by the data in Table 8.3. All compounds listed are highly toxic to a range of species of fish at concentrations much less than 1 mg/l. Also, these compounds are generally quite toxic to other aquatic species such as Daphnia (Table 8.3). The cyclodiene subgroup, e.g., dieldrin and aldrin, tend to have higher mammalian toxicity than other members of the chlorohydrocarbon family (Table 8.3).

The mode of action of the compounds in this group is not fully clear. They are neurotoxins since they act on the nervous system, producing tremors followed by loss of movement, convulsions, and death. DDT appears to act on the nerve axon. All the members are strong inducers of mixed-function oxidase (MFO) in exposed organisms, which was considered in Chapter 3. Resistance to the chlorohydrocarbons has been observed in pesticide populations that have been repeatedly treated with a particular pesticide. With DDT, it is often due to a dehydrochlorinase removing HCl and producing DDE, so resistance is concerned with the development of this enzymatic capacity. Resistance usually takes several generations to develop and may be specific for a particular pesticide.

8.2.6 Ecological Effects

The major ecological effect that has been observed has been the reduction in the reproductive success of carnivorous birds, such as the peregrine falcon, with DDT. DDT is the prime substance that enters birds and is not a strong disrupter of breeding success. However, its metabolic product DDE is very powerful in this area. This has been shown to be due to interference with the endocrine system, causing reduction in the development of the thickness of eggshells in eggs produced. The endocrine system is a complex array of glands and organs that control the hormones in the circulatory system. Substances having this biological effect are usually referred to as **endocrine disrupters**. As a general rule, a 20% reduction in shell thickness leads to a reduction in the population due to the damage and destruction of eggs during the brooding process. A wide range of other ecological effects have been observed in aquatic and terrestrial ecosystems.

8.2.7 Long-Range Transport: The Persistent Organic Pollutants

Persistent organic pollutants (POPs) are a group of substances found in many areas throughout the world in trace concentrations in air, water, sediments, and biota. Most of the POPs are chlorohydrocarbon pesticides such as DDT, hexachlorocyclohexane (HCH; including lindane), and toxaphene. Toxaphene is a complex mixture produced by chlorination of camphene. The particular property of POPs that is of interest is their capacity for long-range movement across continents and oceans. Thus, use can occur in a region from which the POPs evaporate into the atmosphere or dissolve in ocean water and then are transported elsewhere. In the new location, the POPs bioaccumulate to significant concentrations and exhibit the adverse effects previously noted.

The POPs have specific environmental properties from which these characteristics are derived. First, they must be persistent, since the transport process takes a reasonable period to occur. This can be measured as a half-life in soil of greater than 6 months (see Section 3.4). They must be capable of being vaporized, perhaps at a slow rate, and have a Henry's law constant between 10^{-5} and 10^{-2} atoms m^3 mol^{-1} (see Section 15.2), with a vapor pressure greater than 1000 Pa (STP). Also, they must be bioaccumulative with log K_{OW} values between 2 and 5.5.

The POPs have been under scrutiny by international agencies for some time, and currently international agreements are in force that are expected to reduce usage throughout the world.

8.3 THE ORGANOPHOSPHATE INSECTICIDES

This group of substances came under intense investigation during World War II for use as military gases. Initially these compounds were considered quite unsuitable for agricultural use due to their high mammalian toxicity. This was particularly so since the chlorohydrocarbon pesticides were available, which generally have relatively low mammalian toxicity, but with the various environmental problems that have become apparent with the chlorohydrocarbon pesticides, a great deal of attention has been focused on the organophosphate group for development as commercial pesticides. In recent years, a wide range of organophosphate insecticides have been developed that are acceptable for agricultural use.

8.3.1 Chemical Form and Synthesis

The organophosphate pesticides have the following general formula:

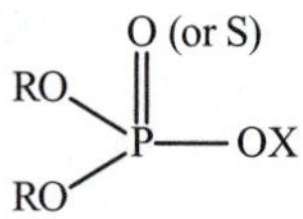

The two R groups are usually methyl or ethyl groups. The oxygen atom in the OX group can be replaced by S with some compounds. Some examples of the chemical structures of this group are shown in Figure 8.5. Although the group has a common core structure, there is still considerable diversity due to variations in the attached chemical groupings.

8.3.2 Physical-Chemical Properties

The defining chemical structure of the organophosphate pesticides contains one P=O and three P–O bonds. From Table 2.3 the oxygen atom has an electronegativity of 3.5 and phosphorus of 2.1. The phosphorus electronegativity is comparable with hydrogen, which is also 2.1. Thus, the O–P bond would be expected to have similar polarity to the O–H bond and be polar. At the same time, the molecule usually

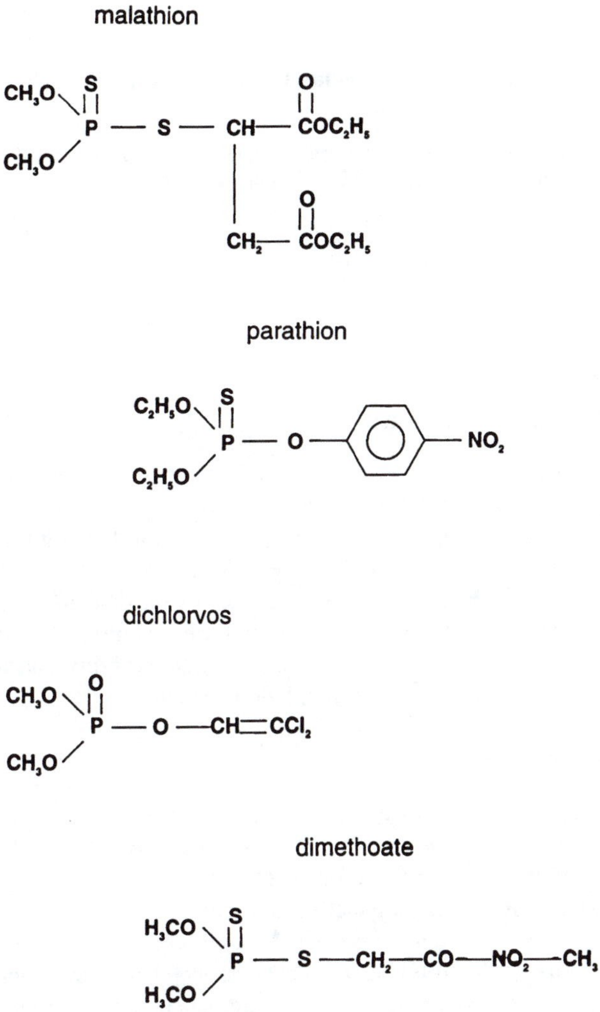

FIGURE 8.5 Chemical structures of some common organophosphate pesticides.

contains a range of other bond types, including O-alkyl, which is of relatively low polarity. These compounds would generally be expected to have greater water solubility than the chlorohydrocarbons and lower lipophilicity. For example, the compounds in Table 8.4 have water solubility in the range of 25 to 10,000 mg/l, whereas the chlorohydrocarbons (Table 8.2) range from 6 to 200 μg/l. The K_{OW} values lie in the range of 2.71 to 3.81 (Table 8.4), which is in the lower range of the lipophilic compounds. Thus, it would be expected that, depending on the identity of the R and particularly the X groups in the molecule, the organophosphate pesticides can have a range of physicochemical properties.

TABLE 8.4
Environmental Properties of Various Organophosphate Pesticides

Compound	Water Solubility (mg/l)	Log K_{OW}	Half-Life Range (soil, days)	LC_{50} (mg/kg of body weight)	LC_{50} (Estuarine fish, µg/l; 96 h)	EC_{50} (Daphnia, µg/l; 48 h)
Malathion	145	2.89	3–7	2500	27–3250	1.8
Parathion	24	3.81	7–10	—	—	0.60
Dichlorvos	10,000 (approximately)	—	—	63	0.05–3.1	0.07
Dimethoate	25	2.71	11–37	500	—	—

8.3.3 Environmental Properties

The organophosphate pesticides are a chemically reactive group of compounds. For example, they are susceptible to hydrolysis when they come into contact with water. This reaction is dependent on the pH, with higher pH values giving more rapid rates of reaction. The half-lives in soil are considerably less than the chlorohydrocarbons and usually range up to about 40 days, as illustrated by the data in Table 8.4 (Howard et al., 1991). Their lack of persistence in soil indicates a similar lack of persistence in biota, and this, together with their moderate water solubility and relatively low lipophilicity, leads to a lack of bioaccumulation capacity.

8.3.4 Toxic Activity

The organophosphate pesticides exhibit strong toxic activity with a wide range of biota. The biota affected range from mammals to insects. All chemicals in this group act by inhibiting the action of several ester-splitting enzymes present in living organisms, and they are particularly active in inhibiting acetyl cholinesterase (Ach). Cholinesterase is an important enzyme that facilitates the transmission of nerve impulses. It operates by hydrolyzing the substance acetyl choline, which is generated in the transmission of nerve impulses. Acetyl choline contains an ester grouping that is the focus of the action of the enzyme. The structure of this substance is shown below:

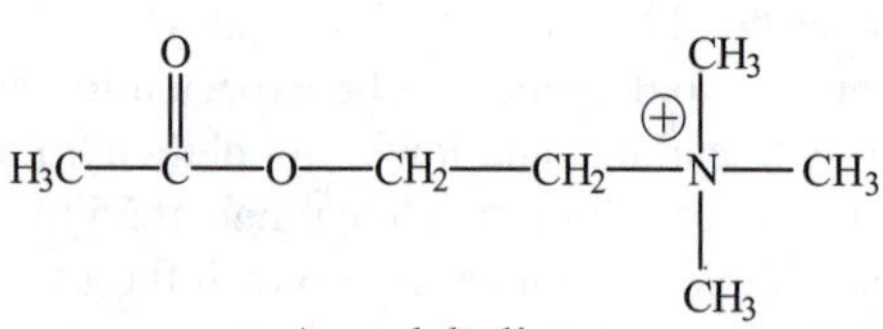

Acetylcholine

Acetyl choline normally has a fraction of a second to interact before it is hydrolyzed by the acetyl choline esterase. The receptor site is thus cleared for the next incoming signal. If this substance is not removed, the acetyl choline accumulates and interferes with the coordination of muscle response. This interference with muscular function of vital organs produces serious symptoms and eventually death.

8.4 THE CARBAMATES

Carbamate insecticides are frequently employed in situations where insects do not respond to the organophosphate compounds. They tend to be more expensive than the organophosphate compounds to produce. The carbamate pesticides have the following general formula:

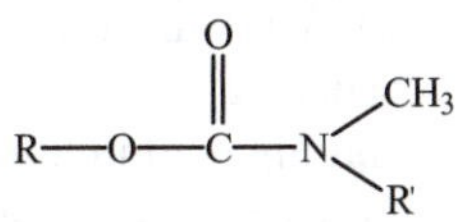

where R can be a variety of groups, often containing aromatic rings, and R′ can be hydrogen or other groups.

Examples of carbamate insecticides include propoxur, also known as Baygon, and carbaryl, the structures of which are illustrated below:

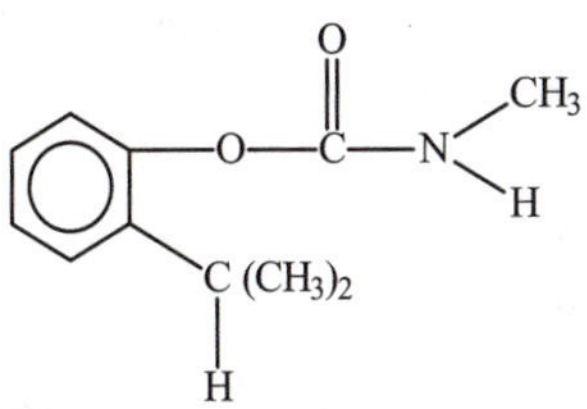

Propoxur

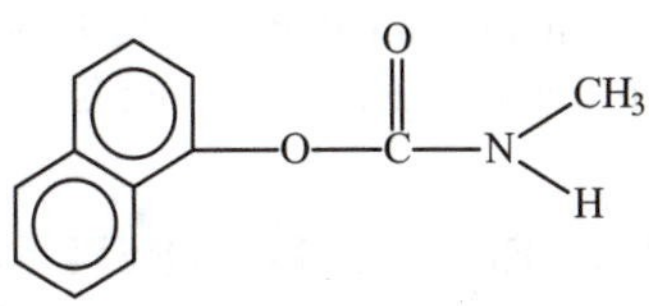

Carbaryl

8.4.1 ENVIRONMENTAL PROPERTIES

The general properties of the carbamates are somewhat similar to those of the organophosphate compounds and are summarized below:

1. The carbamate functional group can be regarded as a hybrid between ester and amide groups, both of which are polar (Table 2.4), leading to a reasonably polar molecule.
2. These substances are relatively water soluble in comparison with the chlorohydrocarbons. For example, propoxur dissolves to the extent of about 2 g/l and carbaryl to the extent of about 120 mg/l.
3. They have limited persistence in the environment, which is reflected by the presence of reactive groups in the molecule. For example, the ester group is easily hydrolyzed when this substance comes in contact with water in the environment. The group, in general, is extremely susceptible to hydrolysis, particularly in alkaline water. This means that the carbamates will exhibit low persistence in the aquatic environment, which has

been found to range from about 1.6 days to 4 weeks. The hydrolysis reaction is illustrated below:

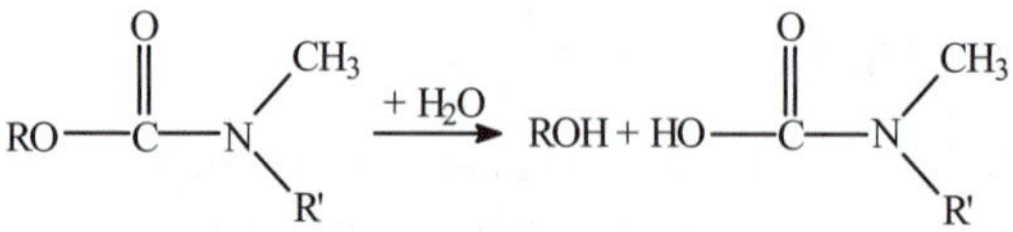

The carbamate insecticides also have generally short lives in soils, with similar half-lives to those observed in the aquatic environment.

4. Some are extremely toxic to mammals and are absorbed readily through the skin, but tend not to bioaccumulate as would be expected due to their relatively low persistence and high solubility in water.
5. The principal biodegradation pathways are hydrolysis and oxidation.

8.4.2 Mechanism of Toxic Action

Carbamates act in a manner analogous to the organophosphate compounds. This means that they are anticholinesterase compounds that inhibit the removal or breakdown of acetyl choline. An important distinction from the organophosphate compounds is that the enzyme cholinesterase is not attached or deactivated for such a long period. This means there are little delayed or long-term effects on the nervous system with this group of compounds.

8.5 PYRETHRINS AND PYRETHROIDS

Pyrethrum is a natural insecticide found in the flowers of certain plants belonging to the genus *Chrysanthemum* (*C. cinerariaefolium* and *C. coccineum*). This group of substances forms another major group of insecticides, which are in wide usage. The *Chrysanthemum* species are daisy-like plants originating from the area now known as Iran. Today pyrethrin flowers mainly come from the highland areas of Kenya, where they contain up to 3% pyrethrins. There are about six principal active components in the pyrethrin flowers. These components have the general structure shown in Figure 8.6. These substances are esters, and the constituent acids and alcohols are practically inactive toward insects.

8.5.1 Physical Properties

The chemical structures of these substances contain some polar groups, such as C=O, but have large nonpolar groups present. As a result, they are oily liquids that are soluble in alcohol and acetone but poorly soluble in water. In this respect, they resemble the chlorohydrocarbon insecticides, differ from most of the organophosphate and carbamate insecticides, and could be classified as **lipophilic**. Most **pyrethroids** are relatively high boiling point, viscous liquids with low vapor pressure due to their relatively high molecular weight. Only a few, for example, allethrin (Figure 8.7), are sufficiently volatile to be useful constituents of mosquito coils.

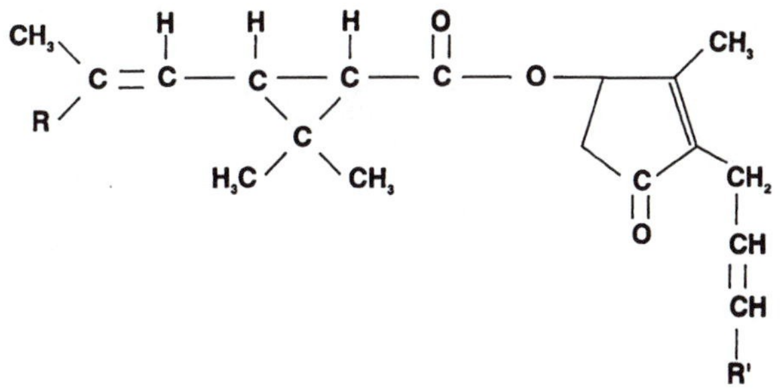

COMPOUND	R	R'
Pyrithrin I	CH_3	$CH_2 = CH$ -
Pyrithrin II	- $COOCH_3$	H_3C -
Cinerin II	- $COOCH_3$	H_3C - CH_2 -

FIGURE 8.6 Chemical structures of some active pyrethrum constituents.

These coils are lit and work on the principle that the pesticide can be vaporized into the atmosphere, thereby repelling insects. Natural pyrethroids are unstable toward moisture, largely due to the presence of the ester group, which is particularly susceptible to hydrolysis. In water, natural pyrethroids are hydrolyzed, with the reaction being both acid and base catalyzed. The structure also contains a conjugated group, i.e., C=C–C=O, which strongly absorbs the high-energy ultraviolet radiation, leading to light-induced photochemical reactions. These factors lead to about a 20% loss of insecticidal activity per year, even within the dried flower heads.

8.5.2 Mechanism of Action and Toxicology

Pyrethroids are neurotoxic, which means that they are poisons that attack the nerve ends. They are also contact insecticides as opposed to systemic insecticides and fumigants. This means that they enter the insect body by absorption through the cuticle following direct contact with the insecticide through droplets, dust, or contaminated surfaces. This mechanism of absorption operates because of pyrethroids' lipophilic properties. The symptoms of pyrethrin poisoning follow the typical pattern of nerve poisoning. First, excitation of the organism occurs, followed by convulsions, after which paralysis and death occur. These substances do not interact with acetyl cholinesterase, as do the organophosphate insecticides and carbamate insecticides. The toxic action is due to a loose chemical binding of the pyrethrin to a neural receptor, which alters the sodium and potassium ion conductance.

Natural pyrethrins are rapidly degraded by mammals and also insects, and usually have low mammalian toxicity. The ester linkage of natural pyrethrins is

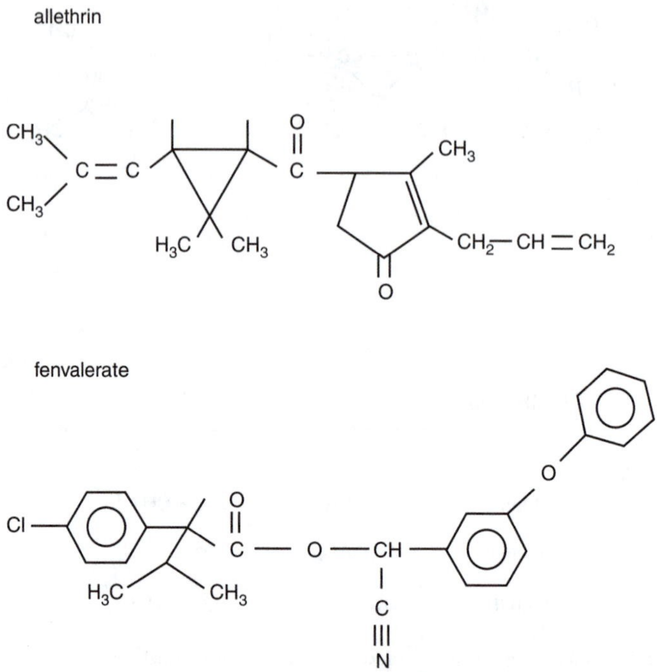

FIGURE 8.7 Chemical structures of some synthetic pyrethroids.

unstable when exposed to enzymes in the insect gut. Once pyrethrins have gained access to the tissues, they appear to be oxidized by a system involving MFO induction, as indicated in Figure 8.8.

In relatively small doses, the pyrethrins typically exhibit a **knockdown effect**, which is the induction of temporary paralysis. There is normally a large difference in the dosage required to cause death over that required to cause a knockdown effect. The knockdown and lethal effects of pyrethrins may be due to two different mechanisms or two different sites of action. This is suggested by the observation that knockdown tends to be associated with the more polar pyrethroids, while lethality seems related to the more lipophilic pyrethroids. As a general rule, the pyrethrins are quite toxic to fish, despite their ready hydrolysis, and also to bees.

In domestic formulations, the pyrethrins are usually used in conjunction with a synergist. The ability of insects to degrade and detoxify pyrethrins is largely due to mixed-function oxidases, and the synergistic compound typically inhibits the MFO induction. Thus, while the synergistic compound alone is relatively innocuous in combination with the pyrethrin, a relatively toxic mixture is formed. Synergistic compounds include piperonyl butoxide and several other substances. The ratio of synergistic compound to pyrethrin is usually about 10 to 1.

8.5.3 Synthetic Pyrethroids

The elucidation of the structure of the natural pyrethrins has made possible the synthesis of related compounds, synthetic pyrethroids, possessing similar insecticidal activity but being more stable to moisture and light. Allethrin (Figure 8.7) was the

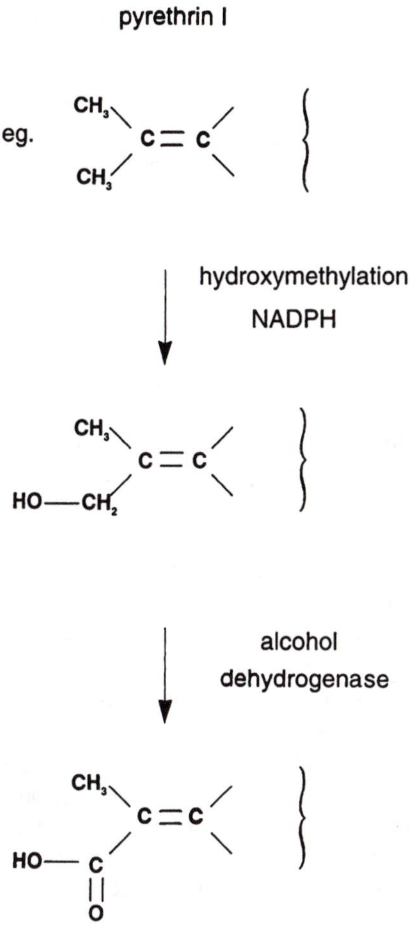

FIGURE 8.8 Oxidative degradation of pyrethrin I.

first synthetic pyrethroid, developed in 1949. The synthesis of pyrethroids generally involves the maintenance or enhancement of the insecticidal activity while increasing the environmental stability.

Synthesis has been based on the principle of copying the molecular geometry of the natural compounds rather than mimicking their structural chemistry, as illustrated by Fenvalerate in Figure 8.7. The low mammalian toxicity associated with the natural pyrethrins extends to some, but not all, of the synthetic pyrethroids. As a general rule, the synthetic compounds have high toxicity to fish.

8.6 PHENOXYACETIC ACID HERBICIDES

The **phenoxyacetic acid herbicides** were developed during World War II, based on the structure of the natural plant hormones, the **auxins**. An example of an auxin, indole-3-acetic acid, is shown in Figure 8.9, and as a result, these substances are

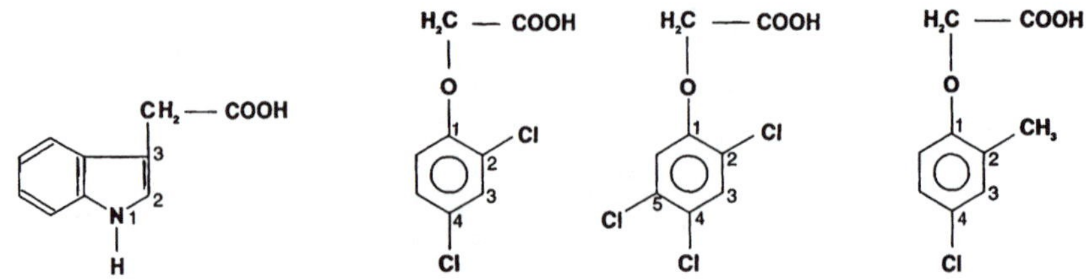

indoyl - 3 - acetic acid(IAA) | 24D | 245T | MCPA

FIGURE 8.9 The chemical structures of some commonly used herbicides and IAA.

often described as the **hormone** weed killers. Examples of common phenoxyacetic acid herbicides are shown in Figure 8.9. These substances can be synthesized by the route shown previously in Section 6.3.2. This involves the use of the sodium salt of 2,4,5-trichlorophenol and chloracetic acid.

These substances are polar and exhibit relatively high solubility in water. As a rule, they have comparatively low toxicity to aquatic animals with LC_{50} values from several hundred to several thousand parts per million. Persistence in the environment is also comparatively low. A particular environmental problem is concerned with trace contaminants, the **dioxins**, which are produced during the manufacturing process and are considered in detail in Section 6.3.2. The dioxins produced include the particularly toxic compound TCDD. If the conditions of temperature and pressure are not carefully controlled during the manufacturing process, the formation of TCDD is enhanced as well as the production of other dioxins. It is believed that TCDD can possibly be formed from the phenoxyacetic acid herbicides after application by processes in the environment (Figure 8.10). TCDD is one of the most toxic substances ever made synthetically, and usually concentrations have been restricted to below 1 ppm in the final product. TCDD is moderately persistent, having a half-life in soil of about 1 year. This substance has been implicated in a range of environmental problems, and as a result, the use of the phenoxyacetic acid herbicides is banned in many countries. In addition, 24D and 245T were used extensively during the military conflict in Vietnam where many adverse effects have been reported.

8.7 KEY POINTS

1. In chemical terms, the pesticides are an extremely diverse group of substances, having only toxicity to pests as a common characteristic. They can be classified chemically as carbamates, phenoxyacetic acids, organophosphates, organochlorines, and so on. Often they are classified according to function as herbicides, insecticides, and fungicides, as well as other names according to the pest targeted.
2. Most pesticides are synthetic chemical compounds that are widely used since costs are relatively low. However, the pyrethrins are natural

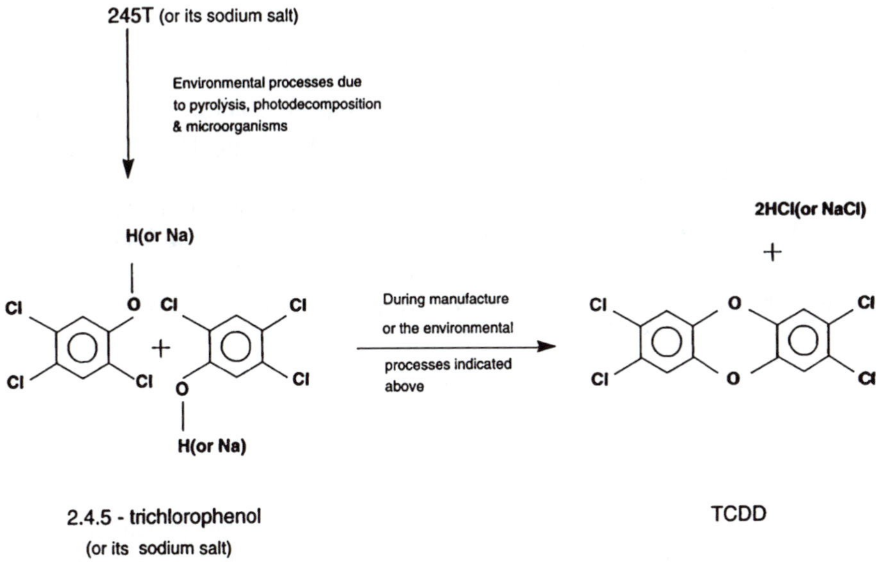

FIGURE 8.10 Possible mechanisms for the formation of TCDD.

pesticides isolated from the *Chrysanthemum* species that have achieved commercial success.

3. The chlorinated hydrocarbon insecticides achieved outstanding commercial success during the 1960s and 1970s, but have been banned in many countries due to their persistence and other unsatisfactory environmental effects. This group is typified by DDT and dieldrin.
4. The chlorinated hydrocarbons contain a limited range of bond types, principally C⋯C (aromatic), C=C, C–H, and C–Cl, with some C–C. The bonds have low dipole moments and are relatively resistant to attack by chemical agents in the environment. Thus, the members are lipophilic with log K_{OW} values ranging from 2.67 to 6.36 and correspondingly low aqueous solubility.
5. The chlorohydrocarbons are strongly bioconcentrated, with 4,4′-DDT exhibiting a log K_{OW} value of 6.36. They are toxic to a wide range of biota, particularly aquatic biota and insects. The cyclodiene subgroup tends to have higher mammalian toxicity than the other members. The major ecological effect has been observed with DDT as a reduction of reproductive success with some bird species.
6. The organophosphate pesticides have the following general structure:

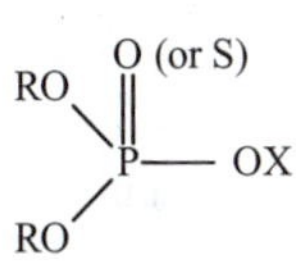

The two R groups are usually methyl or ethyl groups, the O in the OX group is replaced with S in some compounds, and the X group can take a wide diversity of forms. This general structure indicates polar molecules with greater water solubility than the chlorohydrocarbons, but lower lipid solubility. The group is reactive and generally susceptible to hydrolysis in the environment.

7. The organophosphate pesticides are toxic to a wide range of biota and act by inhibiting acetyl cholinesterase. Their lack of environmental persistence leads to a lack of bioaccumulation capacity.
8. The carbamate pesticides have the following general structure:

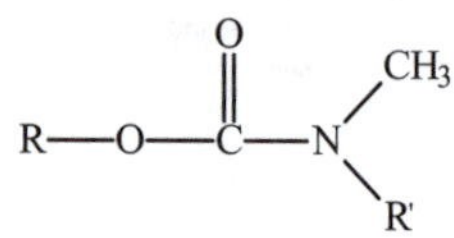

R and R′ can be a variety of different groups. The compounds in this group are relatively water soluble with limited environmental persistence due to their ready hydrolysis. They are toxic to a wide range of biota and act by inhibiting cholinesterase.

9. The pyrethrins are a group of naturally occurring pesticides that are present in certain *Chrysanthemum* species. Their chemical structure is complex but contains the following grouping:

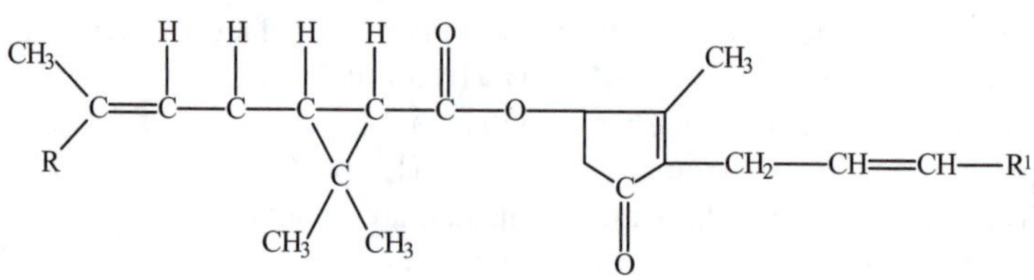

where R and R′ are ester and other groups. In addition, there is a group of structurally related synthetic pyrethroids. The molecules are relatively large and contain large hydrocarbon groupings with a limited number of polar groups. Thus, they are poorly soluble in water and tend to be lipophilic. They are readily degraded in the environment by hydrolysis of the ester linkage to inactive products. They do not interact with acetyl:cholinesterase, but by attachment to a receptor that alters sodium and potassium ion conductance. The group has low mammalian toxicity.

10. The phenoxyacetic acid herbicides are manufactured synthetically, and during this process the polychlorodibenzodioxins, particularly tetrachlorodibenzodioxin (TCDD), can be formed in trace amounts. TCDD is persistent, highly toxic, and bioaccumulative.

REFERENCES

Connell, D.W. and Miller, G.J., *Chemistry and Ecotoxicology of Pollution*, John Wiley & Sons, New York, 1984.

Green, M.B., Hartley, G.S., and West, T.F., *Chemicals for Crop Improvement and Pest Management*, 3rd ed., Pergamon Press, Oxford, 1989.

Howard, H.P., Boethling, R.S., Jarvis, W.F., Heylan, W.M., and Michalenko, E.M., *Handbook of Environmental Degradation Rates*, Lewis Publishers, Chelsea, MI, 1991.

QUESTIONS

1. DDT has been banned in many areas and, in some cases, replaced with endosulfan. Compare the chemical structures and related physical-chemical properties, as well as expected bioaccumulation and persistence of these two substances.
2. Malathion (Figure 8.5) is a typical organophosphate pesticide and is not very persistent in the environment. Write equations for an initial set of reactions that this compound could undergo in the environment.
3. The pyrethrins are lipophilic compounds. Explain why they do not accumulate in soils or biota in terms of their chemical structure and physicochemical properties.
4. The phenoxyacetic acid herbicides contain an acidic group (–COOH), which can be converted into the sodium salt. Compare and contrast the chemical structure, physicochemical properties, and expected environmental behavior of 2,4-dichlorophenoxyacetic acid (2,4-D) and its sodium salt.

ANSWERS

1.

Chemical Structure and Property	DDT	Endosulfan
Chemical structure	Chlorohydrocarbon with only C⋯C (aromatic), C–H, C–Cl, and C–C bonds present	Similar bonds to DDT, but with a S-containing group
Physicochemical properties	A relatively nonpolar compound since nonpolar bonds are present; this means the log K_{OW} would be high and water solubility low	The S-containing group is polar and would be expected to make endosulfan much more water soluble, with a much lower log K_{OW} than DDT
Persistence	High due to unreactive bonds present	Much less than DDT due to a S-containing group, which would be expected to be susceptible to hydrolysis
Bioaccumulation	Persistent and lipophilic, and thus highly bioaccumulative	Much less bioaccumulative than DDT due to its relatively low persistence and log K_{OW} value

2. The following hydrolysis reactions could occur with malathion at the indicated bonds:

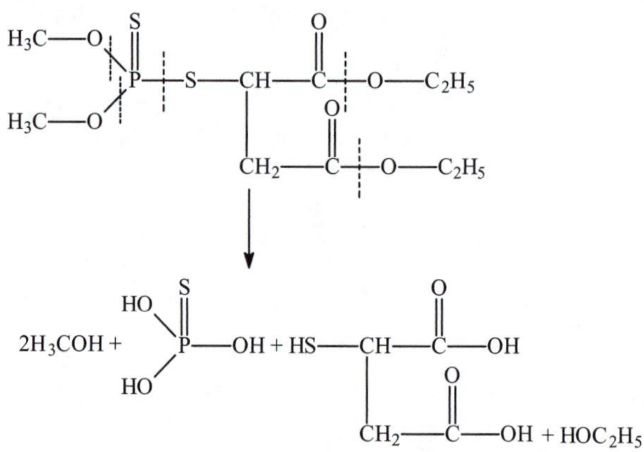

3. The chemical structures contain a high proportion of nonpolar bonds such as C–C, C–H, and C=C, which are nonpolar in character, making the compounds lipophilic. But the structures also contain the ester group, which is susceptible to hydrolysis. In addition, the structures contain a conjugated carbonyl group (C=C–C=O), which absorbs strongly in the ultraviolet range of the spectrum. This leads to activation of the molecule

and degradation. So, while the compounds are lipophilic and thus could accumulate in biota lipid, they lack sufficient persistence to bioaccumulate.

4.

Chemical Structure or Property	2,4-D	Na Salt of 2,4-D
Chemical structure	Weakly ionic	More strongly ionic
Physicochemical properties	Weakly ionic, so solubility in water would be low and vapor pressure low	Greater ionic strength would give greater water solubility but lower vapor pressure due to greater ionic character
Environmental behavior	Moderate accumulation in soil — low evaporation into the atmosphere	Poor accumulation in soil due to high water solubility — very low volatilization in the atmosphere

9 Polycyclic Aromatic Hydrocarbons (PAHs)

9.1 INTRODUCTION

The **polycyclic aromatic hydrocarbons** (PAHs) have been contaminants of the human environment ever since human life first evolved because they are ubiquitous environmental contaminants originating from combustion in cooking fires and fires for warmth. The scientific investigation of the PAHs and their effects started in 1775 with Sir Percival Pott, who attributed scrotum cancer in chimney sweeps in London to exposure to soot and ash. Later investigations strongly suggest that the causative agents present in the soot and ash were PAHs. Direct evidence of the involvement of PAHs as agents of cancer was produced during the 1930s. During this decade, some PAHs were shown to be powerful carcinogens, and since that time, many of the PAHs have been shown to possess similar properties (Grimmer, 1983).

The PAHs family of hydrocarbons consists of molecules containing two or more fused six-carbon-atom aromatic rings. Two common members of the group are naphthalene and benzo(a)pyrene, with two and five fused rings, respectively. In fact, benzo(a)pyrene is a powerful carcinogen, and often the occurrence of PAHs is reported in terms of this substance. In addition, the members share common properties of relatively low water solubility, with the most important members being lipophilic.

Their ubiquitous environmental occurrence stems from their many sources, both natural and anthropogenic. As a general rule, the PAHs are produced by combustion, which can be natural, for example, forest fires, or anthropogenic, for example, combustion in automobiles. These substances have been shown to be widely distributed in aquatic sediments, water, air, plants, and animals. The interest in this group derives from this wide occurrence and the possible induction of cancer in organisms as a result.

9.2 CHEMICAL NATURE OF PAHS

The PAHs comprise a large family of hydrocarbons and thus contain carbon and hydrogen only. Each member consists of a number of benzene rings fused together through two or more carbon atoms. The possible number of different PAHs is enormous, but there are common PAHs that occur throughout the environment (Bjorseth, 1983), some of which are listed in Table 9.1. Naphthalene is considered to be the simplest member of the family, and its structure is among those shown in Figure 9.1. On the other hand, coronene is considered to be the highest-molecular-weight PAH of environmental significance; its structure is also shown in Figure 9.1.

TABLE 9.1
Some Physical Properties of PAH Compounds of Environmental Interest

PAH	Molecular Formula	Molecular Weight	Melting Point (°C)	Boiling Point (°C)	Vapor Pressure (kPa)	Aqueous Solubility (mol/l)	Log K_{OW}
Naphthalene	$C_{10}H_8$	128.2	81	218	1.09×10^{-2}	2.48×10^{-4}	3.36
Acenaphthene	$C_{12}H_{10}$	154.2	93	279	5.96×10^{-4}	2.55×10^{-5}	3.92
Phenanthrene	$C_{14}H_{10}$	178.2	100	340	2.67×10^{-5}	7.25×10^{-6}	4.57
Anthracene	$C_{14}H_{10}$	178.2	218	342	1.44×10^{-6}	4.10×10^{-7}	4.54
Fluoranthene	$C_{16}H_{10}$	202.3	107	384	2.54×10^{-4}	1.29×10^{-6}	4.90
Pyrene	$C_{16}H_{10}$	202.3	149	404	8.86×10^{-7}	6.68×10^{-7}	5.18
Benzo(a)anthracene	$C_{18}H_{12}$	228.3	157	438	—	6.14×10^{-8}	5.61
Benzo(a)pyrene	$C_{20}H_{12}$	252.3	178	495	6.67×10^{-13}	1.51×10^{-8}	6.04
Perylene	$C_{20}H_{12}$	253.3	277	503	—	1.59×10^{-9}	6.04
Coronene	$CH_{24}H_{12}$	300.4	438	590	—	4.67×10^{-10}	6.90

While PAHs usually contain fused benzene rings, there are some that can contain five-membered rings as well, for example, acenaphthene and fluoranthene, as shown in Figure 9.1. It is also worthwhile to keep in mind that there are other PAHs based on these structures as a parent structure and that contain attached alkyl and other groups. In addition, there can be closely related compounds that contain oxygen, nitrogen, and sulfur atoms, as well as fused benzene rings similar to the PAHs.

Naphthalene has a molecular formula of $C_{10}H_8$ and a molecular weight of 128, as shown in Table 9.1. Also, it has a carbon content of 94% and a hydrogen content of 6%. As the PAHs become larger and more complex (moving down in Table 9.1), the percentage of carbon tends to increase and that of hydrogen to decrease. For example, pyrene (Figure 9.1, Table 9.1) has a molecular formula of $C_{20}H_{12}$ and therefore contains 95% carbon and 5% hydrogen. While coronene is considered to be the highest-molecular-weight PAH of environmental interest, the ultimate PAH may be considered to be graphite. Graphite consists of layers of fused benzene rings in which the different layers are held together by comparatively weak forces so that they can slide over one another. Because of this property, graphite is a valuable lubricant.

9.3 ENVIRONMENTAL PROPERTIES

The PAHs are usually solids with naphthalene, the lowest molecular weight member, having a melting point of 81°C. The melting point increases with molecular weight to coronene, with a melting point of 400°C. Similarly, the boiling point increases from naphthalene at 200°C to coronene at 590°C (Table 9.1). As expected with increasing molecular weight and molecular size, the aqueous solubility and vapor pressure decline. Both aqueous solubility and vapor pressure are comparatively low, even for low-molecular-weight compounds, but are very low for the high-molecular-weight compounds (Table 9.1). This suggests that aqueous solubility and vapor

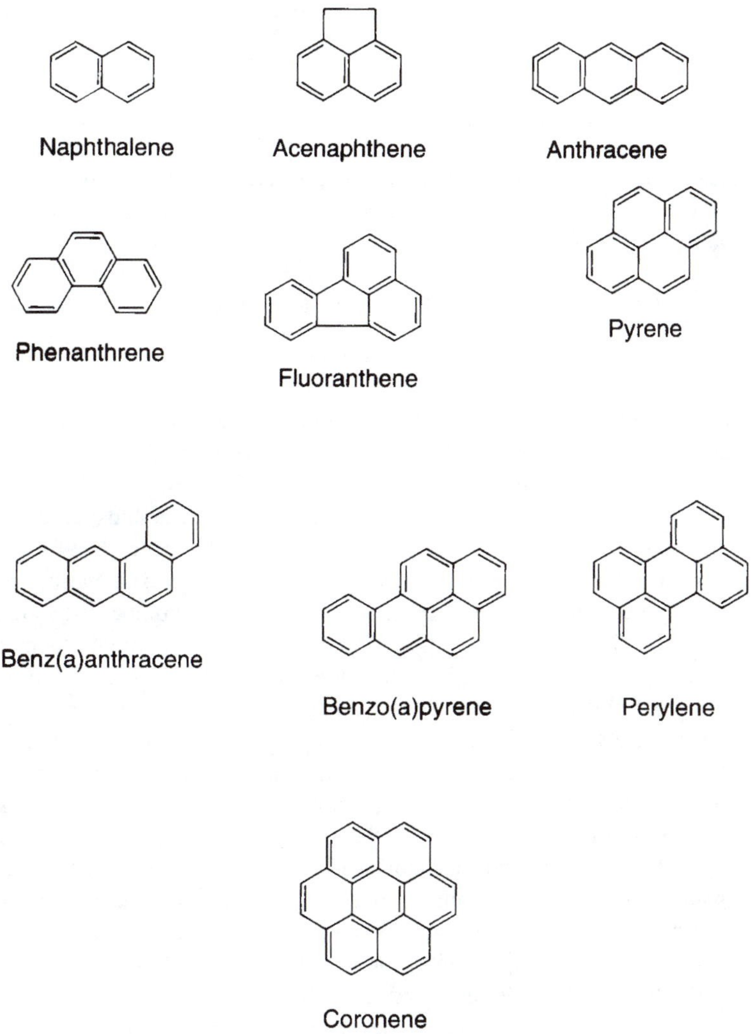

FIGURE 9.1 Chemical structures of some typical PAHs frequently encountered in the environment.

pressure may be important factors influencing environmental behavior for the lower-molecular-weight PAHs, but the influence of these properties will decline with increasing molecular weight.

The octanol–water partition coefficient (K_{OW}) values are a measure of the tendency of a compound to dissolve in biota fat (see also Chapter 2) and other lipoidal substances, such as humic acid in soil and sediments (see Chapter 14). The PAHs significant from an environmental perspective are listed in Table 9.1 and all lie in the range of log K_{OW} 2 to 6, which is generally considered the range for lipophilic compounds. Thus, the PAHs would be expected to bioaccumulate and concentrate in sediments and soils in the environment to an extent depending on their persistence in these media as well.

9.4 FORMATION OF PAHS

Complete combustion of hydrocarbons in oxygen results in the complete oxidation of the carbon and hydrogen present to carbon dioxide and water. As shown below with naphthalene,

$$C_{10}H_8 + 12O_2 \rightarrow 10CO_2 + 4H_2O$$

Naphthalene

Combustion is not as simple as is expressed in this equation, particularly with the combustion of organic substances in the natural environment. At the high temperatures of a flame, greater than 500°C, some of the C–C, C–H, and other bonds are broken to ultimately form free radicals. Depending on the abundance of oxygen present, many of these fragments will react with oxygen to form carbon dioxide and water vapor. But usually oxygen is not sufficiently well dispersed and mixed with the fragments to be able to react efficiently to form carbon dioxide and water. As a result, many organic fragments will react with other fragments close to them, which may be other free radicals formed from the initial hydrocarbon. As the mixture cools, it forms more complex fragments, often leading to PAHs, as shown in Figure 9.2. In this way, a variety of PAHs can be formed depending on the conditions that exist at the time. For example, the formation of PAHs will tend to be more prevalent in an atmosphere where insufficient oxygen is available for complete combustion. The amount of PAHs formed also depends on the combustion temperature and the nature of the organic material combusted. However, irrespective of the type of material burned, which could be coal, cellulose, tobacco, polyethylene, and other polymeric materials, similar ratios of PAHs are formed at a defined temperature. The PAHs listed in Table 9.1 are among the most common PAHs formed through the combustion process. It should be remembered that alkyl-substituted PAHs will also be formed. As a general rule, the higher the combustion temperature, the less alkyl-substituted PAHs will be produced.

During the geological formation of fossil fuels, PAHs can be formed by somewhat different processes. In the formation of coal and petroleum, biological material is broken down by pressure and modest temperatures (less than 200°C). Under these conditions, PAHs can be formed by mechanisms similar to those involved in incomplete combustion. Because of the lower temperatures involved, the transformations occur at a much slower rate. In addition, there may be differences in the types of PAHs that are formed in coal and petroleum. Relatively large amounts of alkyl-substituted PAHs are present in crude oil and coal-derived material, compared with the amounts formed in combustion processes. On combustion of petroleum and coal, some PAHs are released unchanged in emissions and some are transformed into other PAHs. Of course, the combustion process itself will also produce a range of PAHs, as would normally be expected in the incomplete combustion process.

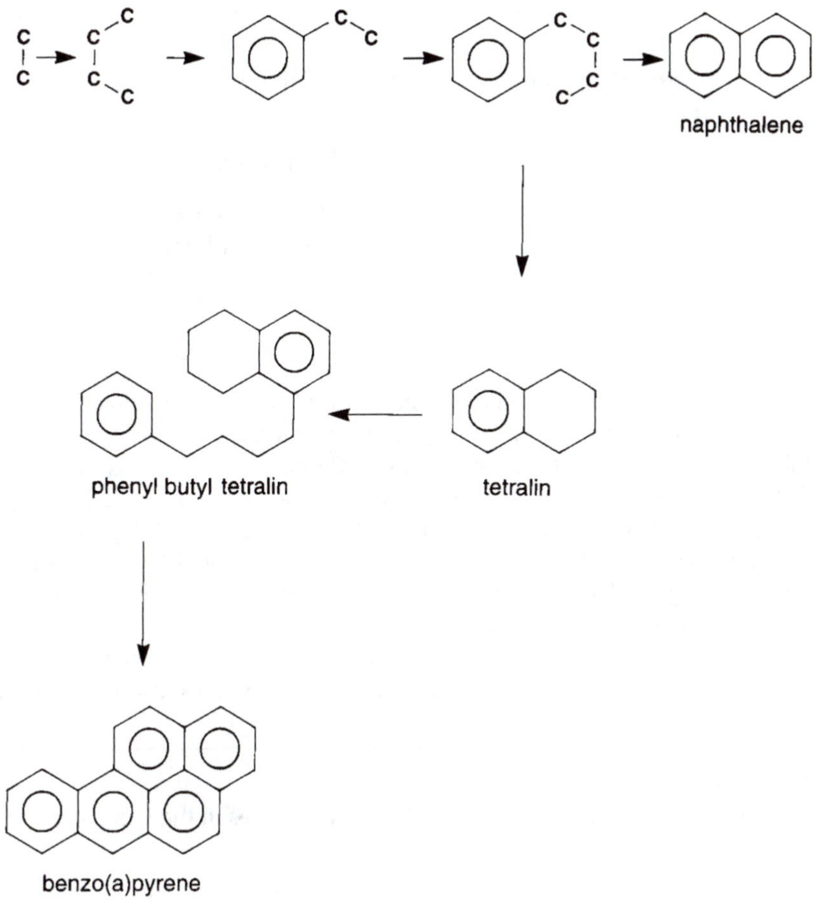

FIGURE 9.2 Possible mechanisms for the formation of PAHs during combustion.

There have been suggestions that there are biological pathways for the formation of PAHs. This is still the subject of considerable scientific debate. In any case, the amount formed in biosynthesis is considered to be relatively low compared to the amounts formed abiotically.

9.5 SOURCES OF THE PAHS

There are a wide range of primarily natural sources of PAHs in the environment. Foremost among these are forest fires and volcanic activity. The actual quantities involved are variable depending on the sporadic nature of these events. The data on production of benzo(a)pyrene shown in Table 9.2 indicate that open burning can produce gaseous emissions that have an extremely high concentration of PAHs. Significant proportions of a range of PAHs, including those in Table 9.1, are produced, although Table 9.2 only reports the PAHs as benzo(a)pyrene, which is considered to be indicative of this group. A large proportion of the PAHs released into

TABLE 9.2
Benzo(a)pyrene Production from Different Processes

Process	Production
Coal-fired power plant	30–930 ng/m^3 emission
Municipal incinerators	17–2700 ng/m^3 emission
Open burning	2800–173,000 ng/m^3 emission
Motor vehicle gasoline combustion	25–700 μg/l gasoline
Tobacco combustion	0.8–2.0 μg/g tobacco

the environment arise from anthropogenic sources such as coal-fired electricity power plants, incinerators, open burning, and motor vehicle exhausts, as indicated in Table 9.2. The relative importance of the anthropogenic sources is indicated in Table 9.3. Industrial processes, residential combustion for heating, mobile transport emissions, and incineration are the most significant sources of PAH production.

Considering the data in Table 9.2 and Table 9.3, it could be concluded that tobacco smoking is relatively insignificant as a source of PAHs. This would be true in overall quantitative terms, but importantly, this source results in direct exposure of humans and, in many cases, close exposure of others who are nonsmokers. Its interesting to note that the cancer-inducing effects of tobacco smoking are not due

TABLE 9.3
Estimated Total PAH Emissions from Different Sources in the U.S., Sweden, and Norway

Source	Quantity[a] (%)		
	U.S.	Sweden	Norway
Industrial processes (aluminum production, iron and steel works, coke manufacturing, ferro-alloy industry, asphalt production, carbon black, petroleum cracking)	3497 (41)	312.3 (62)	202.7 (67)
Residential combustion (wood, coal, oil, gas)	1380 (16)	132 (26)	62.5 (21)
Mobile sources: (gasoline automobiles, diesel automobiles, air traffic)	2170 (25)	47 (9)	20.1 (7)
Incineration: (municipal, open burning, forest fires, agricultural burning)	1150 (13)	3.5 (<1)	13.7 (5)
Power generation: (coal and oil fired, peat, wood, straw, industrial boilers)	401 (5)	13 (3)	1.3 (<1)
Total	**8598**	**507.8**	**300.3**

[a] Metric tons per year.

TABLE 9.4
Concentrations of PAHs in the Ambient Atmosphere Produced by Cigarette Smoking

PAH	Ambient Air Concentration (ng/m^3)
Fluoranthene	99
Pyrene	66
Benzo(a)anthracene	100
Benzo(b,j,k)fluoranthene	35
Benzo(a)pyrene	22
Perylene	11

to exposure to nicotine, although nicotine is toxic, but exposure to PAHs produced by the combustion of tobacco. The range of PAHs identified as being produced by smoking is considerable and includes several established carcinogens, including benzo(a)pyrene. The PAHs identified in ambient air in which cigarettes are smoked are shown in Table 9.4. It can be seen that this includes the well-known carcinogen benzo(a)pyrene.

9.6 OCCURRENCE AND BEHAVIOR OF PAHS IN THE ENVIRONMENT

In accord with the many and varied human and natural sources of PAHs, these substances are very widespread in the environment. The concentrations that occur are highly variable, ranging from very high concentrations in combustion particulates to the very low amounts that generally occur in fresh- and seawater. Some typical examples of the occurrence of PAHs are shown in Table 9.5. The PAHs as a group share the common property that they generally consist of linked six-membered

TABLE 9.5
Typical Examples of Some PAHs in the Environment

PAH	Combustion Particulates (mg/kg)	Air (Typical City Locations) (ng/m^3)	Smoked Fish (μg/kg)	Sewage Sludge (μg/kg, dry weight)
Fluoranthene	4–400	1–15	300–3000	2–7
Benzo(a)anthracene	2–160	0.1–20	20–200	1–4
Perylene	0.1–138	1	1–4	0.1–2
Benzo(a)pyrene	0.2–64	1–1000	4–16	0.5–3
Coronene	0.1–40	2	1–10	0.1–2

aromatic rings. The aromatic bonding within the rings tends to make aromatic compounds resistant to attack by chemical and other agents. So it would be expected that the PAHs would tend to be stabilized by the presence of aromatic rings in conjugated form. On the other hand, the compounds have strong ultraviolet and visible radiation absorption. This means that there is an uptake of energy that can be used to chemically modify and transform these substances. So this characteristic would tend to make the compounds less stable in the environment. In addition, the extensive occurrence of these substances in the natural environment would tend to develop populations of organisms that would have the capacity to degrade them.

The major degradation pathways involve chemical, photolytic, or metabolic processes associated with microorganisms. In many situations all of these processes may occur together, whereas in some circumstances one or various combinations of them may be in operation. These processes are also strongly influenced by environmental conditions such as temperature, availability of oxygen, populations of microorganisms present, and so on.

The chlorination and ozonation of treated water from wastewater treatment plants can result in significant degradation of any PAHs present. Laboratory experiments with these processes have indicated that many PAHs exhibit very low persistence on the order of minutes or hours under these conditions. Somewhat similarly, PAHs in water where oxygen is available, dissolved within the water mass when exposed to sunlight, can exhibit rapid rates of degradation. The half-lives for PAHs in clear water exposed to strong sunlight and oxygen within the water mass can be less than 1 hour, as indicated by the data in Table 9.6. In turbid waters, however, this photolysis would be greatly slowed through the diminution of sunlight and also the partitioning of the PAHs onto particulate matter present and onto bottom sediments. Once sorbed to particulates and bottom sediments, the rate of degradation would be expected to be substantially reduced. The metabolic degradation of PAHs by microorganisms usually occurs through the co-metabolism of the PAHs with normal organic food material. The process usually involves the oxidation of the PAHs to produce oxidized materials that are subsequently further degraded to simple products. The biodegradation of PAHs by bacterial populations in aquatic systems has been shown to

TABLE 9.6
Persistence of PAHs in the Environment

	Half-Lives	
PAH	Clear Water Exposed to Sunlight (h)	Sediment–Water Microcosm (weeks)
Naphthalene	—	2.4–4.4
Benzo(a)anthracene	0.54	—
Phenanthrene	—	4–18
Pyrene	0.75	34–>90
Benzo(a)pyrene	0.034	200–>300

proceed at different rates for different compounds. The half-lives are measured in a sediment–water system, shown in Table 9.6. This indicates that as a general rule, the persistence of the PAHs increases with increasing size of the molecule. The higher-molecular-weight PAHs, such as benzo(a)pyrene, show a half-life on the order of up to about 6 years in this situation.

The PAHs of environmental significance (see Table 9.1) generally lie in the log K_{OW} range from 2 to 6. So these compounds would therefore be classified as lipophilic and would have the potential to bioaccumulate. The relatively low persistence of naphthalene and other low-molecular-weight PAHs indicates that these would have a limited capacity to exhibit bioaccumulation. On the other hand, the high molecular weight compounds, for example, benzo(a)pyrene, are persistent and lipophilic and therefore would be expected to bioaccumulate.

9.7 CARCINOGENICITY AND TOXICITY OF THE PAHS

The major area of environmental concern with the PAHs is their ability to produce cancers in exposed organisms. A range of PAHs has been found to have strong carcinogenic activity with animals, including benzo(a)anthracene and benzo(a)pyrene, as listed in Table 9.7. With many PAHs, evidence establishing carcinogenicity is not available, and they are not classified one way or the other. But importantly, almost all PAHs are suspected to be carcinogens to some degree or another, although this level could be very low.

To understand the process of cancer induction, it is necessary to examine the metabolic fate of the PAHs within organisms. Mammals, for example, have evolved a group of enzymes to convert xenobiotic lipophilic compounds, including PAHs, to polar water-soluble products. The enzymes involved are the mixed-function oxidases (MFOs), which consist of enzymes in the cytochrome P_{450} group (discussed in Chapter 3). This enzyme system is actually stimulated within an organism by

TABLE 9.7
Carcinogenic Activity of Some PAHs

Compound	Overall Evaluation[a]
Phenanthrene	3
Anthracene	3
Fluoranthene	3
Pyrene	3
Benzo(a)anthracene	2A
Benzo(a)pyrene	2A

[a] 1 = carcinogenic to humans; 2A = probably carcinogenic to humans; 2B = possibly carcinogenic to humans; 3 = not classified; 4 = probably noncarcinogenic to humans.

exposure to persistent lipophilic compounds. Repeated exposure results in the induction of increased quantities of the P_{450} type. The capacity to induce these enzymes by organisms is variable, depending on the organism group involved. For example, mammals have quite strong inductive capacity and, as a result, have a good ability to degrade persistent lipophilic compounds. Other organism groups that have much more limited capacity to induce MFO, for example, fish, have a relatively limited capacity.

The initial products of the metabolic degradation of the PAHs involve the insertion of an oxygen atom into the PAH structure to form an epoxide. Hydrolysis followed by further epoxidation leads to the formation of dihydrodiol epoxide. This process is illustrated by the reaction sequence in Figure 9.3. These substances have the capacity to bind to centers in biological molecules such as DNA and hemoglobin. Certain of these complex substances are formed by this process have the capacity to cause tumor formation. However, this later formation can often take a considerable period and is not particularly well understood at present.

The PAHs have acute toxicity to aquatic organisms, which can be measured as the LC_{50} (lethal concentration for 50%) value shown in Table 9.8. As a general rule, toxicity has been found to increase with molecular weight and log K_{OW} value. Its interesting to note that solar radiation has been found to significantly increase the toxicity of PAHs to aquatic organisms. It is believed that this occurs from photoactivation of the PAH molecules present on or within the organisms, and not metabolism or formation of other degradation products.

9.8 EFFECTS ON HUMAN HEALTH AND THE NATURAL ENVIRONMENT

There is a range of information available on the chronic effects of PAHs on aquatic organisms in natural systems as well as in laboratory experiments. Generally exposure to sublethal levels of PAHs, in the lipophilic range of log K_{OW} values from 2 to 6, results in morphological, physiological, and developmental abnormalities in fish. Low concentrations have also been found to cause significant reduction in the hatchability of eggs, as well as larval length and weight. Of particular concern is the occurrence of mutagenic and carcinogenic effects. Benzo(a)pyrene has been shown to induce chromosome aberrations in fish. In addition, the PAHs have been found to cause epidermal hyperplasia and neoplasia. Some species of fish have been shown to develop liver cancer as a result of exposure to PAHs.

Tobacco smoking results in a high mortality in human populations due to the occurrence of cancer and other effects, and the PAHs must be considered to be implicated in this situation. Diet is also particularly noteworthy, and there is a high incidence of mortality associated with agents present in food. Smoked fish, which is relatively high in PAHs, may be among a range of causative factors in some individuals and populations. Also, the low incidence of mortality due to cancer associated with air pollution may be related to the effects of PAHs.

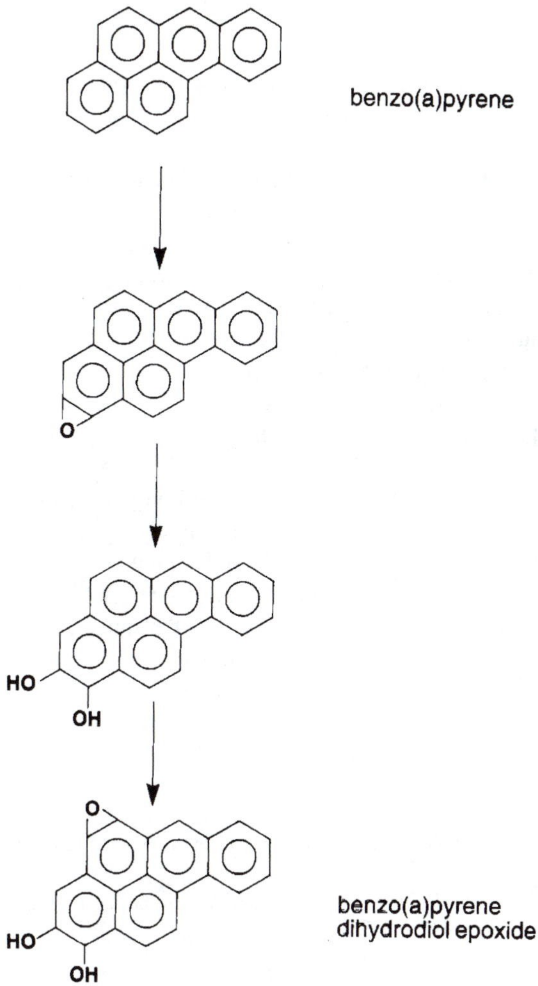

FIGURE 9.3 Metabolic oxidation of benzo(a)pyrene to the dihydrodiol epoxide by the P_{450} enzyme system.

TABLE 9.8
Toxicity of PAHs to Various Aquatic Organisms

PAH	Organism	LC_{50} (96 h) (mg/l)
Naphthalene	Fish	0.1–8.0
	Crustacea	1.0–2.4
Acenaphthene	Fish	0.6–3.0
Phenanthrene	Fish	0.04–0.6

9.9 KEY POINTS

1. The PAHs are produced by a wide range of combustion processes used in human society as well as natural combustion and geological processes.
2. A range of PAHs, including naphthalene, phenanthrene, anthracene, pyrene, benzo(a)pyrene, perylene, and coronene, commonly occur in air, soil, and biota in the natural environment.
3. The PAHs of environmental concern range in molecular weight from naphthalene at 128 to coronene at 300 and have log K_{OW} values from 3.36 to about 6.90, respectively. This means that the common PAHs are lipophilic compounds.
4. The low-molecular-weight PAHs have limited persistence in the environment, but the higher-molecular-weight compounds are more persistent; for example, benzo(a)pyrene can persist in aquatic systems for up to about 300 weeks.
5. The higher-molecular-weight PAHs in the lipophilic group, with log K_{OW} values from about 2 to about 6, are moderately persistent in the environment and so are capable of being bioaccumulated.
6. The common group of environmental PAHs are relatively toxic to aquatic organisms and have LC_{50} values to fish in the range of 0.1 to 8 mg/l.
7. Many PAHs are carcinogenic, with human exposure occurring through tobacco smoking as well as compounds in the diet.

REFERENCES

Bjorseth, A., *Handbook of Polycyclic Aromatic Hydrocarbons*, Marcel Dekker, New York, 1983.

Grimmer, G., *Environmental Carcinogens: Polycyclic Aromatic Hydrocarbons*, CRC Press, Boca Raton, FL, 1983.

QUESTIONS

1. Using the data in Table 9.1, make estimates of the log K_{OW} of the following compounds:

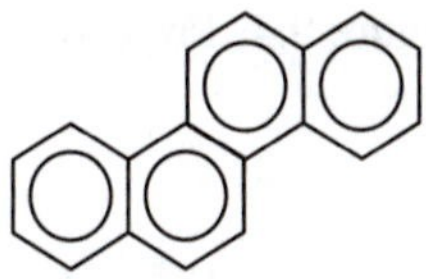

Chrysene

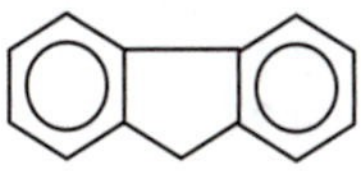

Fluorene

2. PAHs can be formed during pyrolysis by a variety of processes. Suggest possible fragmentation combination patterns, starting with a two-carbon fragment, for anthracene, acenaphthene, and pyrene.
3. An environmental investigation of PAHs has commenced. Data on the persistence in natural aquatic systems are required for fluoranthene and perylene. Make estimates of these characteristics using the information available in this chapter.

ANSWERS

1. a. The molecular weights can be calculated from the structural formula: chrysene, $C_{18}H_{12}$, 228; fluorene, $C_{13}H_{10}$, 166.
   b. Using the data in Table 9.1, plots can be made of molecular weight against log K_{OW}, as shown in Figure 9.4. Projections can be made to give the following log K_{OW} values:
   Chrysene 5.53
   Fluorene 4.24
2. Possible combination pathways are shown in Figure 9.5.
3. The half-life data for the PAHs in Table 9.6 can be plotted against log K_{OW} or molecular weight as shown in Figure 9.6. The maximum and minimum half-lives can be joined by an eye-fitted line of best fit. Fluoranthene and perylene have molecular weights of 202 and 252, respectively, and log K_{OW} values of 4.96 and 6.04, respectively. Using these values, the following half-lives can be obtained by interpolation:

	MW Estimate (weeks)	**Log K_{OW} Estimate (weeks)**
Fluoranthene	30–90	20–50
Perylene	200–300	200–300

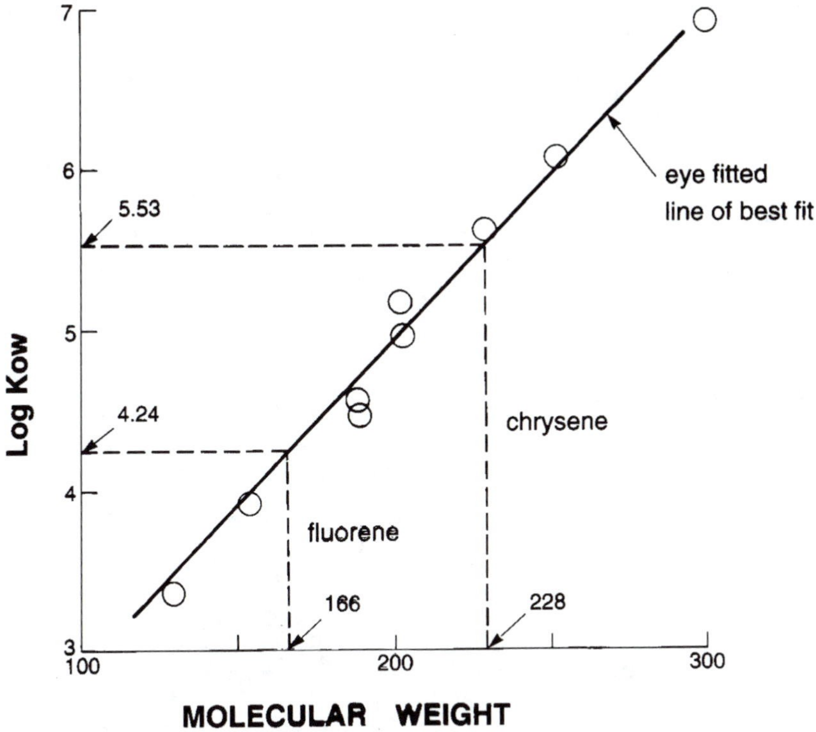

FIGURE 9.4 Plot of the log K_{OW} against molecular weight using the data in Table 9.1 with projections to obtain log K_{OW} for chrysene and fluorene (answer to question 1).

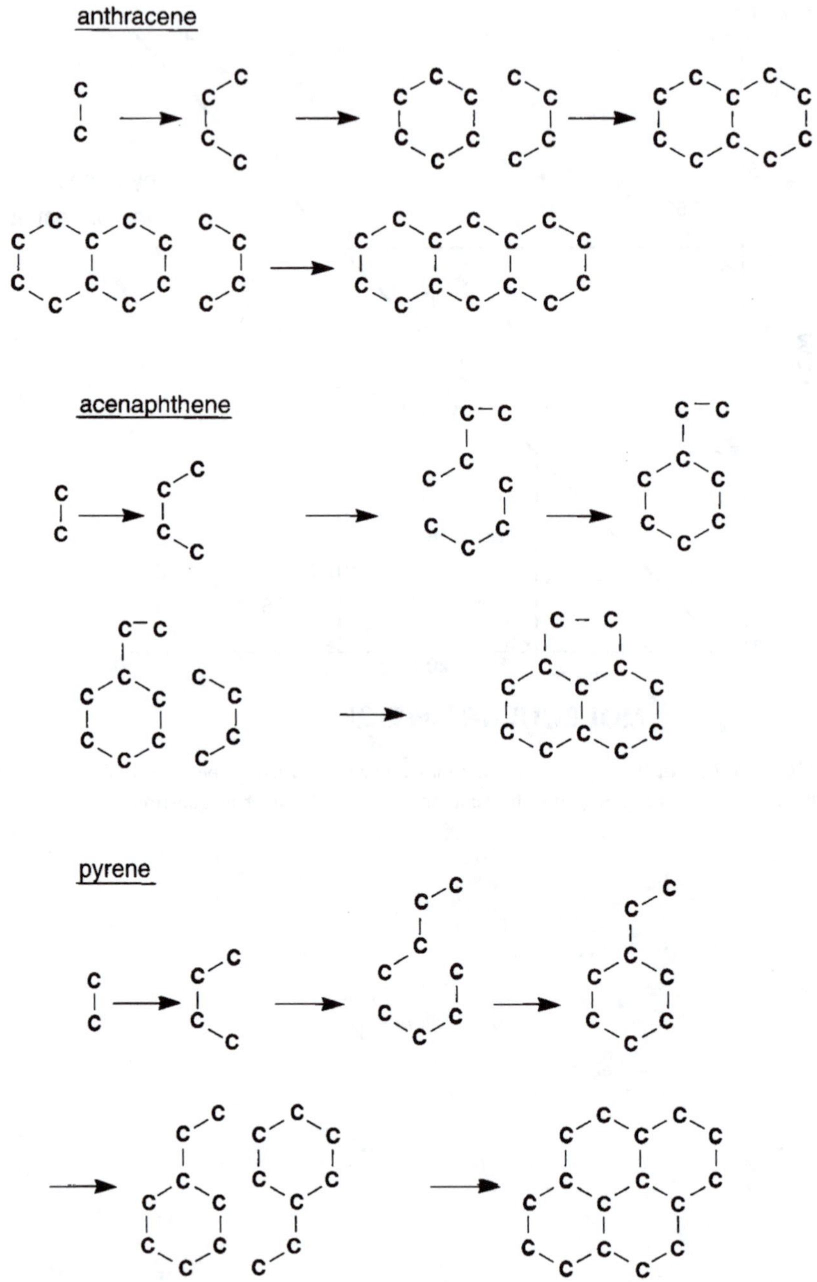

FIGURE 9.5 Possible two-carbon fragment combination processes that can lead to the formation of anthracene, acenaphthene, and pyrene (answer to question 2).

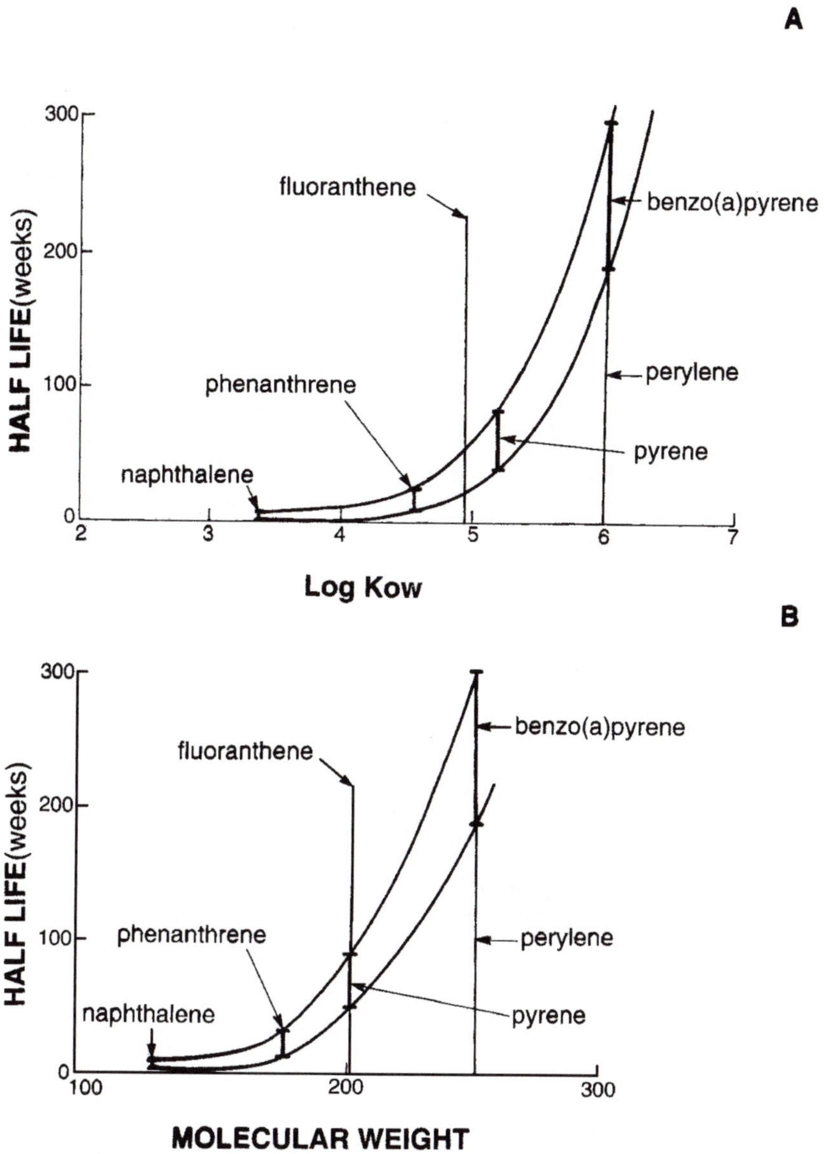

FIGURE 9.6 A: Plot of half-life against log K_{OW}. B: Plot of half-life against molecular weight using the data in Table 9.6 (answer to question 3).

10 Soaps and Detergents

10.1 INTRODUCTION

Cleaning ourselves, our clothing, our homes, our eating utensils, and so on, has been a concern of human beings since the beginning of human society. This serves an aesthetic purpose, but it is also important in the control of disease and maintenance of good health. The cleaning process almost always involves the use of water as a solvent and a carrier of the removed contaminants. This wastewater often is discarded into waterways, where it can cause adverse effects.

Soap is the oldest and best-known chemical agent to assist cleaning. Soap has been prepared since ancient times by recipes handed down from generation to generation. Little was known in the past regarding the chemistry of its preparation and mode of action. Now we have a detailed knowledge of the chemistry of soap, which has been used to develop more effective cleaning agents commonly called **detergents** (Hutzinger, 1992).

The development of new cleaning agents started in Germany during World War I when blockades prevented the importation of raw materials for soap manufacture. This steadily increased in both range of types and overall quantities until today; the annual worldwide production exceeds 15 million metric tons and grows about 3% per year. About half of this is soap, but this proportion is declining. In 1994, consumption in the U.S. totaled roughly 4×10^9 kg, only about 5% being soap. Some of the broad categories are shown in Figure 10.1. The quantities are roughly

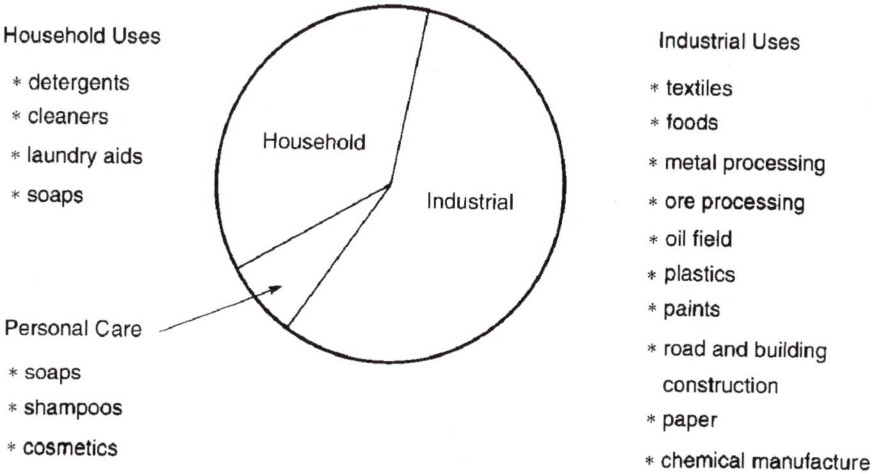

FIGURE 10.1 Surfactant use categories with approximate proportions.

split between domestic and industrial use. Domestic laundry detergents alone account for one quarter of the total amount of surfactants made.

During the 1950s and 1960s, large masses of foam could be seen on treatment ponds in sewage plants and many rivers and streams. This was caused by detergents in wastewaters and aroused concern regarding possible harmful effects on the natural environment. The occurrence of detergent components in natural water bodies has been reduced in recent decades by the use of substances that are more readily degraded. Soap has limited persistence in waterways and has not generally caused environmental problems.

10.2 SURFACTANTS: THE ACTIVE CLEANING AGENTS

Soaps and detergents contain substances that are described as **surfactants** (surface-active agents). They act at the surface, or interface, between polar and nonpolar phases to modify the properties of the phases (Clint, 1992). The surface of water, for example, can be seen as the interface between the air and water phases, with water having polar properties and air having nonpolar properties. The presence of surfactants modifies the surface properties of both phases. They have a key molecular property, illustrated in Figure 10.2 by sodium myristate, a component of soap. The molecule has two parts with very different characteristics. The long hydrocarbon chain forms a nonpolar tail, and the carboxylate group forms a polar head. The head group is hydrophilic (water loving) and imparts water solubility, but the hydrophobic (water-hating) tail would prefer to dissolve in nonpolar materials such as grease. Surfactants are often termed amphiphiles (from the Greek, meaning having a love of two kinds). For this reason, surfactants have low solubilities in water and other solvents.

The surfactant molecules are most stable at interfaces such as air-–water or oil–water surfaces (Figure 10.3). They gather together, orienting themselves with the hydrophilic part in the water and the hydrophobic tail projecting into the air or grease. This aggregation is so efficient that the surfaces can be completely covered with a surfactant **monolayer**, even though the surfactant concentration in the bulk water may be as low as 10^{-4} molar. As a result, the surfactant monolayers greatly reduce the **surface tension** because the negatively charged hydrophobic heads repel one another. Surface tension is a measure of the energy required to form a surface (about 74 and 50 mJ m^{-2} for water–air and water–hydrocarbon surfaces, respectively). Surfactants can reduce the water–air tension to about 25% of its usual value, and the effect is much more dramatic at the water–grease surfaces, where the tension

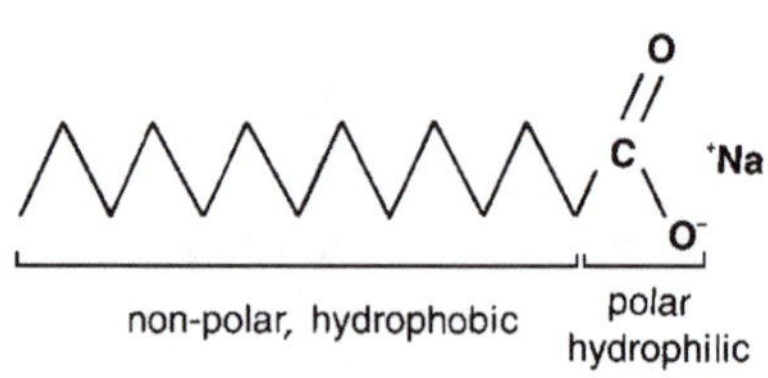

FIGURE 10.2 Structure of a soap component, sodium myristate, $CH_3(CH_2)_{12}COONa^+$. The hydrophobic alkyl tail is flexible.

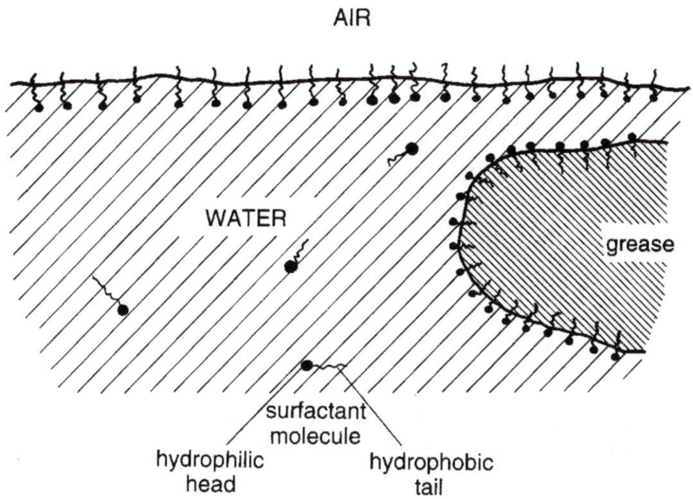

FIGURE 10.3 Surfactant molecules aggregating at water–air and water–grease interfaces. The polar heads are shown as filled circles.

can fall more than 10^5-fold. This effect promotes the formation of structures with very large surface areas, such as foams and emulsions. Soap bubbles are short-lived since they are both kinetically and thermodynamically unstable. On the other hand, oil and water and a little surfactant can form stable, permanent emulsions, largely due to the extremely low tensions produced.

When surfactants reach their solubility limit, they often do not precipitate as do conventional solutes, but display another effect. The molecules begin to clump together forming micelles, tiny aggregates containing about 50 to 100 individuals. More surfactant can seem to dissolve by forming more micelles, while the actual concentration of single molecules remains about the same. Quite concentrated apparent solutions can result. The thick, clear dishwashing liquids that we use daily are not true solutions, but are instead concentrated suspensions of invisibly small surfactant micelles.

Micelles adopt various shapes and sizes. They may be spheres, discs, rods, sheets, or vesicules (hollow spheres), depending on the type of surfactant, concentration, and other conditions, such as temperature. Sphere and sheet-like micelles are shown in Figure 10.4. The polar heads face outward toward the water, while nonpolar tails form a loosely tangled core. The hydrophobic interior of the micelles behaves much like a tiny drop of oil — a good medium for similar hydrophobic compounds such as petroleum or polychlorinated biphenyls (PCBs). These normally water-insoluble compounds can be apparently solubilized in surfactant solutions by partitioning into the oil-like interior of micelles.

Surfactants are broadly categorized according to the charged nature of the hydrophilic part of the molecule:

- **Anionic** — negatively charged
- **Cationic** — positively charged
- **Nonionic** — neutral, though highly polar
- **Amphoteric** — a zwitterion containing positive and negative charges

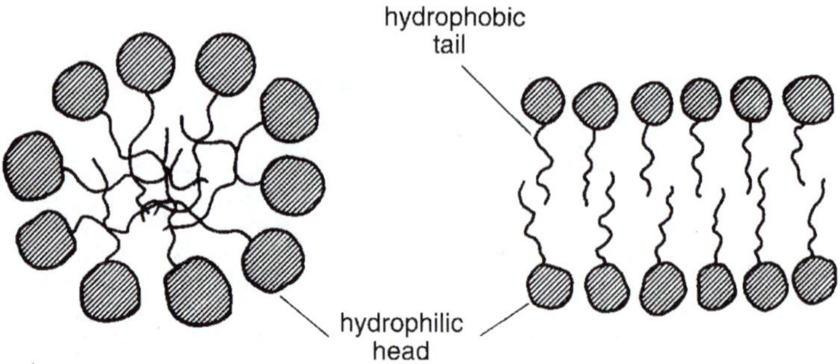

FIGURE 10.4 Cross sections of spherical and bilayer (sheet-like) micelles of surfactant molecules.

Some examples of surfactants in each class are illustrated in Table 10.1, and the broad proportions used are shown in Figure 10.5. The hydrophobic fragments (R groups) are organic groups, usually alkyl chains.

10.3 SYNTHESIS

10.3.1 Soaps

Soaps are salts of long-chain fatty acids. Traditionally, soaps are made by the saponification of natural fats or oils, which are triesters of fatty acids and glycerol.

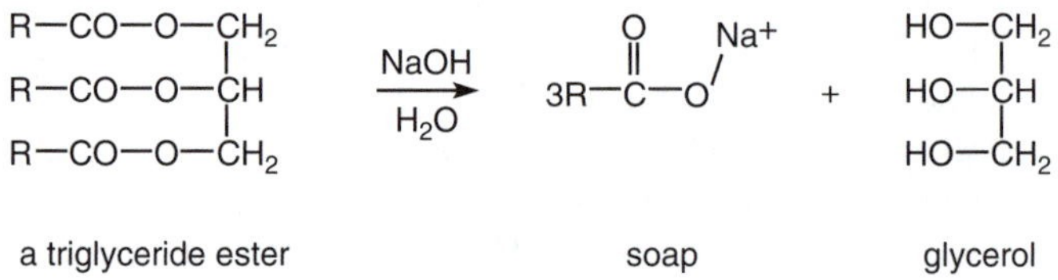

The fat sources are mainly tallow (animal fat) and coconut and palm seed oils. The R groups are linear with variable chain lengths that depend on the source fats. Coconut and palm oil soaps are mainly C_{12} and C_{14}, while tallow soaps are mainly C_{16} and C_{18}.

Today, soaps are usually synthesized industrially by reacting together vegetable oil and water at high temperatures and pressures (where the ingredients are soluble), splitting the ester and forming free fatty acids. The fatty acids can be subsequently neutralized to form soaps, or refined and used for other purposes, including surfactant synthesis.

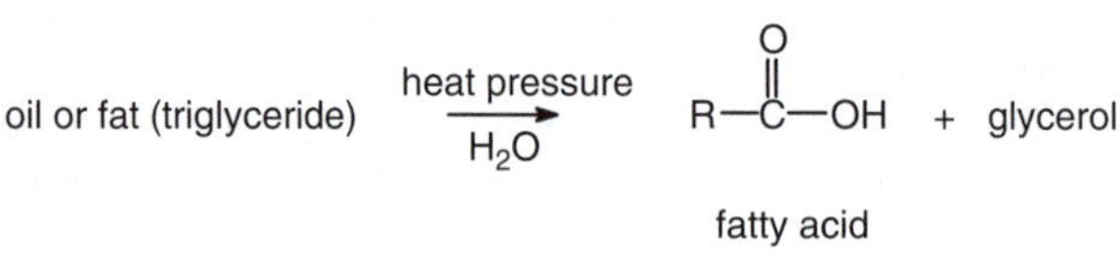

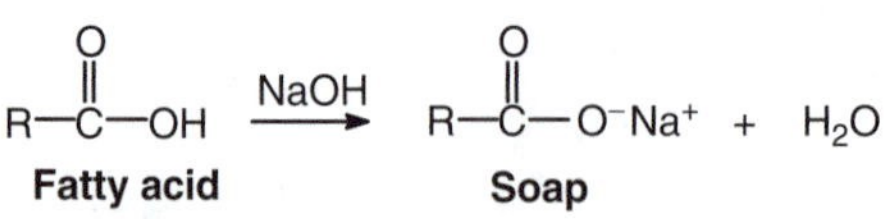

TABLE 10.1

Some Synthetic Surfactants in the Four Surfactant Classes

Common name	Structure	
Anionics		
Carboxylates (soaps)	$R—CH_2—COO^-Na^+$	$R = C_{10\text{-}18}$
Linear alkylbenzene Sulphonates (LAS)	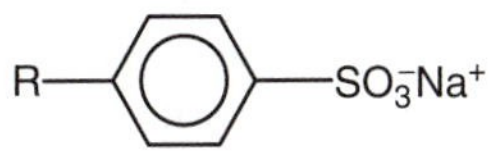	$R = C_{10\text{-}13}$
Alcohol sulfates	$R—CH_2—OSO_3^-Na^+$	$R = C_{11\text{-}17}$
Alcohol ethersulfates	$R—O(CH_2CH_2O)_nSO_3^-Na^+$	
Nonionics		
Fatty alcohol ethoxylates	$R—(OCH_2CH_2)_nOH$	$R = C_{8\text{-}18}$, n = 9-70
Alkyl glucosides	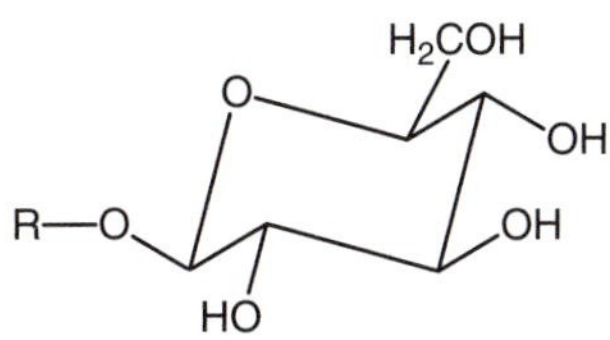	$R = C_{12\text{-}14}$, n = 1-4
Cationics		
A quaternary ammonium chloride	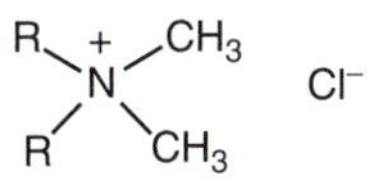	$R = C_{16\text{-}18}$
Amphoterics		
A sulfobetaine	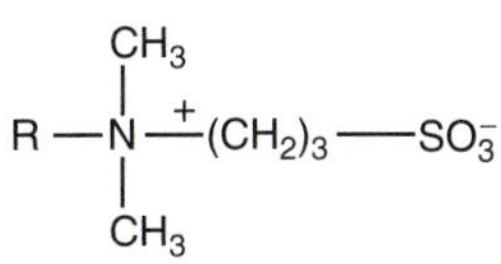	$R = C_{12\text{-}18}$

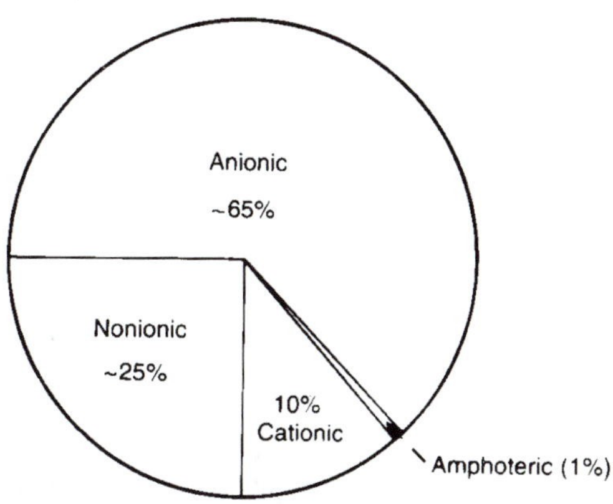

FIGURE 10.5 Broad categories of consumption of surfactants in the U.S. in 1994.

10.3.2 Alkylbenzene Sulfonates and Alkyl Sulfates

The synthetic paths for most other surfactants are varied, but basically involve chemically linking hydrophobic and hydrophilic groups. The raw materials for the hydrophobes can be sourced from natural fats, but most come from the petrochemical industry (~70% worldwide). Today, surfactants consume about 0.4% of the world's petroleum supply.

Alkanes are too unreactive for direct surfactant synthesis. Intermediate compounds containing active centers, such as double bonds, benzene rings, or hydroxyl groups, are used. **Linear alkylbenzene sulfonates** (LASs) are usually made by attaching a benzene ring to a long-chain olefin, an unsaturated hydrocarbon, followed by sulfonation, which provides the hydrophile. Neutralization with sodium hydroxide again finishes the sequence, yielding the surfactant.

The starting alkene (olefin) usually has a range of chain lengths, and the intermediate alkylbenzene can have the phenyl group attached at any position except the terminal carbon. Accordingly, the LAS surfactant synthesized is a **complex mixture** of homologs (different chain lengths) and isomers (different phenyl group attachments). Only the C_{12} homolog and 2-isomer is shown below:

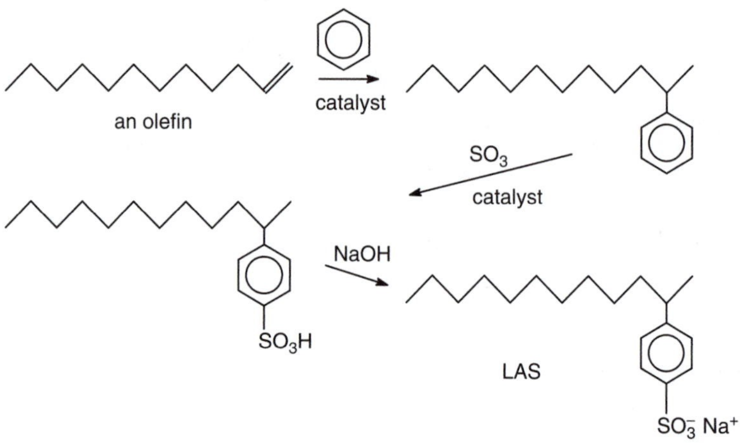

Sulfur trioxide (a product of the sulfuric acid industry) is used in the manufacture of several types of anionic surfactants. **Alkyl sulfates**, for example, are formed by the sulfation of fatty alcohols.

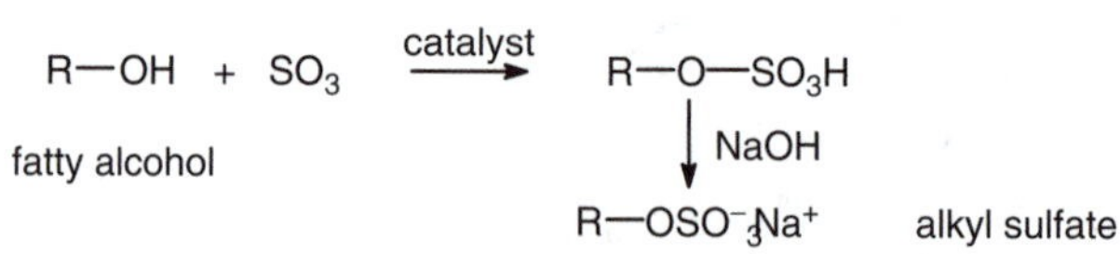

Sulfates differ from sulfonates in having an extra oxygen, contained in the **C–O–S** link.

10.3.3 Alkyl Ethoxylates

The nonionic **alkyl ethoxylates** are formed by reacting ethylene oxide with a long-chain fatty alcohol. The ethylene oxide molecules react with the terminal OH in a short polymerization sequence, building up the water-soluble portion of the molecule:

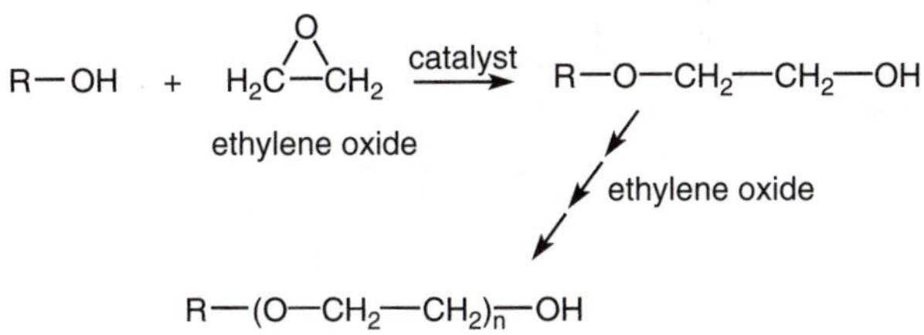

The extent of polymerization can be controlled, depending on the end use of the surfactant. Detergent-grade ethoxylates average about 15 ethylene oxide units per molecule, but the mixture can have a range spanning n = 10 to 20.

10.4 DETERGENTS

10.4.1 Surfactant Components

The major household use of surfactants is in commercial mixtures, such as cleaners or detergents. Laundry detergents are about one-third surfactants, and a breakdown of the main ingredients of a modern compact detergent is given in Table 10.2.

TABLE 10.2
Typical Formulation of a Compact Powder Laundry Detergent

Ingredient	%
Surfactants	30–40
Anionic	30
Nonionic	5
Soap	2
Builders	~35
Polyphosphate/zeolite	
Sodium carbonate/silicate	
Polycarboxylate	
Enzymes	0–1.5
Bleaching agents	0.25
Auxiliaries	~5
Stabilizers	
Brighteners, etc.	

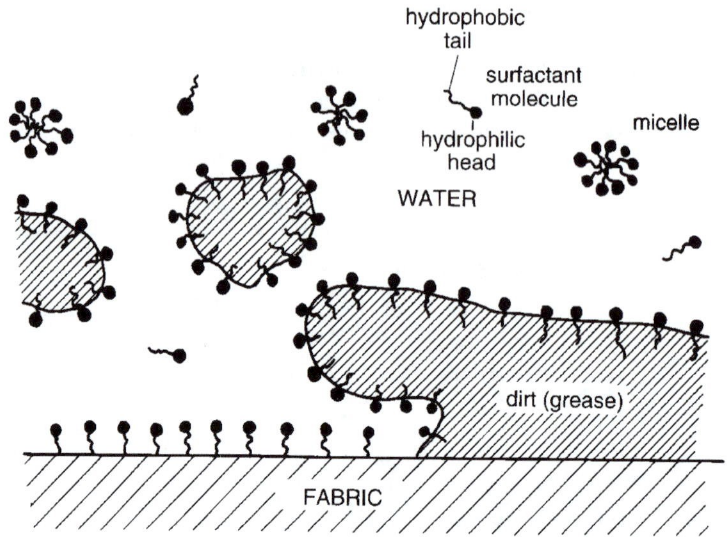

FIGURE 10.6 Detergent action of a surfactant. The cleaned surface fabric is on the left.

Detergent action is quite complex, but basically involves the dislodgment of dirt particles and their suspension in water (Figure 10.6). Dirt particles (food, grease, soil stains) are relatively hydrophobic and firmly attach to weakly polar fabrics. Surfactant molecules attach to the dirt and substrate surfaces, reducing the interfacial tension and promoting wetting. Heat and agitation tend to tear the dirt particles apart and from the substrate. More surfactant absorbs to and stabilizes the freshly exposed surfaces. Micelles play an essential role here in providing a large reservoir of surfactant molecules. Eventually, the dirt particles break free and are dispersed as an emulsion. The new particles have hydrophilic surfaces that have the same charge, which prevents recombination or redeposition by mutual repulsion.

The surfactant components of detergents are anionic, nonionic, or more commonly mixtures of the two. The mixtures exploit the synergistic properties of the two types. Cloth fibers, soil, and organic particles usually carry **negative surface charges**. Sorbed anionic surfactant enhances this negative charge, increasing mutual repulsions and assisting the detersive action. Nonionics promote surface activity at lower surfactant concentrations, since the absence of a head charge permits easier packing into surface regions and micelles. Anionics are better at cleaning cottons, and nonionics are better on synthetics.

Soaps now have declined in their role as cleaning agents. A major factor in this reduced usage is that soaps are sensitive to hard water, forming very water insoluble precipitates with polyvalent ions such as calcium and magnesium.

$$2\,R{-}\overset{\displaystyle O}{\overset{\|}{C}}{-}O^{-}Na^{+}\ \text{(aqueous)} + Ca^{2+}\ \text{(aqueous)} \longrightarrow Ca(R{-}\overset{\displaystyle O}{\overset{\|}{C}}{-}O)_2\ \text{(solid)} + 2\,Na^{+}\ \text{(aqueous)}$$

These greasy solids form the familiar bathtub rings. They can irreparably stain and give clothing a bad odor.

The surfactants do not suffer these precipitation reactions, but their action is still hardness-impaired to some extent. Not surprisingly, the nonionic surfactants are much less hardness affected than their anionic counterparts.

Small quantities of soap are included in laundry detergents to **retard foaming**, by actually exploiting the hard-water precipitation reaction. The precipitated soap is still strongly surface active, and spreads rapidly over the bubble film, displacing other surfactant molecules. But the solid lowers the elasticity of the foam, making it more prone to rupture and collapse (bubbles, like rubber balloons, need to be elastic to resist damage). Foams do not contribute to washing performance and are tolerated to appease consumer preference.

10.4.2 Builders

The second major component of detergents is the **builders**, so called since they increase the action of the surfactants. The builders have several modes of action, including water softening, maintaining an alkaline pH, and improving dirt dispersion.

The major builder in use worldwide is **sodium tripolyphosphate**, $Na_5P_3O_{10}$. The anion of this water-soluble salt is a powerful sequestrant, binding firmly to hardness-causing ions such as Ca^{2+}, Mg^{2+}, and other polyvalent ions.

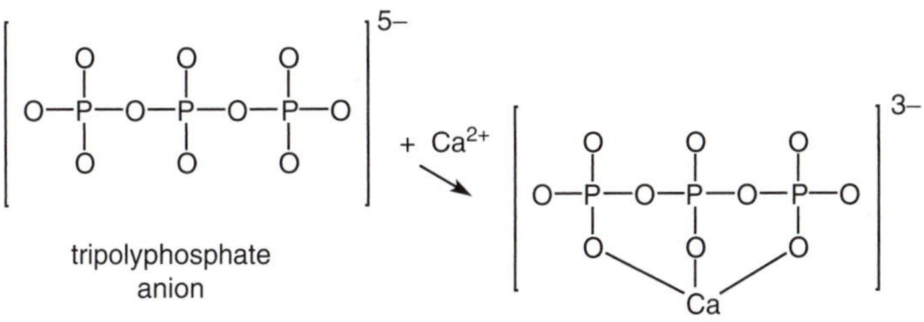

The sequestered metal ion still remains in solution, but in a form that is not available to cause hardness interference with surfactants. Tripolyphosphate is also alkaline and binds to soil particles. Both effects serve to increase the negative surface charges, improving performance of the surfactant.

Tripolyphosphate, however, is a ready source of nutrient phosphorus, and its discharge in large tonnages to waterways contributes to excessive algal growth or eutrophication. For this reason, many government authorities have restricted its use in detergents.

Zeolites have been used as a phosphate builder substitute for 20 years. They behave as a water softener, removing hardness-causing ions by ion exchange. Zeolites are insoluble solid salts called **aluminosilicates**. They occur naturally as minor rock-forming minerals. Detergent zeolite is a synthetic zeolite made from sand, sodium chloride, and bauxite, and it has the formula $Na_{12}(AlO_2)_{12}(SiO_2)_{12}{\cdot}27H_2O$. The anion is a polymer made up of alternating AlO_4 and SiO_4 tetrahedra.

This giant anion is porous, with water molecules and sodium ions occupying the pore spaces. Calcium ions from hard water enter the pores, displacing the sodium and preferentially binding to the surrounding anion.

Zeolites are environmentally benign substances. They have a very low toxicity to mammals and aquatic organisms. Being mineral solids, zeolites deposit as sediments in sewage treatment plants and waterways.

10.4.3 Enzymes and Bleaches

Enzymes and bleaches are two ingredients that augment the cleaning process. Protease and lipase enzyme additives catalyze the hydrolysis of protein and lipid stains, forming hydrophilic products more readily removed by the detergent. Recently, cellulase enzymes have been developed that have a more novel action — they hydrolyze cellulose in the **underlying** cotton fibers. The cellulases, however, selectively attack damaged, nonstructural parts of the fiber, parts that hold stains and cause the cotton to look prematurely worn and jaded. Cellulase washes apparently give a cleaner, newer-looking cloth.

The development of bleaching agents has involved some novel surfactant technology. Bleaches compatible with enzymes are based on peroxygen compounds such as sodium perborate, which contains the active –O–O– group.

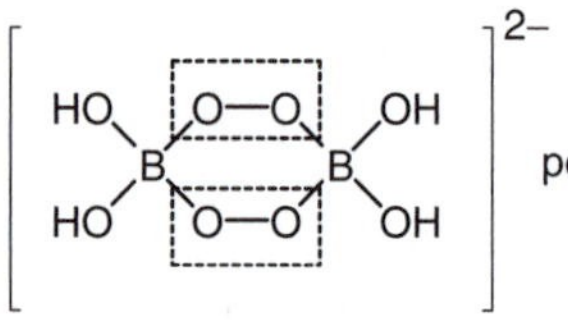

Perborate releases hydrogen peroxide, H_2O_2, when dissolved in water. However, dilute hydrogen peroxide is not an effective bleach in the cool wash conditions popular today. Bleach boosters are needed, such as sodium nonanoyloxybenzene sulfonate (SNOBS). SNOBS is a surfactant that readily exchanges its hydrophile portion for the peroxo group:

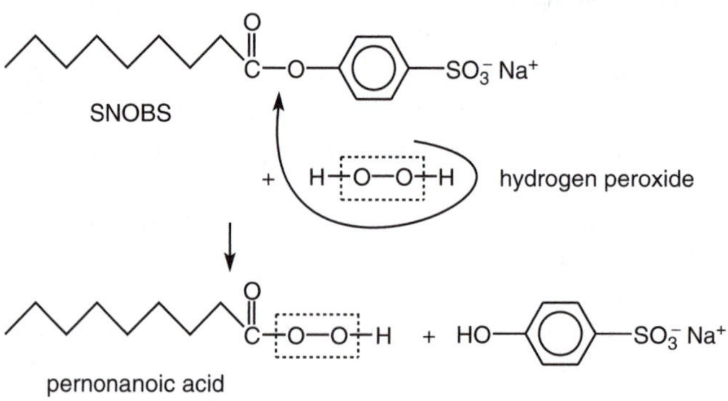

The end result is an in-the-wash synthesis of a surface-active bleach (pernonanoic acid) that has a much higher affinity (and bleaching ability) for the stained fabric.

10.5 CATIONIC SURFACTANTS

Cationic surfactants are generally ineffective as detergents. Their major household use is as fabric and hair conditioners and as disinfectants. Because most natural organic and mineral particles usually carry a negative surface charge, cationic surfactants adopt a head-down position on the surface.

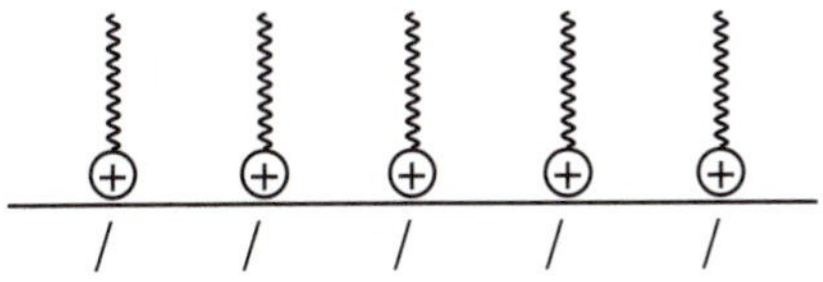

This orientation tends to neutralize the electric charge and offers a hydrophobic surface to the water, both effects discouraging detergency. Just this alignment of surfactant molecules, however, is exploited in fabric and hair conditioners. The fiber surfaces acquire an oil-like film that separates and lubricates the strands, giving a softer feel and reducing static electricity.

Most cationic surfactants behave as **germicides**. Molecules with one long alkyl group are most effective, such as cetylpyridinium chloride, an ingredient of popular mouthwashes.

cetylpyridinium chloride

Cationic and nonionic surfactants are compatible, but cationics and anionics generally are not. The positive and negative molecules combine to form neutral 1:1 complexes that are very insoluble and have much less surface activity. For this reason, fabric conditioners are usually added after the wash cycle, either in the rinse or (as solid-sheet formulations) in the drier.

10.6 TOXICITY

The oral toxicity of surfactants to mammals is low, ranging from about the same as sodium chloride to exhibiting no observable lethal effect at 1000 to >10,000 mg/kg body weight. Many surfactants are classified as edible and used in foods. Sodium dodecylsulfate, $[CH_3(CH_2)_{11}OSO_3^-Na^+]$, for example, is a detergent-grade surfactant used also as an emulsifier and whipping agent. Surfactants are not accumulated by mammals but are very rapidly metabolized and excreted as shorter molecules.

TABLE 10.3
Surfactant Toxicity to Some Aquatic Organisms

Surfactant	Fish LC_{50} (mg/l)	Daphnia LC_{50} (mg/l)	Algae growth inhibition EC_{50} (mg/l)
Anionic			
Linear alkylbenzene sulfonates	3–10	1–70	1–100
C_{12} fatty alcohol sulfates	3–10	5–70	~60
Nonionic			
Fatty alcohol ethoxylates	1–10	1–10[a]	4–50
Alkylglucosides	3.7	—	6.5
Cationic			
Distearyl dimethyl	0.62–3[b]	0.16–0.48[b]	0.1–10
Ammonium chloride	10–24[c]	3.1[c]	

[a] Various invertebrates.
[b] Clarified water.
[c] River water.

Aquatic organisms are much more sensitive to surfactants than mammals, having LC_{50} values in the milligram-per-liter range (Table 10.3). Cationics, in particular, are potent algicides and bacteriocides, with responses below 1 mg/l. The mode of toxicity is not clear, but surfactants seem to interfere with sensitive **external tissues** such as gill membranes.

Toxicities increase with increasing hydrophobic character of the molecule. As the chain length in the linear alkylbenzene sulfates increases, the hydrophobic nature of the compound increases; thus, homologs that have longer alkyl chains become more toxic. The same trend is observed by shortening the polyether chain in ethoxylates (essentially making the hydrophile fragment less effective). This pattern of toxicity fits surfactants into the **general narcosis model**, where partition into a lipid phase is a prime requirement. Surfactants have toxicities not unlike nonsurfactant organic compounds, with similar K_{OW} values and nonspecific modes of action.

The toxicity of cationics is mitigated by the presence of particles or anionic surfactants. The toxicity of the clarified river water without particles is higher (lower LC_{50}) than the unclarified river water (Table 10.3). Sorption to the negative surfaces or the formation of neutral complexes probably reduces the effective numbers of toxic molecules in the water. Commercial mixtures of surfactants decline in toxicity as they degrade forming intermediates that have lost their surface activity.

10.7 BIODEGRADATION

Surfactant biodegradation first became an issue in the 1950s, when alkylbenzene sulfonates (ABSs) largely replaced soaps, as mentioned in the Introduction. Surfactant residues accumulated in sewage treatment plants and waterways, causing spectacular foaming episodes and tastes in recycled water, and reaching levels potentially toxic to aquatic life.

The early surfactants contained highly branched alkyl hydrophobes that are resistant to biodegradation. Today these recalcitrant surfactants are largely obsolete, having been replaced by the linear alkylbenzene sulfonates (LASs) and other biodegradable surfactants.

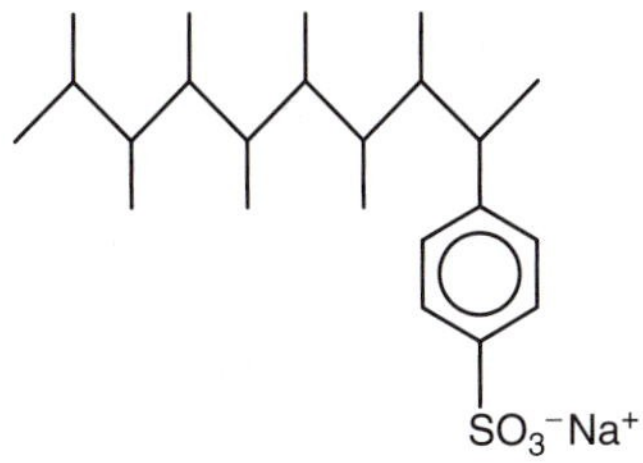

a branched alkylbenzene sulfonate

The variety of surfactant types, coupled with the great diversity of bacterial populations, means that no one degradation process dominates. Enzymatic attack can occur at the hydrophobe, the hydrophile, the connecting link, or all three. Linear alkylbenzene sulfonates degrade via **oxidation** of the hydrophobe chain. Oxidation begins at the terminal carbon forming a carboxyl group — this is the slow step and causes the molecule to lose its surfactant properties. Successive oxidations shorten the chain two carbons at a time (a process that mirrors the way all living organisms metabolize the fatty acids in their cells). Ring opening and mineralization complete the reaction sequence as shown below.

The remainder of the anionic and nonionic surfactants listed in Table 10.1 are regarded as highly biodegradable, often exceeding 98% breakdown when the test criteria demand a minimum of 80 or 90%. Most of the molecules contain chemical groups that exist in natural materials. These groups are prone to attack by hydrolytic enzymes that are common in bacterial assemblages — etherases, sulfatases, esterases, etc.

Cationic surfactants are biodegradable, as long as toxicity thresholds are not exceeded. At concentrations below 1 mg/l, breakdown is rapid and virtually complete, but much higher and their biocidal effect can slow or stop bacterial action. Cationic surfactant biodegradation is improved in the presence of the anionic variety, probably a consequence of the neutralization effect, which reduces toxicity.

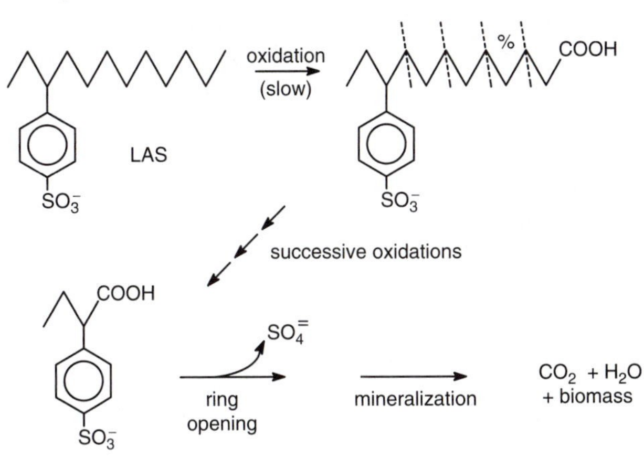

10.8 SORPTION AND BIOACCUMULATION

Cationic surfactants bind strongly to soils and sediments, sorbing to the particles and complexing with humic matter. The organic soil matter, humic, and fatty acids behave as a negative surfactant, forming micellular aggregates. Sediment–water partition coefficients (concentration in soil–concentration in water) can exceed 10^4 for cationics, but are orders of magnitude lower for anionic and nonionic surfactants.

Surfactants show little facility for bioaccumulation, with bioconcentration factors spanning from 3 to 500. The high rates of metabolism tend to limit biological concentrations. Cationics are only slowly taken up by fish since they cling to the gill membranes, but are not transported across this tissue.

10.9 NATURAL SURFACTANTS

Surfactants occur widely in plants and animals, performing many essential functions (Table 10.4). Lecithins, for example, are phospholipids that occur in the walls of living cells. Lecithins have long been commercially extracted from plants (e.g., soya bean) and used in foods. Some bacteria, yeasts, and fungi are prolific surfactant producers, apparently to act as extracellular emulsifiers of water-insoluble nutrients such as hydrocarbons. Surfactant biosynthesis is an emerging technology to provide surfactants designed for specific purposes.

10.10 KEY POINTS

1. During the 1950s and 1960s, large masses of foam on sewage plants and waterways indicated the presence of alkylbenzene sulfonate (ABS) surfactants, which were resistant to degradation. These were later replaced with the linear alkylbenzene sulfonate (LAS) surfactants, with less persistence in water.
2. Soaps and detergents contain surfactants that assemble at the interfaces (e.g., air–water) and modify the properties of the phases.
3. The molecules of surface-active agents (surfactants) usually contain two parts with different characteristics: a hydrophilic part (a polar head) that orients at the two-phase interface to the polar phase and a hydrophobic part (a nonpolar tail) that orients to the nonpolar phase. When this occurs, a monolayer forms at the interface with like charges in proximity. The repulsion of these groups tends to reduce the surface tension. This allows the phases to form mixtures.
4. Surfactants are broadly categorized according to the charged nature of the hydrophilic part of the molecule:
   - **Anionic** — negatively charged
   - **Cationic** — positively charged
   - **Nonionic** — neutral, though highly polar
   - **Amphoteric** — a zwitterion containing positive and negative charges

TABLE 10.4

Some Natural Surfactants

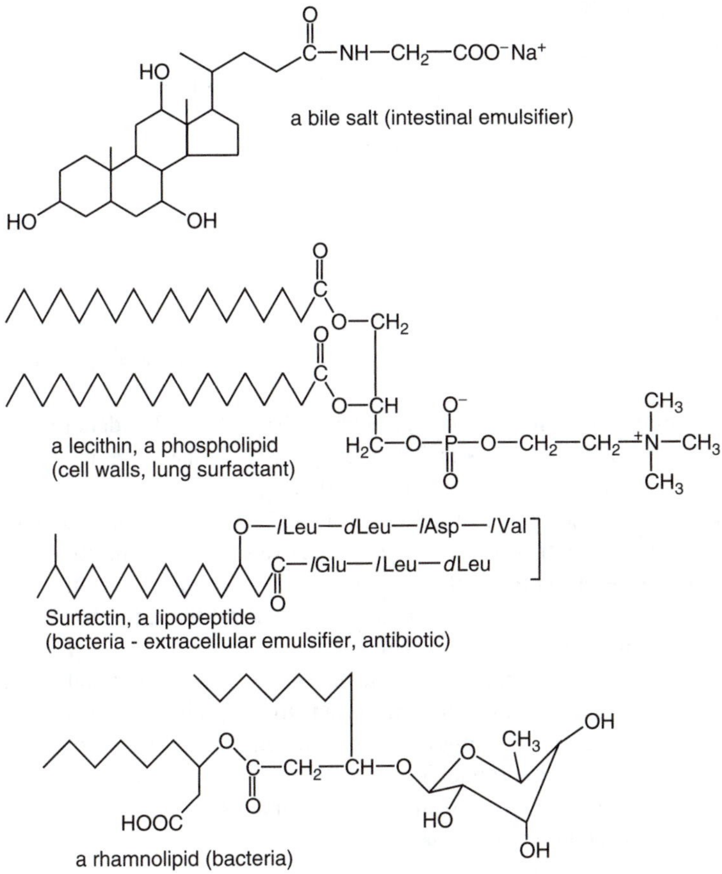

5. Soaps are the metal salts of long-chain fatty acids and traditionally are made by saponification of natural fats or oils that are triesters of fatty acids and glycerol.

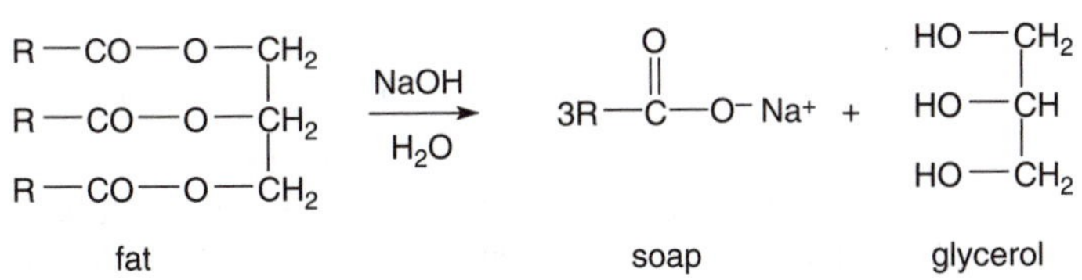

6. The linear alkylbenzene sulfonates are a major group of synthetic anionic surfactants and are synthesized by the following reaction sequence:

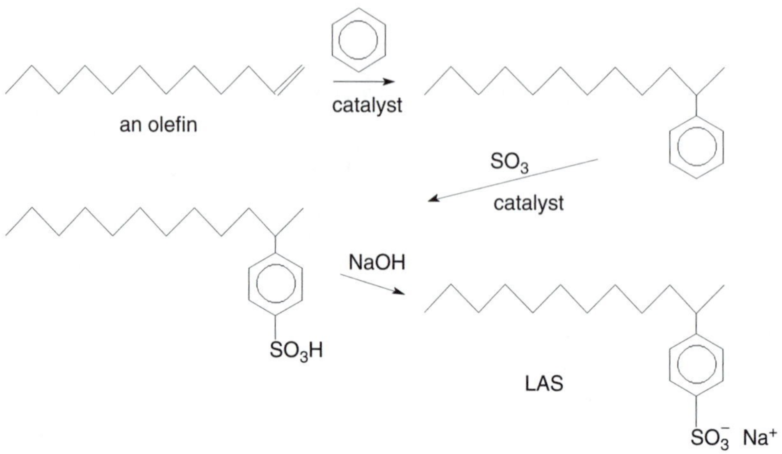

Linear alkylbenzene sulfonate

7. Washing detergents contain surfactants (30 to 40%), builders to enhance the detersive action (about 35%), enzymes to assist in removing stains, etc., from fabrics, bleaching agents (0.25%), and auxiliaries (about 5%).
8. Cationic surfactants are used as fabric and hair conditioners and as disinfectants. The surfactant molecules tend to adopt a head (positive charge)-down position on the surface, which usually contains a negative charge. The fabric and hair surface acquire a neutral oil-like film that lubricates movements and gives a soft feel.
9. The toxicity to mammals of surfactants is generally low and many are used in foods. However, toxicity to aquatic organisms is relatively high (LC_{50} of about 0.1 to 70 mg/l). Toxicity increases with carbon chain length, which increases hydrophobicity.
10. The ABS surfactants used in the 1950s and 1960s had highly branched chains as shown below:

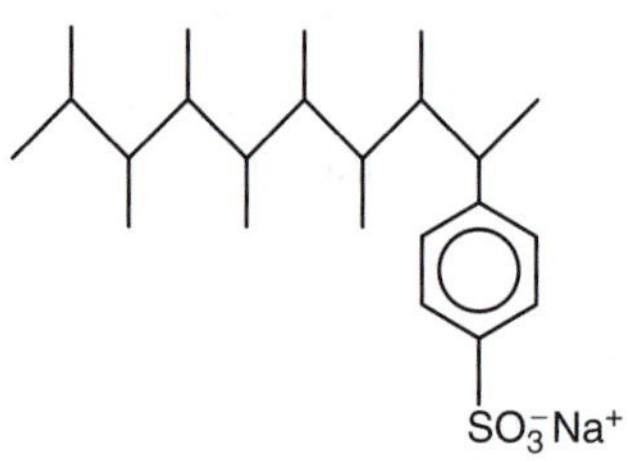

This branching causes resistance to degradation. The LAS surfactants do not have this branched structure and are relatively rapidly degraded in the environment.

REFERENCES

Clint, J.H., *Surfactant Aggregation*, Chapman & Hall, New York, 1992.
Falbe, J., Ed., *Surfactants in Consumer Products*, Falbe, J., Ed., Springer-Verlag, Berlin, 1987.
Hutzinger, O., Ed., Detergents, in *The Handbook of Environmental Chemistry*, Vol. 3, Part F:152–260, Springer-Verlag, Berlin, 1992.

QUESTIONS

1. Which of the following compounds are likely to be surface active?

(a) 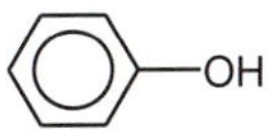

(b) $CH_3(CH_2)_{10}Cl$

(c) Sucrose

(d) $CH_3(CH_2)_8SO_3^-\ Na^+$

(e) $CH_3\ {}^{+}NH_3\ Br^-$

(f) $CH_3(CH_2)_{10}(OCH_2CH_2)_6OH$

2. Classify (a) lecithin and (b) the rhamnolipid (Table 10.4) according to their surfactant type.
3. The cleaning action of soaps is adversely effected by hardness in water. Explain why the alkyl ethoxylates are not so effected.
4. Compare the toxicity and biodegradability that would be expected with the following two surfactants:

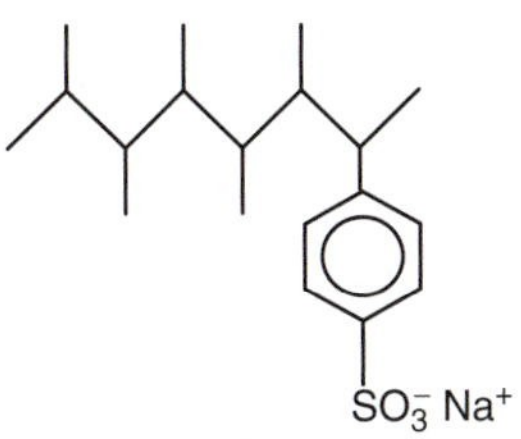

Compound (a)

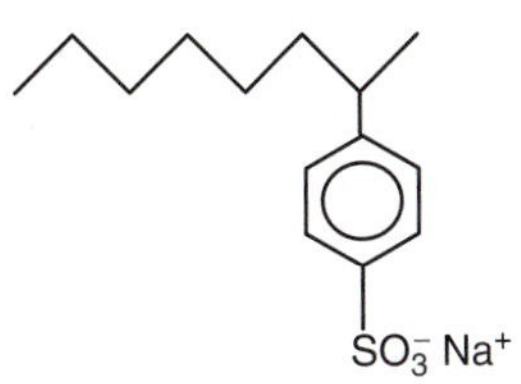

Compound (b)

ANSWERS

1. For a compound to be a surfactant, it must contain within the molecule a clearly differentiated highly polar and highly nonpolar group. Compound a (phenol) has some of these characteristics with the weakly polar benzene ring and the polar hydroxyl group, but these characteristics are not strong enough to produce a surfactant. Compound b does not have the required groups. Compound c (sucrose) has some polar groups, but it does not have a nonpolar group. Compound d has a nonpolar group, $CH_3(CH_2)_8$–, and a polar group, SO_3–, which are well-differentiated groups and would be expected to be a surfactant. Compound e lacks a significant nonpolar part of the molecule and thus does not have the required groups. Compound f is a nonionic fatty acid ethoxylate with a nonpolar group, $CH_3(CH_2)_{10}$–, and a polar group, $(OCH_2CH_2)_6OH$, which give it surfactant properties.
   This means that compounds d and f would be expected to be surfactants.
2. Lecithin contains both positive and negative charges within the molecule and cannot form free ions. It would be classified as amphoteric.
   Rhamnolipid contains nonpolar and polar groups within the molecule but does not form ions. The acid groups would be expected to have a very low ionization capacity. It would be classified as nonionic.
3. Soaps are commonly sodium salts of long-chain fatty acids and react with calcium and other ions in water to form calcium and other salts. These salts do not have the surfactant actions of the sodium salts that constitute soap, so soap loses its activity. The alkyl ethoxylates have the general formula

   $$R—(O—CH_2— CH_2)_n—OH$$

   where R is an alkyl group.
   These molecules gain their surfactant activity due to the polarity distributions present rather than formation of ions, as with the soaps. Since there are no ions present in the alkyl ethoxylates, they do not react with the calcium ions present in the hard water. Thus, the alkyl ethoxylates are resistant to the adverse effects of hard water.
4.

Characteristic	Compound (a)	Compound (b)
Surfactant classification	Anionic	Anionic
Surfactant type	Alkylbenzene sulfonate (ABS)	Linear alkylbenzene sulfonate (LAS)
Toxicity[a]	Mammals — low	Mammals — low
	Aquatic organisms — >compound (b)	Aquatic organisms — <compound (a)
Biodegradability[b]	<Compound (b)	>Compound (a)

[a]Toxicity increases with the size of the carbon chain; compound (a) has 11 carbons and compound (b) has 7 carbons.

[b]Biodegradability increases with branching; compound (a) has more branching than compound (b).

11 Metals in the Environment

11.1 INTRODUCTION

Natural processes, such as chemical weathering and geochemical activities, release the various elements in the Earth's crust into the atmosphere, water, and soil. Transport and transformations of these elements, including the metals, usually involve geochemical as well as biological processes, thereby forming the Earth's biogeochemical cycles.

The influence of pollution on the global environment, through activities such as mining operations, burning of fossil fuels, agriculture, and urbanization, has accelerated the fluxes of some trace metals (Manahan, 2001). The present rate of the global input of trace metals such as mercury, lead, zinc, and cadmium is in excess of the natural rate of biogeochemical cycling. For some metals, global contributions from anthropogenic sources may be small compared to natural fluxes; nevertheless, these may cause localized pollution, such as with mining wastes and leachates.

The term *metal* typically describes an element that is a good conductor of electricity and has high thermal conductivity, density, malleability, ductibility, and electropositivity. However, some elements (boron, silicon, germanium, arsenic, and tellurium), referred to as metalloids, possess one or more of these properties but are not sufficiently distinctive in their characteristics to allow a precise delineation as a metal or nonmetal. Arsenic, in particular, is an important environmental contaminant and will be considered in this chapter. A part of the periodic table is shown in Figure 11.1, where the metals, metalloids, and nonmetals are shown. Throughout this book, the term *metals* has been used broadly to refer to metallic elements and may include metalloids of environmental significance.

Metals react as electron-pair acceptors (Lewis acids) with electron-pair donors (Lewis bases) to form various chemical groups, such as an ion pair, a metal complex, or a donor–acceptor complex. This type of equilibrium reaction can be generalized as follows:

$$M + L \leftrightarrow ML$$

$$K_{ML} = \frac{[ML]}{[M][L]}$$

Note: Charges on chemical species have been omitted for convenience.

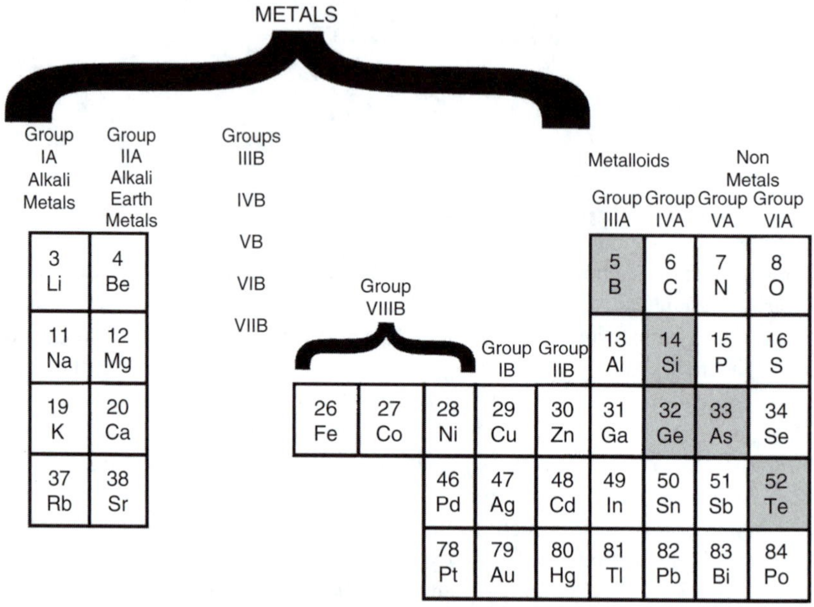

FIGURE 11.1 A part of the periodic table of the elements with atomic numbers and indicating metals, metalloids, and nonmetals.

In the above equation, M represents the metal ion, L the ligand, ML the metal–ligand complex, and K_{ML} the equilibrium (or stability) constant. This equilibrium constant is characteristic of the metal and ligand involved.

Accordingly, a metal cation has a broad general preference for either large, easily polarizable, low electronegativity ions (e.g., sulfide) or smaller, more electronegative anions (e.g., oxides).

Some metals, for example, aluminum, occur widely in the environment, and aluminum is, in fact, one of the most commonly found elements in the Earth's crust. It is found in aluminosilicates, which are insoluble stable complexes from which the aluminum is not mobilized into biological systems. This metal is thus not considered in this chapter. On the other hand, a number of metals are essential for the normal chemical composition of living tissue and are required as essential nutrients. These essential trace metals include iron, magnesium, selenium, and so on.

From the biological point of view, the term *heavy metals* is used widely in the scientific literature to describe toxic metals and has been used in this sense in this book (Bibudhendra, 2000). The definition of heavy metals has been based primarily on (1) the specific gravity of the metals (greater than 4 or 5); (2) location within the periodic table, for example, elements with atomic numbers 22 to 31 and 40 to 51, 78 to 82 (Figure 11.1) and the lanthanides and actinides; and (3) specific biochemical responses in animals and plants. Generally the following heavy metals are of major eenvironmental concerns: Zn, Cu, Cd, Hg, Pb, Ni, Cr, Co, Fe, Mn, Sn, Ag, together with the metalloid arsenic.

11.2 SOURCES OF METALS IN THE ENVIRONMENT

Metals enter the environment from a variety of sources, either natural or human related. On a geological timescale, natural sources such as chemical weathering and volcanic activities have been the major release mechanisms responsible for the chemical composition of freshwater and marine ecosystems.

11.2.1 Natural Sources

In freshwater systems, chemical weathering of igneous and metamorphic rocks and soils in drainage basins is the most important source of background levels of trace metals entering surface waters. Considerable variation in background levels in surface water and bottomed sediment are observed due to the presence of mineralized zones in the drainage basins. Decomposing plant and animal detritus also contribute small, yet biologically significant amounts of metals to surface waters and bottom sediments.

Precipitation and atmospheric fallout, within a drainage basin or directly on the water surface, are the second most important source for trace metals entering water bodies on a global basis. Metals in the atmosphere attributed to natural sources are derived from (1) dusts from volcanic activities, (2) erosion and weathering of rocks and soil, (3) smoke from forest fires, and (4) aerosols and particulates from the surface of the oceans.

Natural inputs into the marine environment can originate from coastal supply, which includes input from erosion produced by wave action and glaciers. There is also a deep-sea supply, which includes metals released by deep-sea volcanism and those removed from particles or sediments by chemical processes. Metals can be transported in the atmosphere as dust particles or aerosols, and also material produced by glacial erosion in polar regions and transported by floating ice.

11.2.2 Pollution Sources

Human activities are also a major source of metal introduction into the aquatic environment. Metal inputs arise from direct discharges of various kinds of contaminated wastes, disturbance of drainage basins, atmospheric precipitation, and fallout. Major inputs are summarized below:

1. **Mining, agricultural, and other operations**. The exploitation of ore deposits invariably exposes fresh rock surfaces and large quantities of waste rock or soil to accelerated weathering conditions. For example, exposure of pyrite and other sulfide minerals to atmospheric oxygen and moisture results in oxidation of this mineral and the formation of acid mine drainage water.
   This is a complex process substantially mediated by *Thiobacillus* and *Ferrobacillus* bacteria (Figure 11.2). Low pH conditions are produced together with various metallic compounds, including ferric hydroxide, which forms yellow-orange deposits in the affected area. Consequently, the release of acid mine drainage from active and abandoned mines,

(1) $FeS_2 + 2H_2O + 7O_2 \longrightarrow 4H + 4SO_4^{2-} + 2Fe^{2+}$
iron pyrites

(2) $4Fe^{2+} + O_2 + 4H^+ \longrightarrow 4Fe^{3+} + 2H_2O$

(3) $FeS_2 + 14Fe^{3+} + 8H_2O \longrightarrow 15Fe^{2+} + 2SO_4^{2-} + 16H^+$

(4) $Fe^{3+} + 3H_2O \longrightarrow Fe(OH)_3 + 3H^+$

FIGURE 11.2 Sequence of reactions involved in the oxidation of pyrites to form acid drainage water.

particularly coal mines, has been widely associated with water quality problems, involving relatively high levels of metals, such as Fe, As, Mn, Cu, Ni, and Co.

This effect is also commonly observed in the agricultural development of swamps and similar wetland, which has necessitated the installation of drains to remove excess water. In many areas, this has exposed pyrite bearing subsoil strata to the atmosphere, resulting in acid drainage formation, as outlined above.

Probably the most serious instance of acid drainage is in Bangladesh. Here an extensive water bore system has been developed to provide drinking water for a population of many millions. However, the use of the bores has exposed the substrata to air, and the development of acid drainage has occurred. The drinking water produced contains excessive levels of arsenic, which has had serious effects on public health.

2. **Domestic effluents and urban storm water runoff**. Appreciable amounts of trace metals are contributed to wastewater effluents by domestic discharges, corrosion of water pipes (Cu, Pb, Zn, and Cd), and consumer products (e.g., detergent formulations containing Fe, Mn, Cr, Ni, Co, Zn, Cr, B, and As). Concentrations are often in the milligram-per-liter range, but vary according to such factors as water usage patterns, time of year, and economic status of consumers. Wastewater treatment by the activated sludge process generally removes less than 50% of influent metals, yielding effluents with significant trace metal loadings. The disposal of sewage sludge can also contribute to metal enrichment (Cu, Pb, Zn, Cd, and Ag) in receiving waters.
3. **Industrial wastes and discharges**. Many trace metals are discharged into the aquatic environment through industrial effluents as well as the dumping and leaching of industrial sludges.
4. **Agricultural runoff**. Enormous quantities of sediment containing trace metals are lost from agricultural regions as a result of soil erosion.

Agricultural soils may become enriched with trace metals from animal and plant residues, phosphatic fertilizers, specific herbicides and fungicides, and through the use of sewage effluent or sludge as a plant nutrient source. On the other hand, trace metals in soils tend to be stabilized through oxidation, formation of insoluble salts, and absorption reactions, dependent on soil characteristics.

11.3 BEHAVIOR AND FATE IN ABIOTIC ENVIRONMENTS

11.3.1 Metals in Aquatic Systems

Natural waters and associated particulate matter are complex heterogeneous electrolyte systems containing numerous inorganic and organic species distributed between aqueous and solid phases. Trace metals entering natural waters become part of this system, and their distribution processes are controlled by a dynamic set of physicochemical interactions and equilibria. The solubility of trace metals in natural waters is principally controlled by (1) pH, (2) type and concentration of ligands and chelating agents, and (3) oxidation state of mineral components and the redox environment of the system. In addition, dynamic interactions at solution–solid interfaces determine the transfer of metals between aqueous and solid phases. Thus, trace metals may be in a suspended, colloidal, or soluble form. In general, suspended particles are considered to be those greater than 100 μm in size, soluble particles are those less than 1 μm in size, and colloidal particles are those in the intermediate range. The suspended particles may consist of (1) organic complexes; (2) precipitated compounds or heterogeneous mixtures of metals in forms such as hydroxides, oxides, silicates, or sulfides; (3) suspended sediments; or (4) clay, silica, or organic matter to which metals are bound by (5) inorganic and organic complexes as outlined in Table 11.1.

Metals in natural waters may exist in the form of free metal ions surrounded by coordinated water molecules. However, the concentrations of anionic species (e.g., OH^-, Cl^-, SO_4^{2-}, HCO_3^-, organic acids, and amino acids) are usually sufficient to form inorganic or organic complexes with the hydrated metal ions by replacing the coordinated water. Other associations occur with colloidal and particulate material, such as clays and hydrous iron, and manganese oxides and organic material. Several types of interactions occur between metal ions and other species in aqueous solutions, as outlined below:

1. **Hydrolysis reactions of metal ions.** Most highly charged metal ions (e.g., Th^{4+}, Fe^{3+}, and Cr^{3+}) are strongly hydrolyzed in aqueous solution and have low pK_1 values.

$$Fe(H_2O)_6^{3+} + H_2O \leftrightarrow Fe(H_2O)_5OH^{2+} + H_3O^+$$

Hydrolysis may also proceed further by the loss of one or more protons from the coordinated water.

TABLE 11.1
Some Forms of Occurrence of Metal Species in Natural Waters

Form	Examples
Dissolved	
Free metal ions (cations)	Cu^{++} (aqueous), Fe^{++} (aqueous), Ca^{++} (aqueous)
Inorganic complexes	$Zn(OH)_4^{++}$, $Cu_2(OH)_2^{++}$, $Pb(CO_3)_2$
Organic complexes	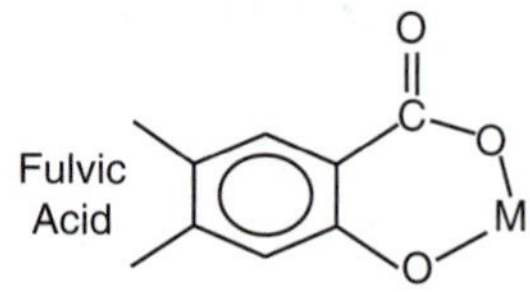
Particulate Insoluble	
Organic complexes	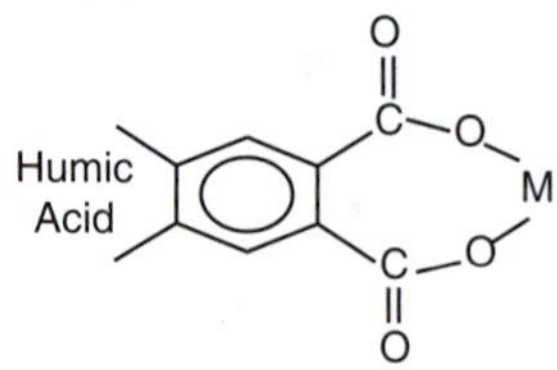
Precipitated oxides, hydroxides, silicates, sulfides, carbonates, etc.	
Suspended sediments with metals in mineral form	
Metals bound to organic matter, colloids, remains of living organisms	

$$Fe(H_2O)_5OH^{2+} + H_2O \leftrightarrow Fe(H_2O)_4(OH)_2^{+} + H_3O^{+}$$

Many divalent metals (e.g., Cu^{2+}, Pb^{2+}, Ni^{2+}, Co^{2+}, and Zn^{2+}) hydrolyze also within the pH range of natural waters.
The hydrolysis of aqueous metal ions can also produce polynuclear complexes containing more than one metal ion, for example,

$$2FeOH^{2+} \leftrightarrow Fe_2(OH)_2^{4+}$$

Polymeric hydroxo-forms of metal ions (e.g., Cr^{3+}) may condense slowly with time to yield insoluble metal oxides or hydroxides. Polymeric species are important in moderate to high concentrations of metal salt solutions.

2. **Complexation of metal ions**. Metal ions also react with inorganic and organic complexing agents present in water from both natural and pollution sources. Dominant inorganic complexing ligands include Cl^-, SO_4^{2-}, HCO_3^-, F^-, sulfide, and phosphate species. These reactions are somewhat similar to the hydrolysis reactions of metal ions in that sequences of

soluble complex ions and insoluble phases may result, depending on the metal and ligand concentrations and pH.

Inorganic ligands are usually present in natural waters at much higher concentrations than the trace metals they tend to complex. Each metal ion has a speciation pattern in simple aqueous solutions that is dependent upon (1) the stability of the hydrolysis products and (2) the tendency of the metal ion to form complexes with other inorganic ligands. For example, Pb(II), Zn(II), Cd(II), and Hg(II) each form a complex series when in the presence of Cl^- or SO_4^{2-} at concentrations similar to those of seawater. The pH at which a significant proportion of hydrolysis products are formed is dependent upon the concentration of the ligand, for example, Cl^- competing with OH^- for the metal ion.

Metals can also bond to natural and synthetic organic substances by way of (1) carbon atoms yielding organometallic compounds, (2) carboxylic groups producing salts of organic acids, (3) electron-donating atoms O, N, S, P, and so on, forming coordination complexes, or (4) -electron donating groups (e.g., olefinic bonds, aromatic ring, etc.).

Of particular importance are organic complexes where two atoms in the ligand can interact with the metal ion as shown for the organic complex of fulvic and humic acids in Table 11.1. These chelation complexes are relatively stable and can be persistent in the environment. In addition, they remove the metal ion from solution and limit its biological activity.

Metal speciation is also controlled by oxidation-reduction conditions. It is affected in two ways: (1) by direct changes in the oxidation state of the metal ions, for example, Fe(II) to Fe(III) and Mn(II) to Mn(IV), and (2) by redox changes in available and competing ligands or chelates.

11.3.2 Metal Speciation in Freshwater and Seawater

Metal speciation in freshwater and seawater differs considerably due mainly to (1) different ionic strengths, (2) lower content of adsorbing surfaces in seawater, (3) different concentrations of trace metals, (4) different concentrations of major cations and anions, and (5) usually higher concentrations of organic ligands in freshwater systems. In freshwater, metals on particulates dominate and soluble metal–ligand complexes are considerably more varied than in seawater; chloro-complexes become the dominant species for Cu, Zn, Hg, and Co, while Ni tends to remain as the free ion and Cr forms hydroxide complexes. Adsorption is negligible for all these metals, because an increase in ionic strength decreases the density of metal ions on particle surfaces due to competitive exchange. Few metals form significant amounts of organic complexes in seawater. In most cases, abundant cations such as Mg^{2+} and Ca^{2+} compete for the organic functional groups.

11.3.3 Bioavailability of Metals

Metals in aqueous pollution can exist in many different chemical forms, as summarized in Table 11.1. For a metal to exert a toxic effect, it must be bioavailable, which means it must be available for uptake by biota. Particulate insoluble matter contains metals that are strongly bound and not in aqueous solution, and so are not bioavailable. However, these particulates can be taken up by biota and assimilated. Free metal ions are generally considered to be most bioavailble and thus have the greatest toxic effects. Other chemical forms of metals have quite different toxic effects, reflecting their chemical constitution.

11.3.4 Interactions between Aqueous and Solid Phases

Dissolved metal ions and complexes can be rapidly removed from solution upon contact with the surfaces of particulate matter through surface bonding phenomena. The formation of metal particulate complexes and subsequent deposition leads to the enrichment of trace metals in sedimentary environments. Enrichment and remobilization of metals in sediments are dependent on factors such as chemical composition of the water constituents (e.g., the amount of dissolved iron and carbonate), salinity, pH, redox values, and the hydrodynamic conditions.

Under suitable conditions, some metals associated with sediments and suspended particles are returned to the overlying water following remobilization and upward diffusion. This process may act as a significant source of trace metal contamination. At least five major processes control the release of metals in this way:

1. **Elevated salt concentrations**. At elevated concentrations, the alkali and alkaline earth cations can compete for adsorption sites on the solid particles, thereby displacing the sorbed trace metal ions.
2. **Changes in redox conditions**. A decrease in the oxygen potential in sediments can occur due to such conditions as advanced eutrophication. This results in a change in the chemical form of the metals, and thus a change in water solubility. Under reducing conditions, trace metals in the interstitial waters occur as (1) sulfide complexes for Cd, Hg, and Pb; (2) organic complexes for Fe and Ni; (3) chloride complexes for Mn; and (4) hydroxide complexes for Cr. With the development of oxidizing conditions, the solubility of metal ions is influenced by a gradual change from metallic sulfides to carbonate hydroxides, oxyhydroxides, oxides, or silicates.
3. **pH changes**. Reduction of pH leads to dissolution of carbonates and hydroxides, as well as to increased desorption of metal cations due to competition with hydrogen ions.
4. **Presence of complexing agents**. Increased use of natural and synthetic complexing agents can form soluble stable metal complexes with trace metals that are otherwise adsorbed to solid particles.
5. **Biochemical transformation**. This can lead to either transfer of metals from sediments into the aqueous phase or their uptake by aquatic organisms and subsequent release via decomposition products.

   After sedimentary deposition, the organic matter, including metals, undergoes diagenesis, which is the chemical modification occurring as a part of the geological deposition process. This involves an increase in molecular weight and loss of some functional groups. A relatively stable and less reactive reservoir for heavy metals in aquatic sediments is formed. However, remobilization may occur through microbial processes.

11.4 TRANSPORT AND TRANSFORMATIONS IN BIOTA

11.4.1 Microbial–Metal Interactions

There are three major microbial processes affecting the environmental transport of metals: (1) degradation of organic matter to lower-molecular-weight compounds, which are more capable of complexing metal ions; (2) alterations to the physicochemical properties of the environment and chemical form of metals by metabolic activities, for example, the oxidation-reduction potential and pH conditions; and (3) conversion of inorganic compounds into organometallic forms by means of oxidative and reductive processes. This third mechanism involves bacterial methylation of a number of elements, for example, Hg, As, Pb, Se, and Sn, in which methyl cobalamin appears to be the primary biological methylating agent. The biological transformations of mercury are shown in Figure 11.3.

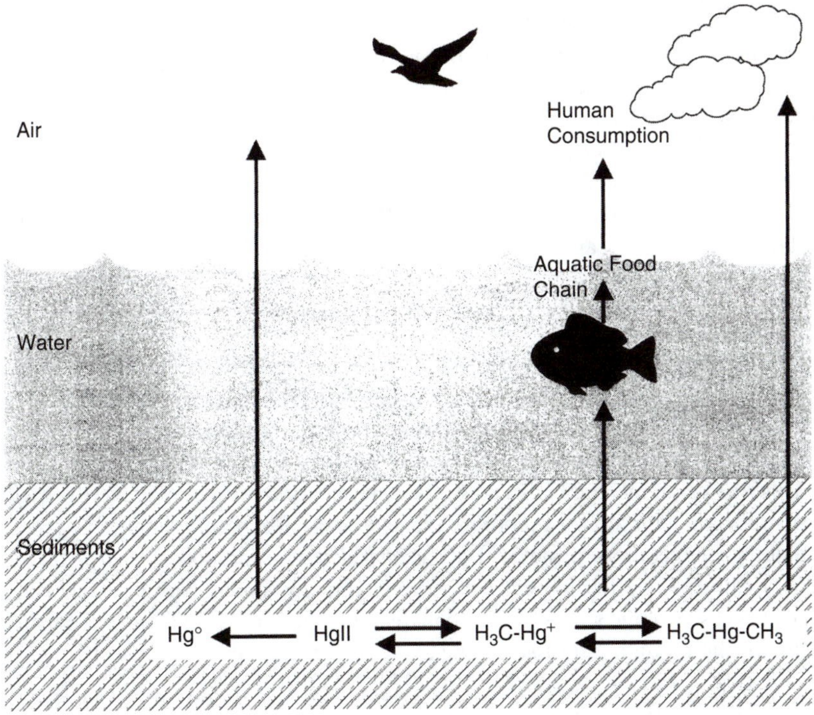

FIGURE 11.3 The biological transformation of mercury in the environment.

11.4.2 Uptake, Excretion, and Regulation Processes

The initial uptake of metals by aquatic organisms can be considered in terms of three main processes: (1) from water through respiratory surfaces (e.g., gills), (2) adsorption from water onto body surfaces, and (3) from ingested food.

Although aquatic organisms readily absorb metals, it is their ability to regulate abnormal concentrations that determines tolerance and is a critical factor in survival. Some animals, such as fish and crustacea, are able to excrete high proportions of abnormal metal intake and consequently regulate the concentration in the body at fairly normal levels. This occurs more commonly with the essential, and relatively abundant, metals, such as Cu, Zn, and Fe, rather than nonessential metals, such as Hg and Cd. Regulation or excretion occurs through the gills, gut, feces, and urine. Nevertheless, there is an upper limit to the amount of metal that can be excreted by animals, above which there is accumulation in body tissues.

11.5 BIOLOGICAL EFFECTS OF METALS

11.5.1 Essential and Nonessential Trace Metals

Many metals, including several heavy metals, are considered to be essential trace elements for normal growth. For example, iron is needed in animal tissue to form hemoglobin in blood, and magnesium is needed to form chlorophyll in plants. Many other metals, including selenium, manganese, zinc, and copper, are required, and in many cases, the biological function is not well understood, but a lack of the metal results in adverse effects. When the essential requirements are met, any further supply of the metal can have toxic effects. In addition, there are several heavy metals, including mercury and cadmium, for which there is no known biological function and levels above zero can have toxic effects.

11.5.2 Metal Tolerance and Biotransformation

Many contaminated organisms are able to tolerate metal concentrations that are in excess of known physiological needs, and in some situations, levels occur at which enzyme inhibition would be expected. Metal-tolerant organisms may contain concentrations of metals two or three orders of magnitude higher than normal.

Detoxification mechanisms can involve storage of metals at inactive sites within organisms on a temporary or more permanent basis. Temporary storage is generally by binding of metals to proteins, polysaccharides, and amino acids in soft tissues or body fluids. Metallothionein, however, effectively stores cadmium in liver and kidney tissues. Storage sites such as bone, feathers, fur, or exoskeleton provide useful means for the elimination of some metals (e.g., Pb and Hg).

11.5.3 Bioaccumulation

The ability of a wide range of aquatic organisms to accumulate both biologically essential and nonessential metals is well established. The bioconcentration factors' ratios of the metal concentration in the animal, $\mu g\ kg^{-1}$, to that in the ambient water, μg l^{-1}, for various aquatic species frequently ranges between 10^2 and 10^6. Bioaccumulation is a significant process in determining the fate of As, Cd, Cr, Cu, Pb, Hg, and Zn, but not Sb and Ni.

Considerable inter- and intraspecies differences exist in the bioaccumulation capacity of individual metals, and different metals exhibit marked variation in kinetics in any one species. In addition, different chemical forms of any one metal may be absorbed and excreted at widely different rates.

Generally, the relative abundance of essential metals in an organism reflects levels necessary to maintain biochemical functions, for example, enzyme systems. Where uptake of essential metals exceeds these levels, homeostatic mechanisms control body levels and tissue distributions.

11.6 TOXIC EFFECTS

The general toxicity mechanism for metal ions can be classified into the following three categories: (1) blocking of the essential biological function groups of biomolecules (e.g., proteins and enzymes), (2) displacing the essential metal ion in biomolecules, and (3) modifying the active conformation of biomolecules. There are similar patterns of toxicity even for dissimilar organisms (Table 11.2). Such similarities in metal toxicity sequences can be explained by the classification of metal ions according to binding preferences, as outlined below.

The most toxic ions exhibit a broad spectrum of toxicity mechanisms:

1. They are most effective at binding to SH groups (e.g., cysteine) and nitrogen-containing groups (e.g., lysine and histidine imidazole) at catalytically active centers in enzymes.

TABLE 11.2
Toxicity Sequences for Metal Ions in a Range of Organisms

Organisms	Sequence[a]
Algae	Hg > Cu > Cd > Fe > Cr > Zn > Ni > Co > Mn
Flowering plants	Hg > Pb > Cu > Cd > Cr > Ni > Zn
Vertebrata	Ag > Hg > Cu > Pb > Cd > Au > Al > Zn > Ni > Cr Co > Mn > K > Ba > Mg > Sr > Ca > Na
Mammalia	Ag, Hg, Cd > Cu, Pb, Co, Sn, Be, Ba, Mn, Zn, Ni Fe, Cr > Li, Al

[a] In this table the atomic symbols represent tripositive ions for In, Al, Cr, La, Y, Sc, and Au; dipositive ions for Ni, Hg, Cu, Pb, Cd, Zn, Fe, Sn, Co, Mn, Mg, Ba, Be, Sr, and Ca; and monopositive ions for Ag, Tl, Cs, Li, H, Na, and K.

TABLE 11.3
Lethal Toxicity of Metals to Marine Organisms LC_{50} (mg l^{-1}) at 96 Hours

Classes of Organisms	Metals						
	Cd	Cr	Cu	Hg	Ni	Pb	Zn
Fish	22–55	91	2.5, 3.2	0.8, 0.23	350	188	60, 60
Crustaceans	0.015–45	10	0.17–100	0.05–0.5	6, 47	—	0.4–50
Mollusks	2.2–35	14–105	0.14, 2.3	0.058–32	72, 320	—	10–50
Polychaetes	2.5–12.1	2.0 to >5.0	0.16–0.5	0.02–0.09	25, 72	7.7–20	1.8–55

2. They can displace endogenous ions (e.g., Zn^{2+}) from metallo-enzymes, causing inactivation of the enzymes through conformational changes.
3. They can form lipid-soluble organometallic substances involving Hg, Sn, and Pb, capable of penetrating biological membranes and accumulating within cells and organelles.
4. Some metals in metallo-proteins exhibit oxidation-reduction activity, for example, Cu^{2+} to Cu^{+}, which can alter structural or functional integrity.

There is a large volume of bioassay data on the acute toxicities of metals to various organisms. Lethal toxicity values for some common metals to selected marine species are shown in Table 11.3.

11.6.1 Effects on Human Health

Many heavy metals in the environment have been implicated in adverse effects on human health. For example, mercury originating from an industrial discharge at Minimata in Japan was discharged into a marine area, resulting in contamination of food fish. Subsequent consumption of the contaminated fish by a local human population led to severe and sometimes fatal diseases.

Lead from industrial sources, motor vehicle discharges, paint, and drinking water has caused contamination of urban environments worldwide. This has adverse effects on human health and also the intellectual development of children. Probably the most severe instance of metal contamination is with the occurrence of arsenic in drinking water in Bangladesh, where severe effects on public health have been observed (see Section 11.2).

11.7 ORGANOMETALLIC COMPOUNDS

11.7.1 Background

An important group of substances, from an environmental perspective, are the metals in combination with organic groups, commonly referred to as **organometallic compounds**. These substances include discharged environmental contaminants, in the organometallic form, as well as chemical environmental transformation products derived from inorganic forms of the metals. A well-known example of this group is the gasoline additive tetraethyllead $(CH_3CH_2)_4Pb$, which is now being phased out of use in many areas due to environmental contamination.

Many metals after discharge into the environment have organic groups attached to them by environmental processes, converting them into organometallic compounds. Mercury, in an inorganic form, can be an important environmental contaminant, but on discharge, most mercury within organisms is converted into dimethyl mercury (H_3C-Hg-CH_3). The properties of this substance are quite different from the original mercury. The dimethyl mercury is more volatile and more toxic than metallic mercury, and its distribution patterns in the environment are different. In general, the organometallic compounds share many properties with organic compounds rather than metals.

Organometallic compounds are commonly defined as compounds having metal–carbon bonds, where the carbon atom is part of an organic group. An example of an organometallic compound is dimethyl mercury (H_3C–Hg–CH_3), mentioned above. Most metals are able to form organometallic derivatives. Also, metalloids (see Figure 11.1 and the introduction to this chapter) form bonds to carbon, but for simplicity, organometalloids will be referred to as organometallics in this chapter. Metals capable of forming stable organometallic compounds in the environment are silicon (Si), germanium (Ge), tin (Sn), lead (Pb), mercury (Hg), antimony (Sb), and tellurium (Te). In some organometallics, the entire bonding capacity of the metal is taken up by bonds to carbon (e.g., $(CH_3)_2Hg$, $(CH_3)_4Pb$). Such compounds are named by first taking the names of the organic groups, in alphabetical order, followed by the name of the metal. The prefixes *di*, *tri*, and *tetra* are used when the same organic group occurs more than once in a molecule. In other organometallics, the metal forms some bonds with carbon and some to other elements (e.g., methyl mercury chloride CH_3HgCl, trimethyllead chloride $(CH_3)_3PbCl$).

The first recorded preparation of an organometallic (an organoarsenic compound) occurred over 200 years ago. Today, organometallic compounds are produced in ever-increasing amounts for a wide range of purposes. These substances are used as catalysts (e.g., organoaluminum compounds), pesticides (e.g., organoarsenic compounds), gasoline additives (e.g., organolead compounds), and polymers (e.g., organosilicon compounds). With reduction in the lead content of gasoline, amounts of organolead compounds manufactured may decline in the future. However, production of organic derivatives of many other metals can be expected to increase.

The relatively toxic nature of some organometallics was first demonstrated over a century ago. A number of poisoning incidents were reported that were ultimately found to be due to a mold producing the volatile trimethylarsenic (Me_3As) from nonvolatile arsenic containing dyes in wallpaper. This also showed that organometallic species can be formed in the environment, from inorganic starting materials. The different properties of organometallics, compared to those of the metal element and inorganic derivatives, influence the environmental mobility and distribution of the metal. Those metals that can form environmentally stable organometallic derivatives have new and added pathways in their biogeochemical cycles.

Overall, environmental stability of organometallics is associated with (1) strong metal–carbon bonds of low polarity, (2) lack of electron lone pairs on the metal atom, (3) lack of energetically low-lying empty orbitals on the metal atom, and (4) organic groups with no hydrogens attached to the -carbon, and no C-to-C unsaturated bonds.

Also, we have largely focused on compounds where the metal is only bonded to carbon. Generally, compounds in which the metal is bonded not only to carbon, but to other nonmetals, e.g., Me_3PbCl, $MeAsBr_2$, are more stable.

11.7.2 Sources

11.7.2.1 Formation in the Environment

Those organometallic compounds that are found in the environment arise from two sources: discharge of synthetic organometallics either deliberately or inadvertently, and formation in the environment through the actions of organisms or by abiotic chemical reactions. Organisms, primarily microorganisms, or compounds produced by them are known to be able to transfer alkyl groups to metal ions. The most common alkyl group transferred is the methyl group. The principal naturally occurring methylating agents are methylcobalamin (a form of vitamin B_{12}), S-adenosyl methionine, and methyl iodide (CH_3I). Methyl iodide is probably formed from the methylation of the iodide anion by S-adenosyl methionine. These methylation reactions take place within organisms, but can also occur in solution if these methylating agents are released in some way.

Within the environment, it also appears that methyl derivatives of metals may be formed in abiotic transalkylation reactions. This type of reaction is well known in the laboratory and involves transfer of a methyl group from one methylmetal species to another, as shown below:

$$Me_3SnCl + HgCl_2 \rightarrow Me_2SnCl_2 + MeHgCl$$

It is suspected that this process may also occur in the environment due to microorganisms acting directly on the metal ion.

11.7.2.2 Environmental Occurrence

Most organometallic derivatives of the group IVA elements (Figure 11.1) found in the environment have alkyl groups (e.g., methyl, ethyl, butyl) attached to the metal. Methyl derivatives of germanium have been found in the oceans, while organometallic forms of tin and lead have significant anthropogenic sources in the environment. Although the mercury–carbon bond is relatively weak, it is not particularly susceptible to oxidation or hydrolysis, and so organomercury compounds are also observed in the environment.

Organoarsenic compounds are relatively widespread in the environment, as a result of both formation in the natural environment and from anthropogenic inputs. The ability to transform inorganic arsenic into organoarsenic compounds is observed with both microorganisms and higher organisms. Apart from simple methylarsenic derivatives, relatively complex molecules such as arsenic-containing lipids and sugars are produced, typically by aquatic organisms.

11.7.3 Behavior

Organometallics may exist in the aquatic environment in a truly dissolved state, i.e., as separate individual molecules surrounded by water molecules. Generally, these molecules resist hydrolysis. For compounds in which one or more of the organic groups have been replaced by a more polar substituent, e.g., Cl, ionization may occur in solution, for example,

$$(CH_3)_3SnCl \rightarrow (CH_3)_3Sn^+ + Cl^- \text{ (aqueous solution)}$$

Furthermore, ligands can bind or coordinate to the central metal atom. There are many different types of potential ligands in natural waters, such as H_2O, Cl^-, NH_3,and humic substances. Coordination of ligands occupies the previously vacant orbitals on the metal atom that promote instability; thus, coordination in solution stabilizes some organometallics.

Organometallics with weakly polar carbon–metal bonds are often hydrophobic. In addition to dissolving in water to a small extent, they are readily sorbed onto particulates and sediments. They also tend to concentrate in a narrow (20-nm to 200-μm) microlayer consisting of microorganisms, organic compounds, and other hydrophobic molecules that exist on the surface of most natural waters. Concentrations of hydrophobic organometallics may be thousands of times greater in this surface microlayer than in the underlying water.

Methylation of metal ions in soils and sediments by agents such as methylcobalamin and S-adenosylmethionine is a widespread phenomenon. Organometallics may also accumulate in these phases by sorption in aqueous systems. For weakly polar, hydrophobic organic molecules, such as chlorinated hydrocarbons, the extent of sorption, to a good approximation, is dependent only on the nature of the compound itself, and the organic matter or organic carbon content of the soil. Thus, the ratio of sorbed to aqueous concentrations at equilibrium (K_D) is given by an expression that is the same for hydrophobic organometallic and organic compounds and takes the form

$$K_D = xK_{OW}\, f_{OM}$$

where x is a constant and f_{OM} is the organic matter fraction in the soil or sediment. The octanol–water partition coefficient (K_{OW}) is a physicochemical characteristic of a compound, and is a measure of its hydrophobicity. Expressions such as this suggest that for a given soil, the more hydrophobic a compound, as evidenced by its K_{OW} value, the greater the sorption. This principle probably holds true for organometallics that tend not to ionize and subsequently form a number of ligand-coordinated species in solution. Organometallics in this category include Me_2Hg and Me_4Sn. For organometallics capable of ionization and speciation, depending on the ligands present, the situation is more complex. The partition coefficient between soil or sediment and water is dependent not only on f_{OM}, but also on other factors, including the concentration of the organometallic entity and the concentration of ligands present, as well as the pH and salinity of the water. As the salinity of the water is increased, it might be expected that the aqueous concentration of the organometallic would decrease and values of K_D increase. In fact, the opposite trend is observed. One reason may be increased competition for sorption sites of soil or sediment between organometallic cations and cations associated with salinity.

11.8 RADIONUCLIDES

11.8.1 Background

All elements in nature are made up of mixtures of isotopes. Most isotopes are stable, but some, because of the nature of the nucleus, are unstable and subject to nuclear disintegration. When these radioactive isotopes — radionuclides — disintegrate, they produce radiation of various kinds, as described in Table 11.4. Radionuclides include metals and nonmetals, but many of environmental significance are metals; thus, they are considered in this chapter. The radiation basically consists of two types. The first type is short-wavelength electromagnetic radiation, which takes the

TABLE 11.4
Types of Radiation Emitted by Radioactive Substances

Type	Nature	Energy and Relative Biological Effectiveness (RBE)	Range	Source
α-Rays	$^4He^{2+}$; charged helium nuclei	High, RBE ≈ 20	Approximately 5 cm in air; stopped by thin sheets of many solids	Disintegration of unstable isotopes of elements with atomic weight of >150
β-Rays	Electrons (or positrons)	Fairly high, RBE ≈ 1	Approximately 3 m in air; stopped by thin sheets of many solids	Light or heavy nuclei
γ-Rays	Electromagnetic radiation of short wavelength (ca. 0.0001–0.01 Å)	Fairly high, RBE ≈ 1	Approximately 4 m in air; several centimeters in lead	Emitted on nuclear disintegration or when a nucleus captures another particle
Neutrons	Particles of unit mass and no charge	Fairly high, RBE ≈ 5	Comparatively long range in air	Liberated when susceptible elements are bombarded with α- or γ-rays from the heaviest elements; no significant natural emitters

form of γ-rays and x-rays, which have shorter wavelengths than ultraviolet radiation and consequently higher energy. The second type of radiation consists of particles and takes three forms: α- and β-rays and neutrons, as described in Table 11.4.

The rate of decay or disintegration is different for each radionuclide. However, for a given radionuclide, the rate of decay is constant and independent of temperature, pressure, physical form, or chemical combination. In all cases, the rate of decay follows first-order kinetics, and thus

$$N = N_o e^{-\lambda t}$$

where N is the number of atoms at any time t, N_O is the initial number of atoms at time 0, and λ is the radioactive decay constant.

The rate of decay of any radionuclide is a characteristic of that element and is commonly measured as the half-life, $t_{1/2}$, which is constant. It can be simply demonstrated that

$$t_{1/2} = \frac{0.693}{\lambda}$$

When an isotope disintegrates, it gives rise to a daughter element. The daughter could be a stable element, and the process of radioactive decay would stop at that point, or it could be an unstable element subject to further disintegration. If unstable, radioactive decay of this daughter element would occur and give rise to a stable or unstable element. If it was an unstable element, a radioactive decay chain of elements would occur.

Biological effects of radiation result from the amount of radiation emitted and also the amount of radiation absorbed by an organism. Of particular importance is the fact that different forms of radiation have different biological effects, and thus the energy of radiation emitted is not relevant, in itself, as a measure of biological effects. The **relative biological effectiveness**, as this factor is known in biology, or **quality factor**, as it is called in human health, is used to measure the amount of a particular type of radiation that is necessary to produce a given biological effect. This, combined with the radiation dose, provides a relevant measure of the potential biological effects of radiation absorbed by a living organism.

11.8.2 Nuclear Energy Usage

Nuclear energy is used for two principal purposes: weapons and electricity generation. It can be derived from two different types of nuclear processes: (1) fission, the splitting of elements with large atoms into smaller fragments with the production of energy, and (2) fusion, the combination of lighter elements to produce heavier elements also with the production of energy. Figure 11.4 shows a simplified set of reaction sequences for some of the principal fission and fusion processes. These

Uranium Fission

$$ {}^{235}_{92}\mathrm{U} + {}^{1}_{0}n \longrightarrow {}^{236}_{92}\mathrm{U} \longrightarrow {}^{136}_{56}\mathrm{Bd} + {}^{85}_{36}\mathrm{Kr} + 3\,{}^{1}_{0}n $$

$$ {}^{238}_{92}\mathrm{U} + {}^{1}_{0}n \longrightarrow {}^{239}_{92}\mathrm{U} \longrightarrow {}^{239}_{93}\mathrm{Np} + {}^{0}_{-1}\mathrm{e} \longrightarrow {}^{239}_{92}\mathrm{Pu} + {}^{0}_{-1}\mathrm{e} $$

Nuclear fusion

$$ {}^{2}_{1}\mathrm{H} + {}^{2}_{1}\mathrm{H} \longrightarrow {}^{3}_{2}\mathrm{He} + {}^{1}_{0}n $$

$$ {}^{1}_{0}n + {}^{6}_{3}\mathrm{Li} \longrightarrow {}^{4}_{2}\mathrm{He} + {}^{3}_{1}\mathrm{H} $$

$$ {}^{3}_{1}\mathrm{H} + {}^{2}_{1}\mathrm{H} \longrightarrow {}^{4}_{2}\mathrm{He} + {}^{1}_{0}n $$

$$ {}^{2}_{1}\mathrm{H} + {}^{2}_{1}\mathrm{H} \longrightarrow {}^{3}_{1}\mathrm{H} + {}^{1}_{1}\mathrm{H} $$

$$ {}^{1}_{1}\mathrm{H} + {}^{7}_{3}\mathrm{Li} \longrightarrow {}^{4}_{2}\mathrm{He} + {}^{4}_{2}\mathrm{He} $$

FIGURE 11.4 Simplified principal reactions used to obtain nuclear energy.

reactions are accompanied by a large release of energy, which can be used in various ways.

Uranium-235 occurs as 0.7% of natural uranium, while the rest of the uranium is principally uranium-238. The uranium-235 is readily fissionable and undergoes the reaction shown in Figure 11.4 on bombardment with neutrons. Uranium-238 is not readily fissionable but can be converted to plutonium by the process shown in Figure 11.4. Plutonium-239 is readily fissionable and serves as an excellent base for weapons and possibly as a fuel for nuclear power generation. With uranium-235 the neutrons produced by the external bombardment produce further neutrons (Figure 11.4), and these impact other uranium-235 nuclei to continue the process and thus give a chain reaction. This chain reaction can occur very rapidly and be used for nuclear weapons.

The rate of reaction of uranium-235 can be controlled by inserting control rods, or moderators, into the fissionable material. These rods absorb neutrons and thus can control the rate of reaction by controlling the availability of neutrons. The concentration of uranium-235 (0.7%) in natural uranium is too low for fuel purposes, and so this material must be enriched with uranium-235 for use as a fuel. The reaction of uranium-235 occurs with the release of large amounts of radiation and heat. This heat can be used to generate steam and run a turbine to generate electricity. This process is the basis of nuclear electricity-generating plants throughout the world.

Currently, nuclear fusion (Figure 11.4) is only used in weapons. There are no electricity-generating plants operating by the fusion process, although research is presently investigating this possibility.

11.8.3 Radioactive Inputs into the Environment

Many natural radionuclides are present in all parts of the natural environment, and consequently, there is a natural background of radioactivity. The distribution of radioactivity in the natural environment is reasonably well understood and is distributed in a reasonably homogeneous fashion throughout the environment. In this way, radioactivity generally does not usually reach particularly high levels in any specific area. In addition, cosmic radiation can result in the production of radionuclides, which add to the natural background activity. Nuclear reactions from weapons testing and electricity generation and related activities have resulted in a new and significant source of radioactivity in the environment.

Nuclear power generation utilizing nuclear fission reactions based on uranium require a cycle of nuclear fuel activities. The nuclear fuel cycle runs from mining and milling through to final storage of spent fuel elements. Radioactive materials can be discharged to the environment at a number of points in this cycle. Mining, milling, refining, and conversion to uranium hexafluoride involve the production of radioactive tailings, which may contaminate terrestrial and aquatic systems. There are a wide range of radionuclides in liquid effluents from power plants. Many of these occur due to neutron activation of impurities in water passing through the plant, although uranium contamination of this water can occur. The most important of these neutron activation products are manganese-56, copper-64, sodium-24, chromium-51, neptunium-239, arsenic-76, silicon-31, cobalt-58, cobalt-60, and zinc-65. Radionuclides distributed to the atmosphere and terrestrial systems consist largely of inert gases and are fairly rapidly diluted to low concentrations that do not interact with biota to any significant extent. On the other hand, radionuclides discharged to aquatic environments do not disperse rapidly and are in fact rapidly accumulated by sediments in the immediate vicinity of the discharge.

11.8.4 Biological Effects of Radiation

Nuclear radiation loses energy by passing through matter ionizing atoms and molecules. This is not selective; however, the major constituent in living matter is usually water, and it is most probably water molecules that interact mainly with radiation. A possible set of reactions are shown in Figure 11.5, and in this sequence there are powerful oxidizing agents produced. These agents produce free radicals and oxidation products on interaction with double bonds, hydrogen bonds, and sulfhydryl groups. The effect of these products on living tissue is to produce damage resulting in death or illness.

11.9 KEY POINTS

1. Metals are elements that are good conductors of electricity and heat, have high density, malleability, ductibility, and electropositivity, and are in groups IA and IB, IIA and IIB, and IIIB to VIIIB of the periodic table.

A. Interactions of radiation with water

$$H_2O + \text{radiation} \longrightarrow H_2O^+ + \text{or } H_2O^-$$

$$H_2O^+ \longrightarrow H^+ + .OH$$

$$H_2O^- \longrightarrow {}^-OH + .H$$

$$OH + OH \longrightarrow H_2O_2$$

$$H + O_2 \longrightarrow HO_2$$

$$HO_3 + e \longrightarrow HO_2^{\cdot}$$

B. Interactions of water products with chemical groups in living organisms

$$RH + HO_2^{\cdot} \longrightarrow R^{\cdot} + H_2O_2$$

$$RH + HO_2^{\cdot} \longrightarrow RO + H_2O$$

C. Interactions of neutrons with biologically active elements

$$^{31}P + n \longrightarrow {}^{28}Al + \alpha + \gamma$$

$$^{28}Al \longrightarrow {}^{28}Si + \beta$$

FIGURE 11.5 Some possible interactions of nuclear radiation with substances, principally water, in living tissue.

2. Metalloids possess one or more of the properties of metals but are not distinctively metals and are in groups IIIA, IVA, and VIA of the periodic table.
3. Metals react as electron-pair acceptors with electron donors to form various groups, such as an ion pair, a metal complex, a coordination complex, or a donor–acceptor complex.
4. Metals can be essential trace elements, including iron, magnesium, and selenium, for the growth lining tissue, but can have adverse effects above the essential levels.
5. *Heavy metals* is a term used to describe toxic metals based on specific gravity (greater than 4 or 5), atomic numbers between 22 to 31 and 40 to 51, 78 to 82, and toxic biological effects.
6. Metals in the environment originate from such sources as weathering of igneous and metamorphic rocks and soil, bottom sediments in aquatic areas, decomposing animal and plant matter, particulates and atmospheric fallout, rivers, coastal erosion, and deep-sea supply.
7. Metals can originate from pollution sources, including mining, agricultural, and other operations, such as acid drainage, domestic effluents, urban storm water runoff, industrial wastes, and discharges.

8. The following simplified set of reactions is involved in forming acid drainage water. This occurs when geological formations containing iron pyrites are exposed to atmospheric oxygen, allowing the development of populations of the bacterial groups *Ferrobacillus* and *Thiobacillus*:

    1. FeS_2 (iron pyrites) $+ 2H_2O + 7_2O \rightarrow + 4H^+ + 4SO_4^{2-} + 2Fe^{2+}$

    2. $4Fe^{2+} + O_2 + 4H^+ \rightarrow 4Fe^{3+} + 2H_2O$

    3. $FeS_2 + 14Fe^{3+}\ 8H_2O \rightarrow 15Fe^{2+} + 2SO_4^{2-} + 16H^+$

    4. $Fe^{3+} + 3H_2O \rightarrow Fe(OH)_3 + 3H^+$

9. Metals in aquatic systems exist in a complex heterogeneous electrolyte system with different chemical species, including soluble forms such as free metal ions, inorganic complexes, organic complexes, and particulate insoluble forms, for example, organic complexes, suspended sediments, and metals bound to organic matter.
10. Most highly charged metal ions (e.g., Th^{4+}, $Fe^{3+} + Cr^{3+}$) in solution are strongly hydrolyzed and have low pK_1 values.

$$Fe(H_2O)_6^{3+} + H_2O \leftrightarrow Fe(H_2O)_5\ OH^{2+} + H_3{}^+O$$

    This hydrolysis reaction may also proceed further by reaction of attached water molecules.
11. Metal speciation in freshwater and seawater differs considerably with metals on particulates and a wide variety of metal–ligand complexes occurring in freshwater, whereas chloro-complexes are dominant in seawater.
12. The bioavailability and toxic effects of the different chemical forms of the metals are different, but as a general rule, the free ions in water are most toxic.
13. Metals discharged to the environment can undergo chemical transformation into different inorganic forms and organometallics. Mercury illustrates these processes with transformation from elemental mercury to HgII to organic forms.
14. Aquatic organisms readily absorb metals but are generally able to regulate abnormal levels. Nevertheless, there is a limit to this process.
15. The metals exhibit similarities in the sequence of levels of toxicity with many diverse biota. Mercury exhibits a high level of toxicity with most biota.
16. Organometallic compounds have metal–carbon bonds where the carbon is part of an organic group. Organometalloids, comprising elements such as silicon and arsenic, share many characteristics with organometallic compounds and can be considered in this group.

17. Organometallic compounds have properties such as toxicity, solubility, and volatility that are quite different from the component elemental metals and more closely related to the properties of organic compounds.
18. The carbon–metal bonds in organometallic compounds of environmental importance are polarized in the direction C^- to M^+. The greater the polarity, the greater the rate of hydrolysis by water in the environment.
19. Organometallic compounds can occur in the environment as a result of anthropogenic discharges, but also as a result of environmental chemical processes acting on naturally occurring metals.
20. Organomercury, organotin, organolead, and organoarsenic compounds are the most important organometallic environmental contaminants.
21. All elements are made up of mixtures of isotopes. Most isotopes are stable, but because of the nature of the nucleus, some are unstable and emit radiation. These are described as radionuclides.
22. The rate of decay of radionuclides is a characteristic of the element and independent of temperature, pressure, chemical form, and physical combination. It is expressed by the equation

$$N = N_o e^{-\lambda t}$$

REFERENCES

Bibudhendra, S., Ed., *Heavy Metals in the Environment*, Marcel Dekker, New York, 2002.
Manahan, S.E., *Fundamentals of Environmental Chemistry*, 2nd ed., Lewis Publishers, Boca Raton, FL, 2001.

QUESTIONS

1. Sodium and potassium are present as free ions in seawater in concentrations of approximately 10.5 and 0.38 g l^{-1}, respectively. Why do these metal ions not exert toxic effects on marine biota at these relatively high concentrations?
2. A freshwater body has a low level of visible brown coloration in the water and exhibits very little plant growth. Measurements of sunlight intensities have shown that there is sufficient light available for photosynthesis and the unfiltered water has sufficient magnesium present to sustain plant growth. How can these observations be explained, and what further chemical analysis can be undertaken to verify any possible explanation?

3. Washing detergents can contain polyphosphate builders in the form of sodium tripolyphosphate:

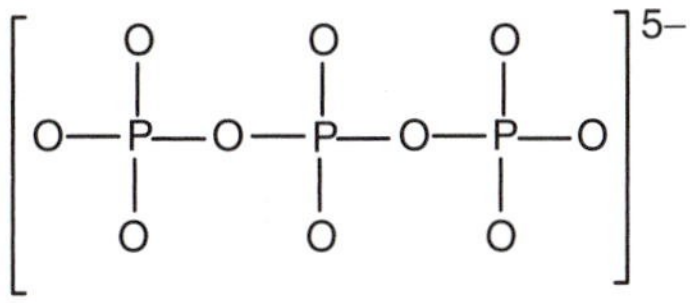

Outline possible environmental effects these substances could have on metals in waterways.

ANSWERS

1. Sodium and potassium in group IA of the periodic table (Figure 11.1) are referred to as the alkali metals. This means they are relatively low in density, having specific gravities of 0.971 and 0.862, respectively, which is considerably lower than the toxic heavy metals, at greater than 4 to 5. The atomic numbers are 11 and 19, also less than the toxic heavy metals, at a minimum of 22. These metals also have important functions in biota, for example, maintaining the ionic balance of bodily fluids. For these reasons, these metal ions do not exert toxic effects on marine biota.
2. The brown coloration is very often due to humic and fulvic acids, which contain chemical groups that could chelate with the magnesium ion (Mg^{2+}) as shown:

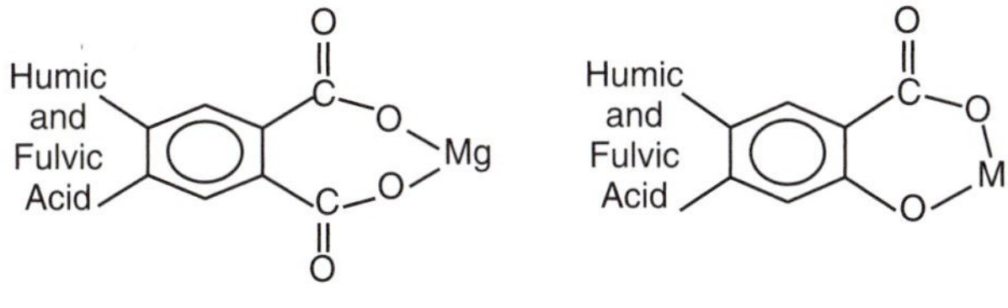

These chelation reactions would remove the Mg^{2+} ion from solution, and it would not be available to be taken up to form plant matter. Plant matter contains chlorophyll, which has magnesium as an essential component. Thus, little plant growth would occur.
It is likely that some of the humic acid complex is present as insoluble particulates in the water. Thus, filtration of the water and analysis of the particulates for magnesium could be used to confirm this hypothesis.

3. The polyphosphates are used in detergents to form chelation complexes principally with Ca^{2+} and Mg^{2+}, thereby reducing the adverse effects of these ions on performance. Similarly, other essential trace metals, including Ca^{2+} and Mg^{2+} mentioned above, could be depleted, reducing their availability to biota. Complexes could take the form

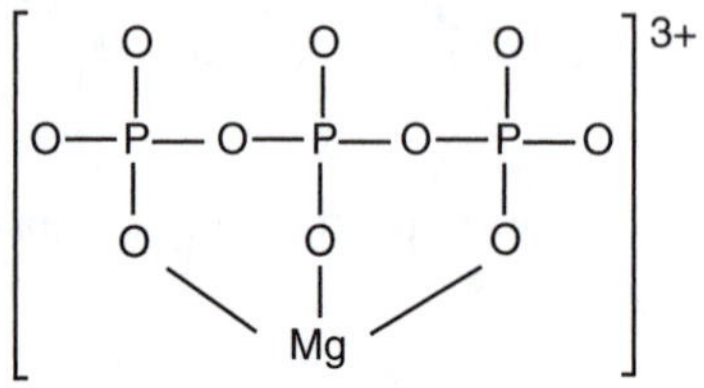

In addition, the effects of toxic metals could be reduced by similar complexation reactions.

Part IV

Processes in the Natural Environment

12 Atmospheric Chemistry and Pollution

12.1 INTRODUCTION

The atmosphere plays a major role in global processes supporting life on Earth. The interaction of photosynthesis and respiration with carbon dioxide and oxygen in the atmosphere maintains the global living system. However, the composition of the atmosphere has changed over geological time and continues to change. The atmosphere also plays an important physical role as a heat reservoir. Heat absorbed by the atmosphere during daylight hours is released overnight, helping to maintain an equitable temperature. Without the atmosphere, the Earth's surface temperature would rise very high when bathed in sunlight, but fall very low during darkness.

The major features of the atmosphere are shown in Figure 12.1. The zone closest to the Earth's surface is described as the **troposphere** and is the zone of most significance to living organisms. In this zone the normal atmospheric composition of nitrogen, oxygen, and other gases is observed as shown in Table 12.1. It is noteworthy that while there are several gases of biogenic origin, including oxygen, carbon dioxide, and methane, there are others that are inert gases and play no apparent role in living systems, including argon, neon, and helium.

As distance from the Earth's surface increases, the composition of the atmosphere changes. The **stratosphere** is the next zone above the troposphere; it consists mainly of N_2, O_2, and water with some O and ozone. From an environmental perspective, the ozone in this layer is of particular importance since it absorbs short-wavelength solar radiation. Ozone is a reactive and unstable substance, but it can persist in the stratosphere due to the low air pressures that exist in this zone (0.1 to 0.001 atm). This means that there is a long mean free path between collisions of molecules that occur relatively infrequently, and so the possibility of intermolecular reactions is reduced. Similarly, the other zones at further distances from the Earth, the mesosphere and thermosphere, contain highly reactive substances, O_2^+, NO^+, and O^+, which could not exist for any significant time at the Earth's surface. These reactive ions in the upper zones of the atmosphere also play a major role in maintaining living systems by absorbing short-wavelength solar radiation. This radiation is relatively high in energy and would be damaging to living systems at the Earth's surface.

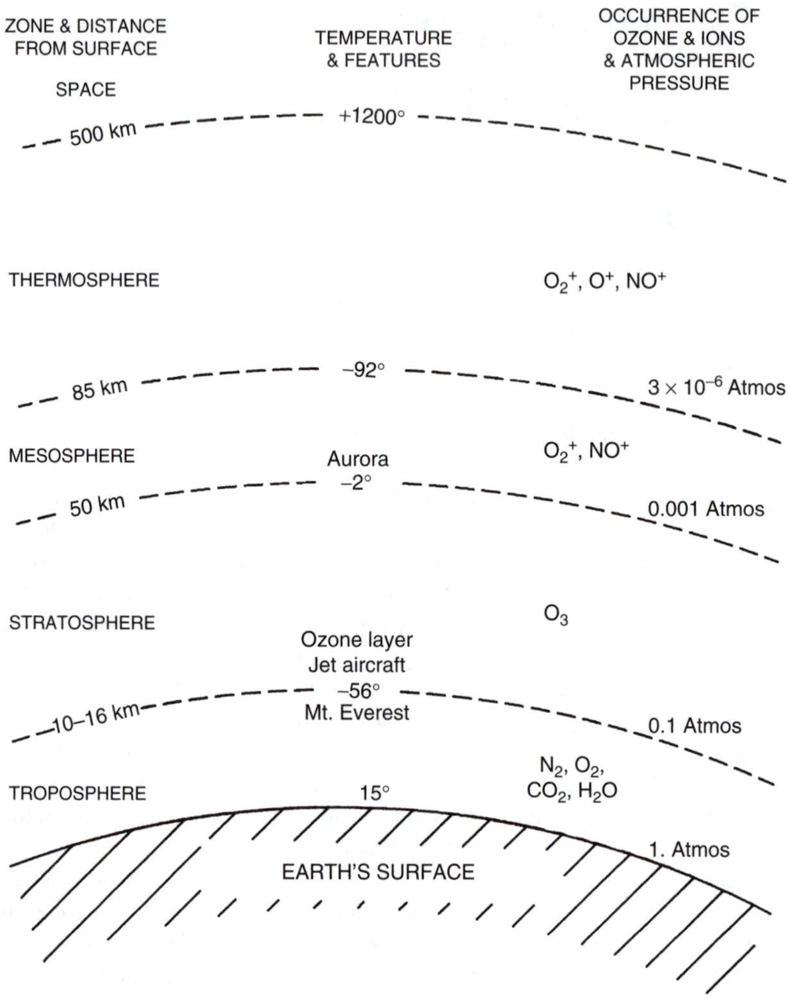

FIGURE 12.1 The major features of the Earth's atmosphere (not to scale).

12.2 OVERVIEW OF PHOTOSYNTHESIS AND RESPIRATION

Photosynthesis is the conversion of energy from sunlight to energy stored in chemicals (carbohydrates) that form the chemical building blocks of plants and are the basic energy input to the Earth's living systems. It requires principally water, sunlight, and carbon dioxide as the inputs to plants to allow photosynthesis to occur. The photosynthetic process and conversion of carbon dioxide to carbohydrate in plants is an endothermic reaction that can be summarized by the following equation

TABLE 12.1
Average Composition of the Atmosphere at the Earth's Surface

Component	Composition (Volume/Volume) in Parts Per Million
Nitrogen (N_2)	780,900 (78.09%)
Oxygen (O_2)	209,500 (20.95%)
Argon (Ar)	9,300 (0.93%)
Carbon dioxide (CO_2)	300 (0.03%)
Neon (Ne)	18
Helium (He)	5.2
Methane (CH_4)	1.7

$$6CO_2 + 6H_2O + \text{Energy} \rightarrow 6CH_2O + 6O_2 \quad (12.1)$$

Plant matter

This equation indicates that photosynthesis forms plant matter, resulting in the release of oxygen to the atmosphere. Photosynthesis can be conveniently divided into two sets of reactions: light and dark reactions. The **light reactions**, as indicated by the name, require sunlight in order to proceed. The light reactions produce two compounds, adenosine triphosphate (ATP) and a reduced form of nicotinamide adenine dinucleotide (NADPH + H^+), which are a means of transferring energy and reducing power to the dark reactions. The **dark reactions** do not require light; however, they can proceed in its presence. Here the ATP and NADPH + H^+, synthesized in the light reactions, are used to drive the conversion of carbon dioxide to carbohydrates. The series of reactions that form carbohydrates are called the Calvin–Benson cycle.

Respiration is the reverse reaction to photosynthesis (Equation 12.1). It occurs principally as a result of plant matter being utilized as food by microorganisms and animals. It oxidizes the carbohydrates formed in photosynthesis back to carbon dioxide and water and releases the energy that was stored in the chemical bonds. Thus, it releases carbon dioxide to the atmosphere. The released energy from respiration is utilized by synthesizing ATP, which can then be used to drive other chemical reactions necessary for normal plant metabolism. Respiration occurs in three consecutive stages: glycolysis, the tricarboxylic acid cycle, and oxidative phosphorylation.

Photosynthesis and respiration on a global scale must occur at approximately the same rate; otherwise, major changes would occur in the oxygen and carbon dioxide content of the atmosphere. In fact, over geological time these processes have occurred at slightly different rates, with photosynthesis occurring at a slightly faster rate. This has led to an ongoing increase in the oxygen content of the atmosphere and a parallel decrease in the carbon dioxide content, as discussed in the next section.

12.3 THE ENVIRONMENTAL SIGNIFICANCE OF PHOTOSYNTHESIS AND RESPIRATION

12.3.1 Photosynthesis, Respiration, and the Atmosphere over Geological Time

The evolution of the linked processes of photosynthesis and respiration has had and continues to have a highly significant effect on the chemical composition of the atmosphere and environmental conditions on Earth. From the available geochemical evidence for the Earth and other planets, it seems that the very earliest atmosphere of the Earth contained essentially no oxygen. The principal components were nitrogen, carbon dioxide, and water vapor, with trace amounts of nitrogen monoxide, hydrochloric acid, hydrogen, and carbon monoxide. With such a composition, the conditions were mildly reducing.

The only possible sources for the production of oxygen gas were the photolysis of water molecules and carbon dioxide by sunlight. The reactions for water are outlined below:

$$H_2O + \text{Light} \rightarrow OH^- + H^+$$

$$OH^- + OH^- \rightarrow O^{2-} + H_2O$$

$$O^{2-} + OH^- \rightarrow O_2 + H^+$$

The reactions for carbon dioxide are:

$$CO_2 + \text{Light} \rightarrow CO + O^{2-}$$

$$O^{2-} + O^{2-} \rightarrow O_2$$

However, unless either the hydrogen or carbon atoms are removed from the atmosphere, they will eventually react with the released O_2, reforming CO_2 and H_2O, leading to no net gain of oxygen. Hydrogen atoms have a very small mass (1.66×10^{-27} kg) and are therefore very hard to retain in the atmosphere. They simply cannot be held by gravity and are lost into space. Carbon was removed from the atmosphere by forming carbonate minerals. These two processes could have led to the concentration of atmospheric oxygen gradually increasing; however, there were a number of substances, such as reduced forms of iron (Fe^{2+} and Fe^{3+}) and manganese (Mn^{7+}, Mn^{4+}, Mn^{2+}), that would readily remove any oxygen in the atmosphere. This anoxic period, termed the Hadean period, extended from the formation of the Earth approximately 4.6 billion years ago (Bya) until 3.7 Bya.

The Archaean period extended from 3.7 to 2.5 Bya. It was during this period that life first evolved. The earliest forms of life would have been simple single-celled bacteria. These would not have had a nucleus or any other subcellular organelles and are termed **prokaryotes**. Two main mechanisms for obtaining energy were developed. One group of bacteria, chemotrophs, catalyzed the reduction of inorganic atoms such as iron and sulfur and in this manner gained sufficient energy to oxidize organic carbon and form carbohydrates. The second group, autotrophs, use energy from the sun (photosynthesis) or from the oxidation of mineral matter (chemosynthesis) to convert CO_2 to carbohydrates.

The oldest preserved evidence of life on Earth comes from rocks found in a location called North Pole in Western Australia. There, in rocks approximately 3.5 million years old, are the characteristically column- or mushroom-shaped calcium carbonate deposits called **stromatolites**. These are deposited by colonial forms of **cyanobacteria** (also called, although incorrectly, blue-green algae). These bacteria were able to photosynthesize, and the deposits are a direct result of the photosynthesis. Despite being among the earliest forms of life to appear, stromatolites still occur in several locations throughout the world, perhaps the largest and most famous being those at Shark Bay in Western Australia and in the Bahamas, although examples also occur in glacial lakes in Antarctica and volcanic springs in Yellowstone National Park. The number and activity of these photosynthetic organisms must have been low, as there is no evidence to support a marked increase in the levels of atmospheric oxygen, which was probably stable at approximately 1 ppm, during the Archaean period.

The period from 2.5 to 0.7 Bya, the Proterozoic, was marked by great changes in the chemistry of the Earth. The number and activity of photosynthetic organisms increased markedly and resulted in the conversion of the atmosphere from mildly reducing to oxidizing. One of the key pieces of evidence to support the claim of increased oxygen production during the Proterozoic is the change in abundance of banded iron formations over time. As stated previously, reduced metals, particularly iron and manganese, readily absorb any free oxygen:

$$Fe^{2+} + O_2 \rightarrow Fe_2O_2$$

$$4Fe^{3+} + 3O_2 \rightarrow 2Fe_2O_3$$

$$4Mn^{7+} + 7O_2 \rightarrow 2Mn_2O_7$$

$$2Mn^{4+} + 2O_2 \rightarrow Mn_2O_4$$

When reduced forms of iron and manganese ions are oxidized, their aqueous solubility decreases substantially and they precipitate out of solution forming the iron oxide deposits. The increase in oxygen production over time is evident by the parallel increase in the amount of these iron ore deposits. Known reserves deposited during the Hadean and Archaean periods total 0.8×10^{14} tons; between 2.5 and 2.0 Bya the mass increased to 6.4×10^{14} tonnes, while between 2.0 Bya and the present there was approximately 0.4×10^{14} tonnes. The marked decrease in tonnage of iron ore deposits after 2 Bya is not due to a decrease of oxygen, but rather a reduction in the amount of iron present in a reduced state.

The oxidation of iron and other reduced chemicals would have negated the increased synthesis of oxygen, and concentrations of atmospheric oxygen most probably did not increase until essentially all the reduced forms were oxidized. Estimates of the oxygen content at about 2 Bya are typically 1% (10,000 ppm) compared with the current content of about 21%.

It was also during the Proterozoic that **eukaryotic bacteria** developed. **Eukaryotes** are organisms that have subcellular organelles, including a nucleus in which the DNA is located. The development of eukaryotes was not possible before this period, as the replication of eukarotic DNA requires actinomyosin, a protein that requires oxygen for its synthesis.

The Phanerozoic period (0.7 Bya to the present) was marked by the evolution of multicelled organisms. For example, there are fossils of jellyfish that have been dated as being 670 million years old (0.67 billion years old). The presence of sufficient concentrations of oxygen (>1%) was also vital to the evolution of multicellular organisms for two reasons. First, relatively high external atmospheric oxygen concentrations are needed in order to supply oxygen to cells within the organism with sufficient oxygen by passive diffusion processes. Second, multicellular organisms require collagen, the biosynthesis of which requires oxygen.

It was only with the development of large photosynthesizing plants with solid cell walls that the oxygen levels rose above 7 to 10% (70,000 to 100,000 ppm) to the current levels of approximately 21% (210,000 ppm). This happened not only because the plants produced more oxygen, but also because the amount of carbon that could be removed from the atmospheric cycle was greatly increased. Single-celled organisms with their soft cell walls are very rapidly broken down, releasing carbon back to the atmosphere. But with trees, which have hard cellulose structures that are relatively resistant to breakdown, there is a much greater chance of the carbon stored in the plant being preserved in sediments, with a subsequent reduction of carbon reentering the atmosphere.

It can be seen that photosynthesis and respiration have had a very significant effect on the atmospheric environment. They have led to the conversion of the atmosphere from a reducing environment to an oxidizing one, which has allowed the evolution of animals and life as we currently know it.

12.3.2 The Greenhouse Effects

Photosynthesizing plants, algae, bacteria, and animals, in fact the whole living system, perform an integral role in the carbon cycle and the modulation of the Earth's climate. The link between plants and climate is **carbon dioxide**, a greenhouse gas and the gas used to create the carbohydrates in all plants (Baird, 1995). The carbon cycle is illustrated in Figure 12.2.

The carbon cycle consists of a series of linked compartments through which carbon, in various forms, passes in various quantities and rates. It can be subdivided into three sections based on the average residence time of carbon. The biological portion consists of the atmosphere, land and aquatic biota, and surface waters. The exchange between the compartments of this portion of the cycle is very rapid, with an average residence time of 1 to 5 years. The second portion, the soil compartment, retains carbon for a longer period. Typical residence times are 30 to 50 years. The third portion, the geological, consists of the deep-water and rock compartments.

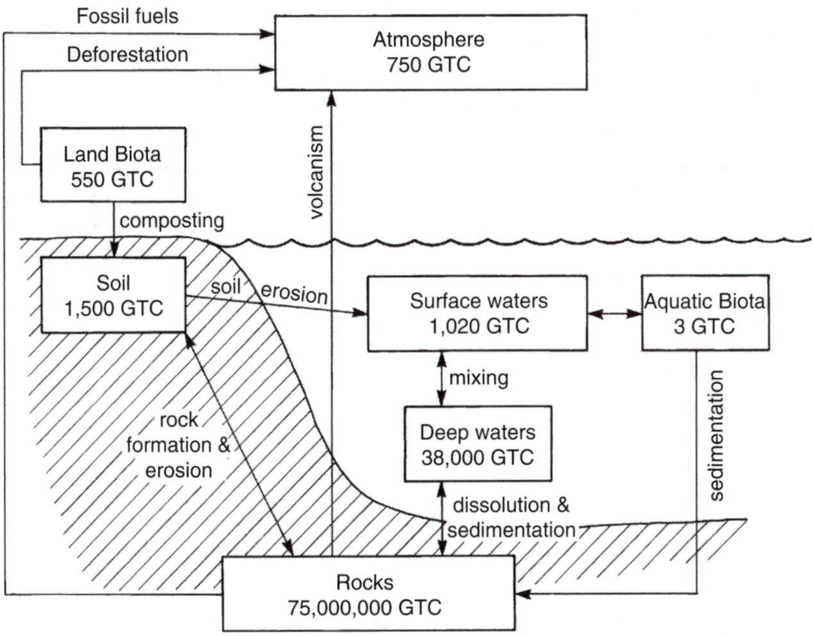

FIGURE 12.2 The carbon cycle with the amounts of carbon in the various compartments indicated as GTC (gigatonnes of carbon, 1 giga = 10^8).

Exchange of carbon between these compartments and the others is extremely slow. Carbon can reside in deep water or rocks for periods of approximately 1000 years and many millions of years, respectively.

As illustrated by Figure 12.2, all of the compartments are reversibly linked to each other either directly or indirectly. Thus, in a sense, an equilibrium is established between the compartments. However, Figure 12.2 also illustrates that the combustion of fossil fuels and deforestation can short-circuit carbon from the cycle and increase the atmospheric concentration of carbon dioxide. It is these sources that have led to the observed increase of carbon dioxide levels and the associated problem termed the **greenhouse effect**. From 900 A.D. to 1750, the atmospheric concentration of carbon dioxide was fairly steady at 280 ppm, with maximum deviations of 10 ppm. Since 1800, when the industrial revolution began, carbon dioxide levels have increased until the present value of 355 ppm, with the rate of change increasing with time.

There are a number of so-called **greenhouse gases**: carbon dioxide, methane, and water vapor are particularly important. Water is the most abundant of the greenhouse gases. However, as its concentrations are controlled by factors such as evaporation from the oceans, and they are not rising, it is not considered further here. In contrast, atmospheric concentrations of CO_2 and methane are increasing. At present, the main concern is with the levels of CO_2, although the methane concentrations are increasing relatively rapidly and each CH_4 molecule has the same effect as 30 CO_2 molecules.

The daily variation in the temperature of the Earth is quite small, yet every day the Earth absorbs large amounts of energy from the sun. The temperature is maintained at a relatively consistent level by the Earth acting in a manner similar to a black body. In other words, it reradiates essentially the same amount of energy that it absorbs. Greenhouse gases do not interfere with the Earth absorbing energy from the sun; however, they do reduce the amount of energy that can be lost by radiation, as shown in Figure 12.3. Solar radiation enters the Earth's atmosphere in a spectrum extending across a wide range of the electromagnetic spectrum. However, on striking

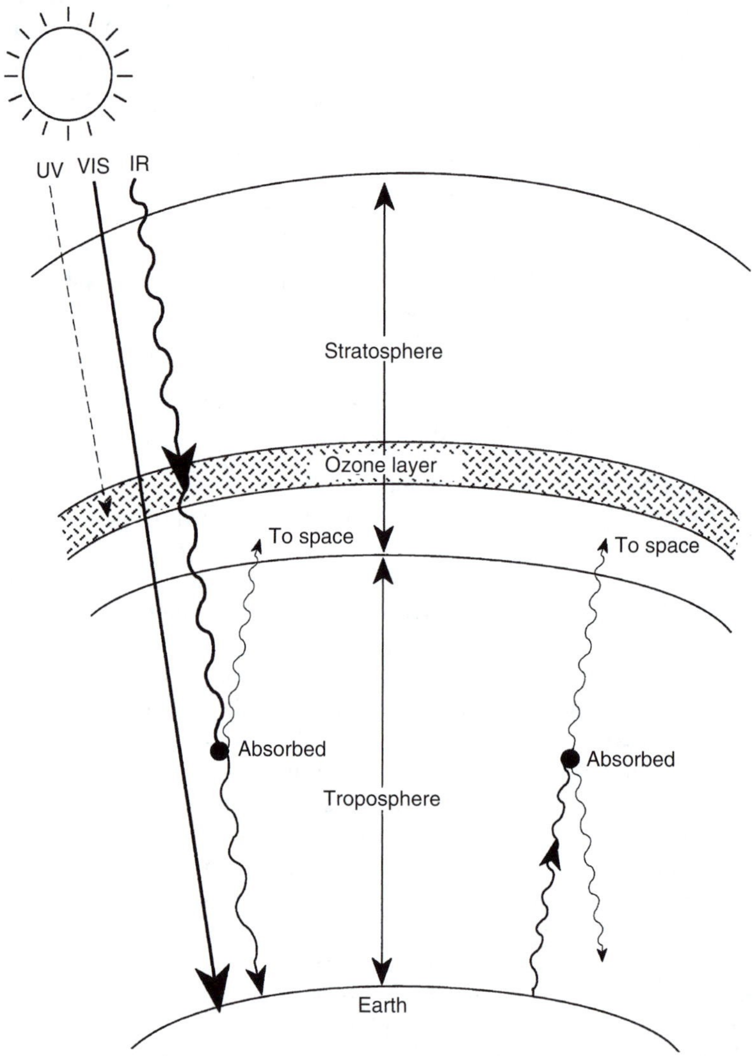

FIGURE 12.3 Simplified diagram of the behavior of solar radiation in the Earth's atmosphere indicating the influence of the ozone layer and the greenhouse effect.

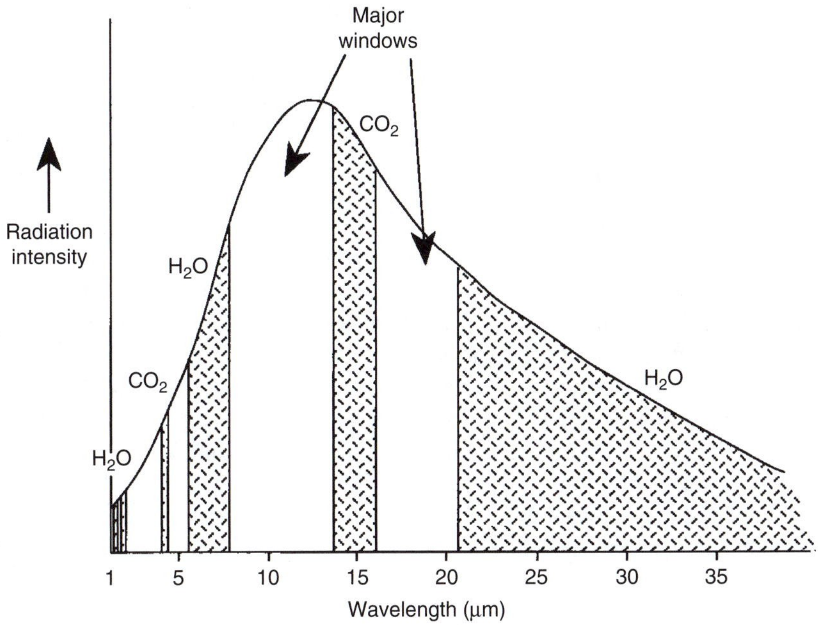

FIGURE 12.4 The relative intensity of emitted radiation from the Earth's surface together with major absorption bands of CO_2 and H_2O vapor. The intensity of absorption for each band is not shown.

the Earth's surface, most of this radiation is reirradiated in the infrared (IR) region, 0.75 to 100 µm.

Water and carbon dioxide have major absorption bands in the infrared spectrum of radiation, leaving the Earth's surface as shown in Figure 12.4. Therefore, it is expected that increases in greenhouse gases will lead to increases in the absorption of the reirradiated energy, which is converted into heat. This could lead to an increase in the Earth's temperature, which could raise sea levels, decrease the temperature gradient between the equator and poles, change weather patterns, and change the amount and distribution of rainfall.

There are many gaps in our knowledge, such as the role of the deep ocean in the carbon cycle and the effects that clouds have on the amount of solar radiation reaching the Earth.

Analysis of gases trapped in ice from Greenland has revealed a similarity in the trends of the concentrations of CO_2 and temperature when viewed over the past 150,000 years. However, when viewed from a longer timescale, a different picture emerges. Since the formation of the Earth there has been an increase of 25 to 30% in the solar energy released daily by the sun and absorbed by the Earth. Yet mean air temperatures have been remarkably stable. For example, there is no evidence to indicate the temperature reached 100°C (water would boil) or 0°C (water would freeze). This maintenance of temperatures suitable for life occurred due to the continuous removal of CO_2 from the atmosphere, with the ensuing increased levels of energy radiated into space. All the evidence indicates that the solar output will continue to rise. So in order to maintain temperatures, atmospheric CO_2 levels should decrease. Instead, the reverse is happening.

12.4 REACTIVE INTERMEDIATES IN THE ATMOSPHERE

12.4.1 Hydroxyl Radical

Apart from possible interaction with solar radiation, much of the chemistry of organic compounds in the troposphere revolves around reaction with certain reactive species or intermediates that are continually being formed and consumed. The most important of these reactive intermediates is the hydroxyl radical (OH·) (Manahan, 2001). Molecules that do not react or react only very slowly with OH· (e.g., chlorofluorocarbons (CFCs)) are unlikely to react with any other reactive intermediates. Sources of OH· in the troposphere are shown in Table 12.2.

It can be seen that solar radiation is critical for the formation of OH·, and so there is considerable diurnal variation in concentrations, which reach their peak during the day. Actual concentrations are somewhat difficult to measure, but it appears that annual average concentrations over a 24-hour period in the troposphere

TABLE 12.2
Sources of Hydroxyl Radicals in the Troposphere

1. Reaction of oxygen atoms, formed from the photolysis of ozone, with water vapor:

$$O_3 \xrightarrow{hv} O_2 + O$$

$$O + H_2O \longrightarrow 2OH^{\bullet}$$

2. Photolysis of nitrous acid:

$$HONO \xrightarrow{hv} HO^{\bullet} + NO$$

3. Photolysis of hydrogen peroxide:

$$H_2O_2 \xrightarrow{hv} 2HO^{\bullet}$$

4. Reaction of hydroperoxy radicals with nitric oxide:

$$HO_2^{\bullet} + NO \rightarrow HO^{\bullet} + NO_2$$

Note: Hydroperoxy radicals are generated from the photolysis of aldehydes,

$$H-CHO \xrightarrow{hv} H^{\bullet} + {}^{\bullet}CHO$$

$$H^{\bullet} + O_2 + M \longrightarrow HO_2^{\bullet} + M$$

and also in the oxidation of hydrocarbons.

are 5×10^5 and 6×10^5 radicals cm^{-3} for the Northern and Southern Hemispheres, respectively.

This is seemingly a large number of species in every cubic centimeter, but these concentrations correspond to approximately 0.02 ppt by volume, which shows that OH· is indeed a trace species. It is also a transient species. It is consumed mainly by reaction with hydrocarbons (particularly methane) and also carbon monoxide (CO). Depending on levels of these compounds, the lifetime of an average hydroxyl radical between formation and consumption is on the order of 0.02 s or less.

From a chemical point of view, OH· is a neutral species, containing in total nine protons and nine electrons. It is a combination of a neutral oxygen atom with six valence electrons and a neutral hydrogen atom with one valence electron:

If the hydrogen shares its electron with one of the unpaired electrons on oxygen, a covalent bond is created between the oxygen and hydrogen, and the hydrogen attains the filled-shell electron arrangement of helium. Oxygen, however, does not achieve a filled shell. There is still one unpaired electron. The oxygen atom has residual bonding capacity and OH· would be expected to be quite reactive. It is important to distinguish between the hydroxyl radical (OH·) and the hydroxide anion (OH^-), which is commonly encountered in water chemistry. The hydroxide anion has one more electron than OH· and hence has an overall charge of –1. Both oxygen and hydrogen have achieved filled shells in OH^-.

12.4.2 Ozone

Approximately 90% of Earth's ozone is to be found in the stratosphere, but 10% exists in the troposphere. In the troposphere, with hydrocarbons, ozone reacts mainly with unsaturated and aromatic molecules, but not saturated ones, and a variety of other substances.

It is a product of photochemical air pollution (smog) (see Section 12.5), but also exists in trace levels in relatively clean, unpolluted air. Levels at the surface of the Earth are variable, but are typically around 5×10^{11} molecules cm^{-3}. Concentrations of 1×10^{13} molecules cm^{-3} have been recorded in severe photochemical air pollution episodes.

In the troposphere, concentrations of ozone tend to increase with altitude, suggesting a stratospheric source. It is generally accepted that periodically, meteorological conditions in the upper troposphere and lower stratosphere are such that ozone from the stratosphere is injected into the troposphere. The other source of ozone in the troposphere is nitrogen dioxide (NO_2), which can be formed in the atmosphere from a number of sources, including motor vehicle exhaust gases initially containing mainly NO (see Section 12.6). Photolysis of NO_2 is the only known way by which ozone can be produced in the troposphere:

$$NO_2 \xrightarrow{hv} NO + O$$

$$O + O_2 + M \longrightarrow O_3 + M$$

M is a third chemical species, most likely N_2 or O_2, that dissipates the excess energy of the ozone molecule that is produced. If there is no third chemical species available, the ozone would not form at all or dissociate on formation. Radiation with wavelengths of less than about 400 nm are required for this process to occur.

The NO_2 can be formed by reaction of nitric oxide (NO) with alkylperoxy (ROO·) and hydroperoxy (HOO·) radicals — intermediates formed in the oxidation of hydrocarbons and CO in the troposphere. During the day, in atmospheres with low concentrations of hydrocarbons and CO, however, a dominant mode of reaction for NO is its reaction with ozone. Under such conditions, as shown in Figure 12.5, a cyclic situation known as the Leighton relationship is created, with ozone being continuously formed and consumed. In the absence of any other reactions, steady concentrations of NO_2, NO, and ozone are eventually observed with time. The relationship between these steady-state concentrations is given by

$$\frac{[O_3][NO]}{[NO_2]} = \text{Constant}$$

where the constant is a function of solar radiation intensity.

12.4.3 Nitrate Radical

The nitrate radical or nitrogen trioxide (NO_3 or NO_3·) is another molecule that is unusual, but is an important reactive intermediate in the troposphere. It is an overall

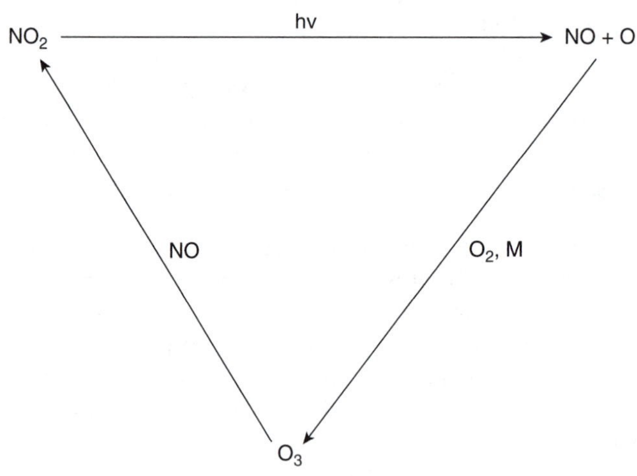

FIGURE 12.5 The Leighton relationship, which describes the cyclic reaction between NO, NO_2, and O_3 in air with little or no hydrocarbons or CO.

neutral species and a radical with one unpaired electron. It should not be confused with the nitrate anion (NO_3^-), which has an additional electron. There is abundant spectroscopic evidence for the existence of NO_3, but it has never been isolated as a pure chemical species. It is formed from oxidation by ozone of NO_2, derived from anthropogenic and natural sources in roughly equal proportion. Motor vehicles contribute about 40% to the anthropogenic sources. The reaction can be represented by the following equation:

$$NO_2 + O_3 \rightarrow NO_3 + O_2$$

During the day, however, NO_3 is rapidly broken down by visible light via two possible pathways:

$$NO_3 \xrightarrow{hv} NO_2 + O \quad \text{or} \quad NO + O_2$$

The relative importance of each pathway depends on the wavelength of the radiations involved. Because of the instability of NO_3 toward light, highest concentrations are observed at night. Even at night, however, NO_3 can be consumed by reaction with NO, or excess NO_2:

$$NO + NO_3 \longrightarrow 2NO_2$$

$$NO_2 + NO_3 \qquad N_2O_5$$

The dinitrogen pentoxide (N_2O_5) can react with water, forming nitric acid:

$$N_2O_5 + H_2O \rightarrow 2HNO_3$$

For this reason, detectable levels of NO_3 are rarely observed if the relative humidity is over 50%. In drier air, at night, the lifetime of NO_3 is on the order of minutes.

The nitrate radical reacts with hydrocarbons in much the same way as the hydroxyl radical, i.e., by abstraction of hydrogen atoms or addition. In a sense, their role as oxidants of hydrocarbons in the troposphere is complementary. During the daytime, when NO_3 is rapidly photolyzed, OH· is the most important. At night, however, levels of OH· diminish because it is photochemically produced and levels of NO_3 increase.

12.5 FORMATION OF PEROXYACETYL NITRATE (PAN) AND RELATED COMPOUNDS

Aldehydes, ketones, ethers, esters, carboxylic acids, and alcohols are all known to exist in the troposphere. Natural sources include plants that emit some of the terpenoids containing oxygen. Primary air pollutants are those that are emitted directly from an identifiable source such as a stack or automobile exhaust. Oxygen-containing compounds typically comprise 5 to 10% of the total organic compound concentration of automobile exhaust.

The aldehyde present in the greatest amount in automobile exhaust, and usually the major aldehyde in ambient air, is formaldehyde. Both formaldehyde and acrolein, another aldehyde typically found in trace levels in the troposphere, are potent irritants of the eyes, nose, throat, and skin. Elevated levels occur in photochemical air pollution events.

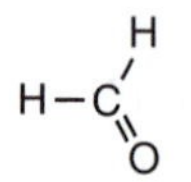

Formaldehyde

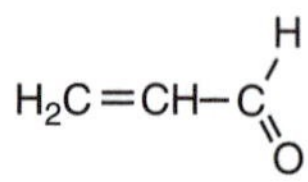

Acrolein

The troposphere is an oxidizing medium, and it is not surprising that hydrocarbons emitted into the troposphere are usually oxidized. Even oxygen-containing compounds can be further oxidized. Secondary air pollutants are those formed in the troposphere as a result of chemical reactions. A significant fraction of oxygen-containing compounds found in the troposphere, particularly in urban areas, are secondary in nature. They are usually less volatile than the compounds from which they were formed and contribute to the haze associated with photochemical air pollution.

An interesting oxygen- and nitrogen-containing compound found in the troposphere is PAN. This compound is a secondary pollutant first found in photochemical air pollution events. It is toxic to plants and extremely irritating to the eyes.

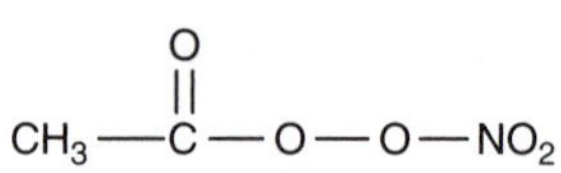

Peroxyacetylnitrate

Trace levels can now be identified all over the world, even in relatively clean air from over the mid-Pacific Ocean, where it is formed from natural precursors.

Petroleum hydrocarbons (PHCs) are discharged into the atmosphere from the incomplete combustion of fossil fuels and evaporation in substantial amounts. The major source is the internal combustion motor used in motor vehicles. Natural sources of hydrocarbons are mainly vegetation, which releases large quantities of terpenoids into the atmosphere. But in addition, methane is generated in substantial quantities by the anaerobic decay of vegetation.

PHCs are not significant primary pollutants since they do not have a strong physiological impact on animals or plants. However, the secondary reactions of partly combusted PHCs with oxides of nitrogen result in the production of important atmospheric pollutants. A simplified sequence of reactions leading to the production of ozone and PAN is shown in Figure 12.6. The primary pollutants involved are nitric oxide and olefins produced by the partial combustion of PHC fuels. As atmospheric pollutants, ozone and PAN can be considered together since they are derived from the same primary pollutants and have similar environmental properties,

FORMATION OF PEROXYACETYLNITRATE (PAN),
OZONE + OTHER OXIDANTS

$$N_2 + O_2 \xrightarrow[\text{internal combustion motor}]{\text{complex reactions}} NO + NO_2$$

where $NO \gg NO_2$,

But NO_2 triggers the following reaction sequence:

$$NO_2 + h\nu \longrightarrow NO + O$$
$$O + O_2 + M \longrightarrow O_3 + M$$
$$O_3 + NO \longrightarrow NO_2 + O_2$$

where M is any third body.

Further reactions now occur with olefins also derived from the internal combustion motor:

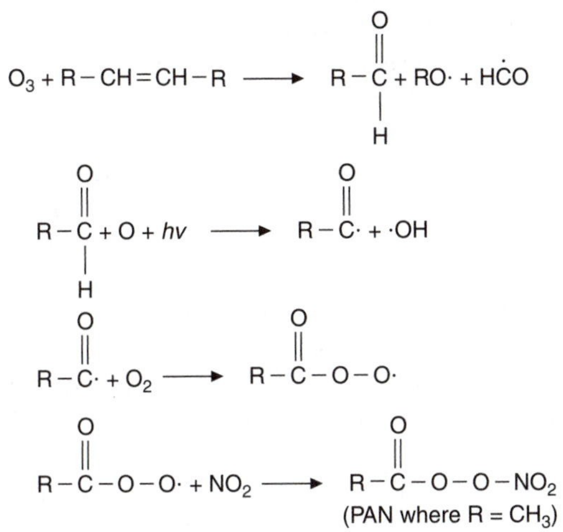

FIGURE 12.6 A simplified reaction sequence in an atmosphere containing products from the internal combustion motor and leading to the formation of peroxyacetyl nitrate (PAN) and ozone.

primarily related to their oxidizing capacity. Ozone and PAN are the major atmospheric pollutants produced by the process described in Figure 12.6, but a great number of other substances with pollutant properties are also produced. Of particular importance are PAN homologs and aldehydes.

The mixture of ozone, PAN, and the other substances is generally described as **photochemical smog**. Photochemical smog is particularly important in urban areas with a large number of cars and also a high incidence of solar radiation. Los Angeles is considered to be a classic example of the development of photochemical smog, causing losses to agricultural crops, ornamental plants, and native vegetation.

Photochemical oxidants are highly reactive and are removed from the atmosphere comparatively rapidly by reactions with vegetation and soil. However, ozone shows low water solubility, so rain-out is not an important factor with this compound.

These substances show a classic diurnal cycle of concentration of the various pollutants as related to the incidence of solar radiation and use of motor vehicles. Clear unpolluted air generally has a concentration of ozone ranging between 0.01 and 0.02 ppm. However, slightly smoggy atmosphere generally ranges from 0.1 to 0.2 ppm, with the highest concentrations ranging from 0.5 to 0.8 ppm. Usually, the concentration of ozone is 10 to 20 times greater than the concentration of PAN.

12.6 HALOGENATED HYDROCARBONS, CHLOROFLUOROCARBONS, AND THE OZONE LAYER

Halogenated hydrocarbons can be considered to be hydrocarbon molecules in which one or more hydrogen atoms have been replaced by a halogen atom, e.g., Cl, Br, or I. The most prominent halogenated hydrocarbons in the troposphere are probably the chlorofluorocarbons (CFCs) and related compounds. These are anthropogenically derived and are of concern because of their effect on the Earth's ozone layer. It would be a mistake, however, to think that all halogenated hydrocarbons found in the atmosphere are the result of human activity, since some are not. The halomethanes CH_3Cl, CH_3Br, and CH_3I are natural components of the troposphere. The source is thought to be biological activity in the oceans. There are abundant halide ions in the oceans that can be incorporated into organic molecules.

The halomethane present in the greatest amount is chloromethane (CH_3Cl). Its concentration, which seems to be fairly constant, is 1.6 ppb by volume. The lifetimes of both CH_3Cl and CH_3Br in the troposphere are sufficiently long, such that some molecules can diffuse into the stratosphere and take a minor part in natural ozone depletion processes.

CFCs were first manufactured on an industrial scale in the 1930s. They have been used as refrigerants, aerosol propellants, in the manufacture of foam plastics, and as cleaning agents for microelectronic apparatus. Related groups of compounds are the hydrochlorofluorocarbons (HCFCs) and the hydrofluorocarbons (HFCs). The HCFCs comprise a central carbon skeleton with hydrogen, fluorine, and chlorine atom substituents; HFCs have only hydrogen and fluorine bonded to carbon. The molecular formula of these groups of compounds can be determined from the Rule of 90. Each compound has a code number, e.g., CFC-11, HCFC-124, or HFC-134a. Adding 90 to the code number results in a three-digit number, which indicates the numbers of carbon, hydrogen, and fluorine atoms present. Any remaining atoms are assumed to be chlorine. This procedure completely characterizes one-carbon and some two-carbon compounds. For other two-carbon compounds, this procedure only results in a molecular formula, e.g., $C_2H_3F_3$. The particular isomer involved cannot be distinguished. To denote a particular isomer, the code number is followed by a lowercase letter, e.g., a, b, or c.

CF_2Cl_2	CHF_2CF_3	$CClF_2CH_3$
CFC-12	HFC-125	HCFC-142b
Dichlorodifluoromethane	Pentafluoroethane	1-Chloro-1,1-difluoroethane

CFCs that have been of greatest importance in terms of production are CFC-11 ($CFCl_3$) and CFC-12 (CF_2Cl_2). Although many uses are closed to the atmosphere, leakage to the air eventually occurs. It has been estimated that 90% of all the CFC-11 and CFC-12 ever produced has been released into the atmosphere. (The total cumulative production of CFC-12 up to 1990 was approximately 10^7 tonnes.) Tropospheric concentrations of a number of CFCs are already greater than that of CH_3Cl. The main problem with CFCs is that they have no reactive tropospheric chemistry — they are effectively inert so that diffusion into the stratosphere is their only significant fate. Their reactions with tropospheric and stratospheric components are considered below. These reactions result in depletion of stratospheric ozone and possibly increased UV wavelength radiation reaching the Earth's surface, as illustrated in Figure 12.3. International agreements, first in Montreal in 1987, followed by revisions in London (1990) and Copenhagen (1992), have aimed to phase out or limit production of CFCs and related compounds.

While resident in the troposphere, CFCs can act as greenhouse gases. Stretching vibrations associated with both C–F and C–Cl bonds lie in the main window for IR radiation transmission (between about 8 and 13.5 μm), as shown in Figure 12.4. As a result, CFCs can absorb outgoing terrestrial radiation very efficiently. As a matter of fact, on a molecule-for-molecule basis, CFCs are more effective absorbers than CO_2 and methane (CH_4).

The primary fate of organic compounds in the troposphere is reaction with hydroxyl radicals (OH·). Although highest concentrations of OH· are only observed during the daytime, this process is usually the most rapid of all the possible fates. With saturated molecules such as alkanes, reaction with OH· occurs by abstraction of a hydrogen atom. The products are water and a hydrocarbon type radical that can undergo further reaction:

$$CH_4 + OH^{\bullet} \rightarrow {}^{\bullet}CH_3 + H_2O$$

To illustrate this, and also the complexity of tropospheric organic chemistry, consider a possible fate of a tetrafluoroethane molecule emitted into the troposphere, as shown in Figure 12.7. Following initial reaction with OH·, a complex sequence of reactions occurs, some involving the UV component of light found in the troposphere. Molecular oxygen is also involved with intermediates formed, including acetaldehyde, formaldehyde, and carbon monoxide.

Acetaldehyde Formaldehyde

Ultimately, for every molecule of tetrafluoroethane, two molecules of CO_2 are formed so the compound has been oxidized.

Because reaction of saturated hydrocarbons with OH· involves breakage of a C–H bond, it follows that the rate constant for such processes depends on the strength of the C–H bond involved. Other factors include which other atoms are bonded to

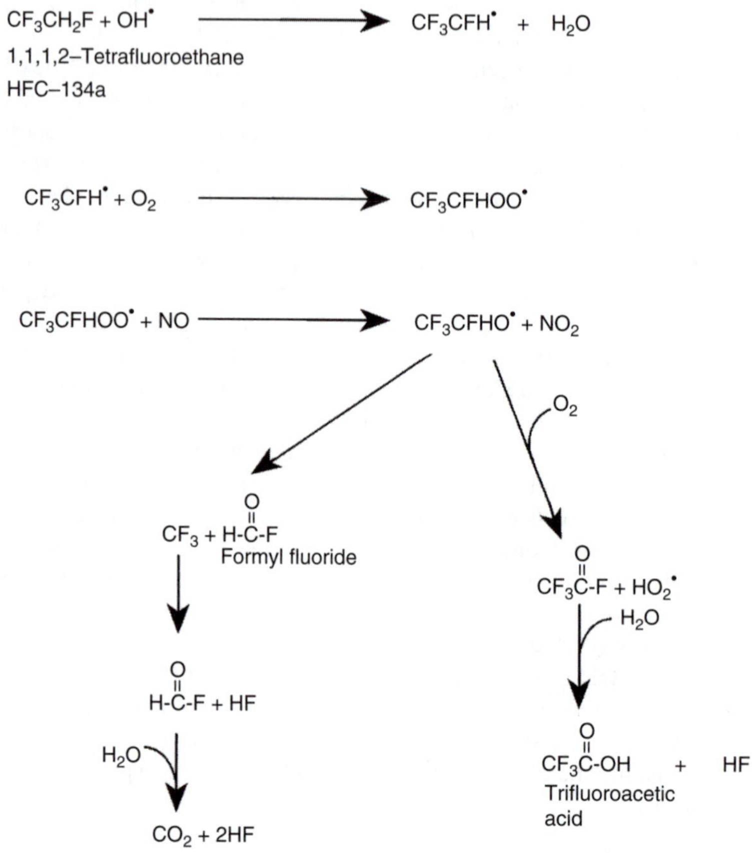

FIGURE 12.7 A diagram of a possible fate for HCF-134a in the troposphere.

that particular carbon. With alkanes, where a total of three hydrogens are attached to a given carbon atom, the hydrogens are denoted as primary (1°) hydrogen atoms. The hydrogen atoms of a methyl group (CH_3–X) can be regarded as primary. Similarly, where a total of two hydrogens are bound to a particular carbon, the hydrogens are secondary (2°), and where a carbon has only one hydrogen attached, this is a tertiary arrangement. For example,

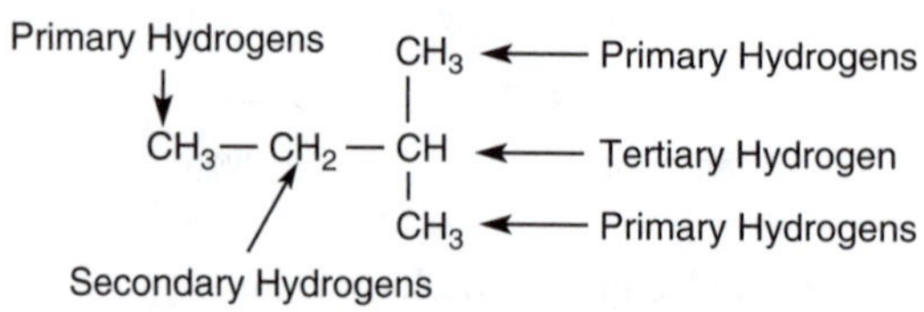

With hydrocarbons, it is generally found that C–H bond strengths are in the order 1°> 2°> 3°. Ease of abstraction of hydrogens tends to have the same order. However, among alkanes, methane itself is relatively unreactive.

TABLE 12.3
Lifetimes for the Reaction of OH• with Methane and a Series of Chlorinated and Fluorinated Methanes in the Troposphere

Compound	Lifetime (years)[a]
Methane (CH_4)	7.5
Chloromethane (CH_3Cl)	1.4
Dichloromethane (CH_2Cl_2)	0.5
Trichloromethane ($CHCl_3$)	0.6
Tetrachloromethane (CCl_4)	>63
Fluoromethane (CH_3F)	3.7
Difluoromethane (CH_2F_2)	5.8
Trifluoromethane (CHF_3)	320

[a] Lifetime at 298K with OH· concentration of 5×10^5 molecules cm^{-3}.

Source: Atkinson, R., *Chem. Rev.*, 85, 69–201, 1985.

Electronegative or electron withdrawing substituents attached to carbon reduce the magnitude of the rate constant thereby increasing the lifetimes. As shown in Table 12.3, the lifetimes resulting from reaction of OH· with a series of fluoro- and chloromethanes are the result of factors such as the number and type of hydrogens and halogens involved. The extremely small rate constants observed leading to the long lifetimes shown are part of the reason CFCs are effectively inert in the troposphere. Rates of reaction with OH· are extremely slow. In addition to this, the equilibrium for the reactions is decidedly unfavorable.

Replacing CFCs with compounds containing C–H bonds increases reactivity in the troposphere and decreases the potential to diffuse into the stratosphere and affect the ozone layer. HCFCs and HFCs are first-generation replacements for CFCs. HFC-134a (CF_2CH_2F) is regarded a viable alternative to the CFC-12 used in some automobile air conditioning systems. Figure 12.7 shows a possible fate for HFC-134a in the troposphere, beginning with reaction with OH·. Oxidation eventually results in the formation of the number of additional polar products that would be susceptible to removal from the troposphere by rain. HCFCs and HFCs still have some ozone depletion potential, however.

Complete replacement of the halogens in CFCs by hydrogen would result in hydrocarbons, which are unsuitable for some CFC applications, as they are flammable. A possible new class of replacement compound comprises hydrofluoroiodoalkanes or fluoroiodoalkanes. The C–I bond can rupture on absorption of UV radiation in the troposphere, enhancing the breakdown of these compounds.

12.7 FORMATION OF OXIDES OF SULFUR AND NITROGEN AND ACID RAIN

In natural systems, decomposing animal and vegetable matter, volcanic action, and wind erosion can result in the release of several gases. These gases usually contain carbon, sulfur, and nitrogen, which are needed in the photosynthetic process for the production of proteins, nucleic acids, and other substances in plants and animals. In addition to obtaining nutrients from the atmosphere, plants can derive a portion of their requirements dissolved in precipitation, as well as from soils. Although the molecular species in the atmosphere and dissolved in precipitation are beneficial at low concentrations, at higher concentrations they may be injurious to plants, animals, or microorganisms.

Combustion of fossil fuels has provided a new source of airborne substances. Thus, significant additions have been made to atmospheric sulfur and nitrogen from fossil fuel combustion. Direct toxicity and toxic effects, produced by solubilization of sulfur and nitrogen gases in precipitation, can have a deleterious effect on natural ecosystems.

The atmosphere includes 0.03% of carbon dioxide, which, in equilibrium with water as precipitation, produces a pH of about 5.7. In rainwater affected by atmospheric pollutants, additional acidity is usually due to three mineral acids: sulfuric, nitric, and hydrochloric acids. Generally, sulfate ions are dominant, with lesser proportions of nitrate ions and comparatively low proportions of chloride ions.

Sulfur occurs in coal in proportions of 1 to 3%, and in petroleum products the proportions can be somewhat higher. Other important sources of sulfur are sulfide ore smelters and volcanoes. Combustion of the fossil fuel produces sulfur dioxide, which can be then oxidized and converted to sulfuric acid:

$$2SO_2 + 2H_2O + O_2 \rightarrow 2H_2SO_4$$

The oxidation of sulfur dioxide in an effluent gas is strongly influenced by the relative humidity. Little oxidation occurs at relative humidities below 70%, but at higher humidities there is relatively rapid oxidation and conversion to sulfuric acid. Manganese and iron salts derived from fly ash catalyze these reactions.

The oxides of nitrogen and nitric acid are naturally produced in the atmosphere by the energy discharged in lightning flashes. However, the major pollution source of these substances is the internal combustion process (Figure 12.8). Acid precipitation can be partially neutralized by the occurrence of bases in the atmosphere, such as ammonia and sea spray.

Sulfur dioxide in the atmosphere can interact with organisms in a number of ways. It can be absorbed into plants directly in the molecular form, or it may be adsorbed onto the moist surfaces of plants, soils, aquatic systems, and so on. Alternatively, it can be converted into sulfuric acid and remain in the atmosphere as aerosol droplets to be removed by precipitation. The use of tall smokestacks, together with suitable atmospheric and meteorological conditions, can lead to transfer and deposition of sulfates thousands of kilometers from the emission source.

FORMATION OF NITROUS AND NITRIC ACID
BY COMBUSTION

Reactions in the Combustion Chamber

1. Generation of atomic oxygen

$$O_2 \longrightarrow O + O$$

$$\text{Step (i)}\quad CO_2 + OH \longrightarrow CO_2 + H$$

$$\text{Step (ii)}\quad H + O_2 \longrightarrow OH + O$$

2. Formation of nitric oxide utilizing atomic oxygen and atmospheric nitrogen

$$\text{Step (i)}\quad O + N_2 \longrightarrow NO + N$$

$$\text{Step (ii)}\quad N + O_2 \longrightarrow NO + O$$

Reactions in the Atmosphere

1. Formation of nitrogen dioxide and nitrogen trioxide

$$2NO + O_2 \longrightarrow 2NO_2$$

$$O_2 + NO \longrightarrow NO_2 + O_2$$

$$NO_2 + O_3 \longrightarrow NO_3 + O_2$$

2. Formation of N_2O_5 and reactions of nitrogen trioxide

$$NO_3 + NO_2 \longrightarrow N_2O_5$$

$$NO_3 + NO \longrightarrow 2NO_2$$

3. Formation of nitrous and nitric acids in the presence of moisture

$$N_2O_5 + H_2O \longrightarrow 2HNO_3$$

$$NO_2 + NO + H_2O \longrightarrow 2HNO_2$$

FIGURE 12.8 A possible reaction sequence resulting in the formation of nitrous and nitric acids from the combustion of fossil fuels.

Significantly, an indirect effect of acid rain and the lowering of the pH of aquatic areas is the release of toxic metals adsorbed on bottom sediments. Thus, effects in natural aquatic ecosystems may be due to these metals as well as low pH.

12.8 KEY POINTS

1. Photosynthesis by plants uses energy from sunlight to convert carbon dioxide and water into plant carbohydrate with the release of oxygen to the atmosphere. This reaction can be summarized by the following equation:

$$6CO_2 + 6H_2O + \text{Energy} \rightarrow 6CH_2O + 6O_2$$

Carbohydrate plant matter

2. The plant matter produced by photosynthesis is utilized as food by animals and other organisms by the respiration process. This process is the reverse of photosynthesis and can be represented by the following equation:

$$6CH_2O + 6O_2 \rightarrow 6CO_2 + 6H_2O + \text{Energy}$$

(Carbohydrate)

3. The photosynthesis and respiration processes must occur at approximately the same rate to maintain the O_2 and CO_2 content of the atmosphere.
4. Over geological times the development of photosynthesis by larger plants has had an effect on the chemistry of the Earth. The production of oxygen gas by plants has changed the composition of the atmosphere and converted the chemical state of the Earth from reducing to oxidizing. Photosynthesis and respiration currently play a major role on the Earth, as the key processes in the carbon cycle and maintaining the existing atmospheric levels of carbon dioxide.
5. The atmosphere plays several key roles in maintaining the global ecosystem. The principal roles are:
   - The external medium for photosynthesis and respiration
   - Source of carbon dioxide and a sink for oxygen
   - A heat reservoir
   - A shield against damaging short-wavelength radiation
6. The atmosphere can be divided into zones, starting with the troposphere at the Earth's surface (depth, about 10 to 16 km; average temperature, 15°; normal atmospheric composition, 1 atmosphere pressure), stratosphere (depth, 15 to 50 km; temperature, –56 to –2°C; 0.1 to 0.001 atmosphere pressure), mesosphere (depth, 50 to 85 km; temperature, –2 to –92°C; 0.001 to 3×10^{-6} atmosphere pressure), thermosphere (depth, 85 to 500 km; temperature, –92 to +1200°C; $<3 \times 10^{-6}$ atmosphere pressure), and space.
7. The composition of the troposphere is 78.09% N_2, 20.95% O_2, 0.93% Ar, 300 ppm CO_2, 18 ppm Ne, 1.5 ppm He, 1.5 ppm CH_4, and many other minor components.
8. The upper atmosphere contains a range of chemical entities that are highly reactive, e.g., O_3, O^+, O_2^+, etc. These substances have a limited life at the Earth's surface, but can exist in the upper atmosphere due to the low pressures, giving long mean free paths and relatively infrequent intermolecular collisions during which reaction could occur.
9. There are several important reactive intermediates that participate in reactions in the atmosphere. These include OH· (hydroxyl radical). The OH· radical is formed from ozone, water, and other reactants in the atmosphere by a series of reactions. The uptake of energy through solar radiation is essential for some of these reactions. Thus, the OH· radical is formed during daylight hours when it reaches its peak concentration.

10. About 90% of the Earth's ozone occurs in the stratosphere, where it plays a major role in absorbing biologically damaging short-wavelength radiation.
11. Ozone in the troposphere is an important component of smog, where it is formed by photolysis of NO_2. A significant proportion of the NO_2 present is formed from motor vehicle exhaust gases:

$$NO_2 \xrightarrow{h\nu} NO + O$$

$$O + O_2 + M \longrightarrow O_3 + M$$

12. The nitrate radical (NO_3) is formed from NO_2, of which a significant proportion in urban atmospheres is derived from motor vehicle exhaust gases. The formation reaction with ozone can be represented by

$$NO_2 + O_3 + M \rightarrow NO_3 + NO_2$$

During daylight hours, NO_3 is rapidly broken down by photolysis reactions.
13. Formaldehyde, acrolein, and peroxylacetyl nitrate (PAN) are all important atmospheric pollutants that have damaging biological effects. PAN is a secondary pollutant formed by a series of reactions involving nitrogen- and oxygen-containing intermediates initiated in the internal combustion motor and continued in the external atmosphere.

$$NO_2 \xrightarrow{\dot{h\upsilon}} NO + O$$

$$O + O_2 + M \longrightarrow O_3 + M$$

14. The chlorofluorohydrocarbons (CFCs) have been manufactured since the 1930s and used as refrigerants, aerosol propellants, and in a range of other applications. The CFCs produced in greatest quantities are CFC-11 ($CFCl_3$) and CFC-12 (CF_2Cl_2).
15. The CFCs are relatively inert in the troposphere, but in the stratosphere they can lose a halogen, which can react with ozone, resulting in depletion of this layer. This may result in a larger amount of short-wavelength radiation reaching the Earth's surface.
16. The sulfur in fuels can be oxidized during combustion to produce sulfur dioxide, which can be further oxidized to sulfuric acid:

$$2SO_2 + 2H_2O + O_2 \rightarrow 2H_2\,SO_4$$

17. The nitrogen in the atmosphere (about 78%) can react with oxygen in the atmosphere (about 21%) through a series of complex reactions in the internal combustion engine and the external atmosphere to give nitrous and nitric acids.

REFERENCES

Baird, C., *Environmental Chemistry*, W.H. Freeman and Co., New York, 1995.
Manahan, S.E., *The Fundamentals of Environmental Chemistry*, 2nd edition, Lewis Publishers, Boca Raton, 2000.
Wayne, R.P., *Chemistry of Atmospheres*, 2nd ed., Clarendon Press, Oxford, 1991.

QUESTIONS

1. The composition of the atmosphere has changed over geological time from predominantly carbon dioxide, nitrogen, and water to predominantly nitrogen, oxygen, and water, with trace amounts of carbon dioxide. Use equations to explain how this has occurred and outline in broad terms which global components have changed in elemental composition as a result.
2. In 1750, the atmospheric concentration of carbon dioxide was about 280 ppm (volume/volume), compared with about 355 ppm (volume/volume) today. Calculate the proportion of carbon present in rocks that has been used as fuels assuming there is 7.5×10 tonnes of carbon in rocks. In carrying out this calculation, assume that air and carbon dioxide have the same density, the total mass of the atmosphere is 5.14×10^{15} tonnes, all of the carbon dioxide produced by burning fossil fuels still remains in the atmosphere, and burning fossil fuels has been the only process producing an increase in atmospheric carbon dioxide.
3. Many smog components are produced by motor vehicle exhaust gases. The general 24-hour cycle of motor vehicle traffic is shown in Figure 12.9. Draw possible patterns for NO (the main nitrogen oxide in vehicle exhaust), NO_2, and PAN and explain the reasons behind these patterns.

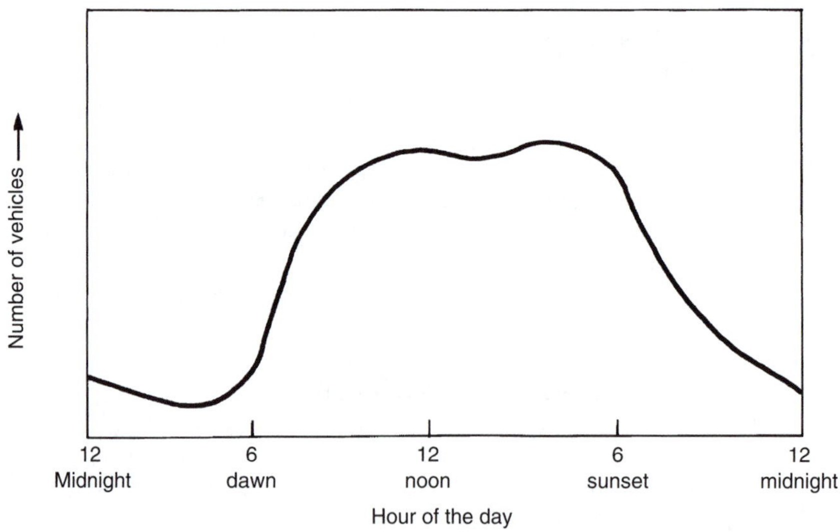

FIGURE 12.9 The general 24-hour cycle of motor vehicle traffic.

4. The more commonly produced CFCs are CFC-11 ($CFCl_3$) and CFC-12 (CF_2Cl_2). Suggest chemical reactions that could explain the reactions of the substances with ozone.

ANSWERS

1. The change in composition has occurred because after plants evolved on Earth, the global rates of photosynthesis and respiration were not the same:

   Photosynthesis

   $$6CO_2 + 6H_2O + \text{Energy} \rightarrow \underset{\text{Plant carbohydrate}}{6[CH_2O]} + O_2$$

   Respiration

   $$\underset{\text{Plant carbohydrate}}{6[CH_2O]} + O_2 \rightarrow 6CO_2 + 6H_2O + \text{Energy}$$

   Photosynthesis has resulted in the production of plant carbohydrate matter, which was not fully consumed by organisms, such as animals and microorganisms. Thus, the rate of photosynthesis exceeded the rate of respiration. The excess plant matter was substantially incorporated into sedimentary strata over geological time. Thus, the upper compartments of the crust of the Earth became enriched with carbohydrate and, of particular importance, carbon. This material then formed large deposits of coal and other fossil fuels. On the other hand, the atmospheric compartment was steadily depleted in carbon dioxide, particularly carbon, and enriched with oxygen. The impact of these processes on nitrogen have not been great, so nitrogen has remained substantially unchanged.
2. The increase in concentration of carbon dioxide is 355 to 280 ppm (volume/volume), i.e., 75 ppm. If it is assumed that the atmosphere and carbon dioxide have the same density, then the mass of carbon dioxide produced is $(75 \times 10^{-6}) \times (5.14 \times 10^{15})$ tonnes of CO_2.

Tonnes of CO_2 produced	=	38.55×10^{10}
Tonnes of carbon produced	=	$12/44\ (38.55 \times 10^{10})$
Therefore carbon produced	=	10.48×10^{10} tonnes
Amount of carbon in rocks (from Figure 12.2)	=	7.5×10^{15} tonnes
Thus the fraction of carbon in rocks used as fossil fuel	=	$(10.48 \times 10^{10})/(7.5 \times 10^{15})$
	=	1.4×10^{-5} or 0.0014%.

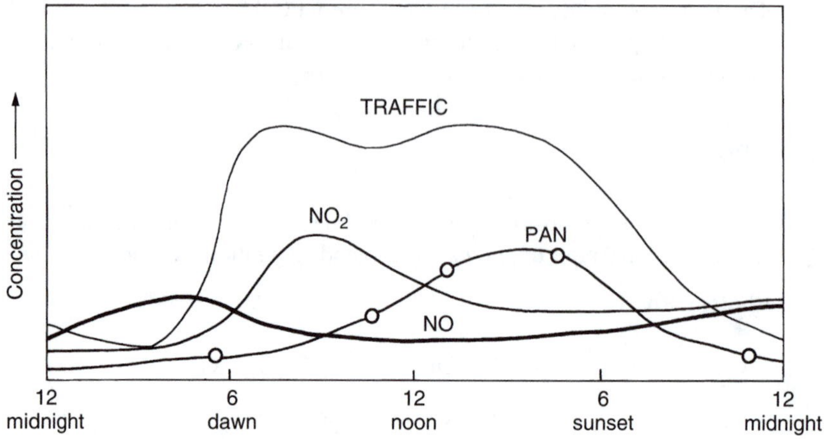

FIGURE 12.10 General pattern of change in traffic of NO, NO_2, and PAN over a 24-hour cycle.

3. An outline of possible changes over the 24-hour cycle of NO, NO_2, and PAN due to motor vehicle exhausts is shown in Figure 12.10.

   NO: A primary component of motor vehicle exhaust gas that is low, due to low vehicle traffic, overnight. It starts to rise at dawn with increasing vehicle traffic, but is converted to NO_2 during the day.

   NO_2: Formed from NO early in the day, but consumed to form PAN later in the day.

   PAN: Formed later in the day by reaction of NO and other components acted on by solar radiation.

   Overnight the input of vehicle exhaust gases declines substantially and the input of solar radiation ceases. Thus, all components tend to decline or remain reasonably constant.

4. The CFCs are very stable compounds that contain only strong carbon halogen bonds.

   CFC-11 ($CFCl_3$): The photoreaction of these compounds leads to the splitting of the CFC molecule at the low pressures in the stratosphere to form the highly reactive $^{\bullet}CFCl_2$ and $Cl^{\bullet}$. Thus,

$$CFCl_3 \xrightarrow{h\nu} {}^{\bullet}CFCl_2 + Cl^{\bullet}$$

Solar radiation in the stratosphere contains short-wavelength, high-energy radiation that can be absorbed by the CFC molecule resulting in this reaction. The chlorine can further react with ozone as shown:

$$Cl^{\bullet} + O_3 \longrightarrow ClO + O_2$$

This results in depletion of ozone present.
CFC-12: A similar process can occur with this compound.

13 Chemistry of Natural Waters and Water Pollution

13.1 INTRODUCTION

Water covers about 71% of the global surface, so the water environment comprises a major part of the human environment. About 97% of this water environment is in the oceans, and the remainder in lakes, rivers, and so on. The study of water bodies has been a major area of scientific activity since the beginning of scientific endeavors.

Limnology is the study of freshwaters. This branch of science is concerned with the physics, chemistry, and biology of lakes, rivers, and other freshwater bodies. In addition, limnologists have always had a strong interest in the effects of pollution on natural systems, particularly sewage. In fact, in the early part of the 20th century some of the basic principles of the pollution ecology of streams were established. This emphasis on the human interaction with freshwater systems continues to the present day.

Oceanography has a similar long history of activity (Meadows, 1978). Some say that the voyages of Captain James Cook, which started in 1768 with the *Endeavour*, mark the start of scientific oceanography. In Cook's voyages, observations were made of the temperature and depth of the oceans as well as biological phenomena. Many oceanographic cruises have been undertaken since the time of Cook, and these continue to the present day. Expeditions following Cook began the study of the ocean currents and winds, principally due to their importance in the movement of sailing ships carrying cargoes for trading purposes.

The chemistry of natural water bodies has developed enormously in parallel with the development of the science of chemistry in general (Manahan, 1999). Currently, we are able to characterize and quantify chemical components of water bodies and evaluate the dynamics of their transformations. This capacity was not available to early investigators. In recent years, with the development of governmental bodies and agencies for aquatic research, the water-based sciences have received a boost in activity. At the present time, considerable resources and effort are employed in the development of an understanding of the functions of natural water bodies as a basis for conservation and management of these important natural environments.

13.2 PHYSICOCHEMICAL PROPERTIES OF WATER

To appreciate the functions of natural water bodies, we must start at the molecular level. The water molecule, H_2O, is planar with an angle of 104.5° between the O–H bonds, as shown in Figure 13.1. Each O–H bond is polar and has a dipole moment of 1.5 Debyes, with the positive charge directed toward the hydrogen atom. This is in agreement with the difference in electronegativity of oxygen (3.5) and hydrogen (2.1) of 1.4. Due to its polarity (1.84 Debyes), water has a high boiling point and vapor pressure, compared with other compounds of similar molecular weight. Thus, water would be expected to be a good solvent for ionic salts, such as sodium chloride. This characteristic is reflected in the high salt content, which is dissolved in seawater. On the other hand, nonpolar substances such as animal and plant fats, which are present in biotic cells, do not dissolve in water to any significant extent. If this were not so, water would penetrate living cells freely and thereby threaten their survival.

Water has a number of remarkable properties. Perhaps one of the most noteworthy is its freezing behavior. Although it freezes at 0°C, it exhibits a maximum density at about 4°C. This means that when ice forms, instead of falling to the bottom, it floats to the surface. In this way, the ice forms a layer on the surface insulating the lower waters and preventing them from freezing.

Most large water bodies, both natural and artificial, exhibit a vertical stratification into layers at different temperatures. For example, in the oceans the surface temperatures show considerable variation, but at the lowest depths the water temperature is usually close to 4°C at all latitudes. This is the temperature at which water exhibits its maximum density. This suggests that whenever water at 4°C is formed in the ocean, it falls to the bottom and accumulates where it is substantially insulated from thermal changes by the overlying water and the sea floor. In lakes and dams, as well as the oceans, solar radiation can heat the upper layers of the water body, producing lower densities. The heating produced by solar radiation penetrates to the depth permitted by the clarity of the upper layers of the water. The upper layer, which is often at a higher temperature and lower density, is referred to as the **epilimnion**, and the lower layer, having higher density and lower temperatures, is referred to as the **hypolimnion**. A distinctive temperature profile is observed with a **thermocline** separating the two layers, as shown in Figure 13.2.

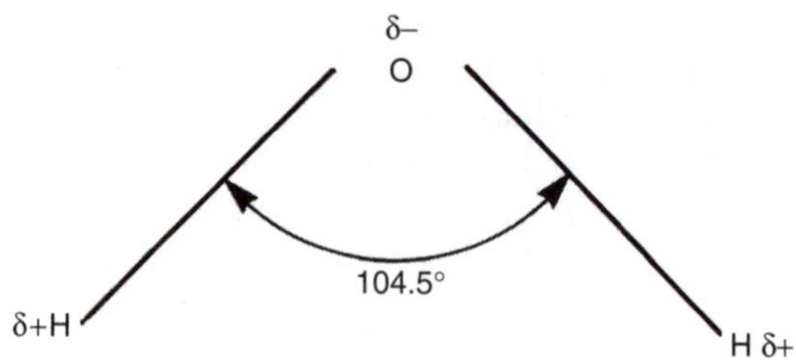

FIGURE 13.1 Diagrammatic representation of the water molecule.

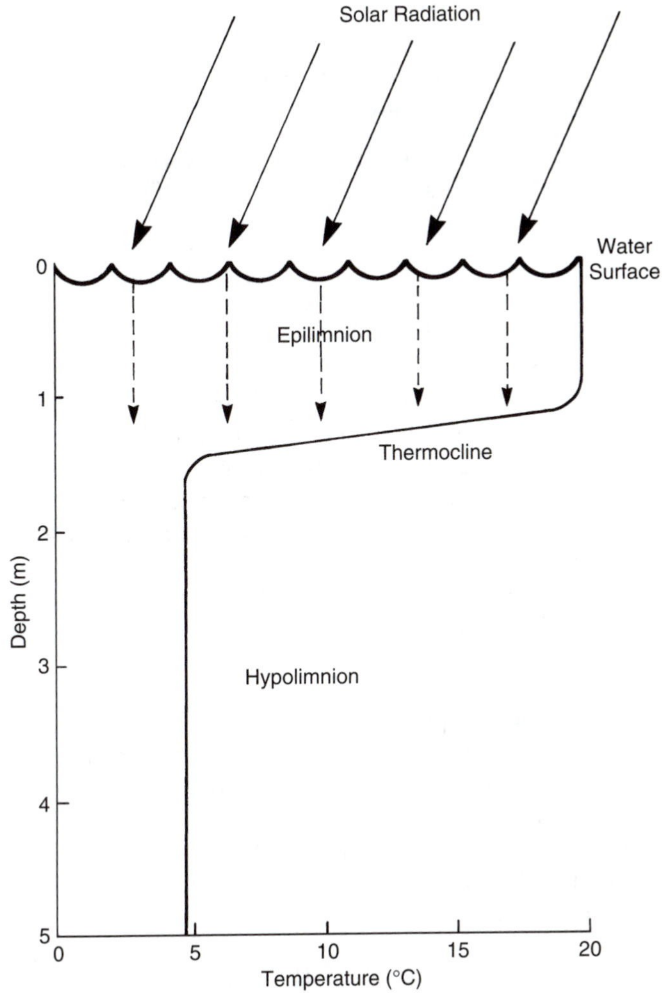

FIGURE 13.2 Stratification of a water body produced by solar radiation.

Gases from the atmosphere dissolve in natural water bodies and have an important influence on chemical and biological systems. Gases such as oxygen, nitrogen, and methane are all symmetrical molecules and would be classified as nonpolar compounds. Thus, it would be expected that these substances would have limited solubility in water that is polar. This is reflected in the low aqueous solubility values shown in Table 13.1. Carbon dioxide also has low solubility in water, but the solubility of this gas is more complex than the others since it dissolves in water and also dissociates to form ions.

A convenient measure of the environmental behavior of gases is the Henry's law constant (H). The Henry's law constant is defined as follows:

$$H = P/C_W$$

TABLE 13.1
Henry's Law Constants for Some Atmospheric Gases (1 atm pressure and 25°C)

Gas	Aqueous Solubility (mg/l)	Henry's Law Constant (H) (atm/mol/l)
O_2	8.11	773
CO_2	0.55	29
N_2	13.4	1610
CH_4	24	746

where P is the partial pressure of the substance in the atmosphere above the water and C_W the corresponding concentration of the substance in water. Some Henry's law constants for atmospheric gases are shown in Table 13.1. It should be noted that the Henry's law constant for carbon dioxide may not be constant, as the other gases are, since carbon dioxide undergoes the dissociation reaction in water that was previously mentioned.

The Henry's law constant is useful for calculating concentrations that would be expected in water and the atmosphere. For example, the solubility of oxygen in pure water can be calculated as follows:

1. **Calculation of partial pressure of oxygen in the atmosphere**: The total pressure of the gases in the atmosphere is 1 atm. But the atmosphere also contains 0.03 atm of water vapor at 25°, as well as 20.95% of oxygen as a permanent atmospheric gas. This means that the partial pressure of oxygen in the atmosphere is 0.2095/(1.00 – 0.03), which is 0.216 atm.
2. **Calculation of the concentration of dissolved oxygen (DO) in water**: From Henry's law,

$$S = P/H$$

and

$$H = 773 \text{ atm mol l}^{-1}$$

Thus,

$$S = \frac{0.216 \text{ atm}}{773 \text{ atm mol l}^{-1}}$$

$$= 2.79 \times 10^{-4} \text{ mol/l}$$

$$= 8.9 \text{ mg/l}$$

This is in reasonable agreement with the directly measured value reported in Table 13.1 at 8.11 mg/l.

The oxygen dissolved within natural waters has a critical role since it is needed for biotic respiration. Without this oxygen supply, biota would not be able to survive within the water mass. Although there is a relatively large supply of oxygen in the atmosphere, which contains about 21% of oxygen, this substance has only limited solubility in water, as reflected in the data in Table 13.1. It is also interesting to note that the solubility of oxygen in water varies with a variety of factors. Of particular importance is temperature, with increases resulting in a major decline in solubility in the range of 0 to 35°, as shown in Table 13.2. The salinity of seawater and estuarine waters also causes a decline in the solubility of oxygen in water.

Pure water weakly dissociates into hydrogen ions and hydroxyl ions with a dissociation constant of 10^{-14}, producing 10^{-7} g of ions l^{-1} of each ion (see also Section 2.5). The pH is defined as

$$pH = -\log [H^+]$$

The p designates the power of the hydrogen ion activity.

Thus, the pH of pure water is 7. In natural waters the pH is governed by a variety of factors. Of particular importance is the solubilization of carbon dioxide from the atmosphere to produce carbonic acid, which dissociates into ions, as shown in Figure 13.3. The solubility of carbon dioxide in natural waters varies with temperature, and the dissociation into ions is strongly influenced by salts and other substances present. The pH of natural freshwaters is usually in the range of 6 to 9.

TABLE 13.2
Variation of Solubility of Oxygen in Water with Temperature (at 760 mm pressure)

Temperature	Oxygen (mg/l)
0	14.16
5	12.06
10	10.92
15	9.76
20	8.84
25	8.11
30	7.53
35	7.04

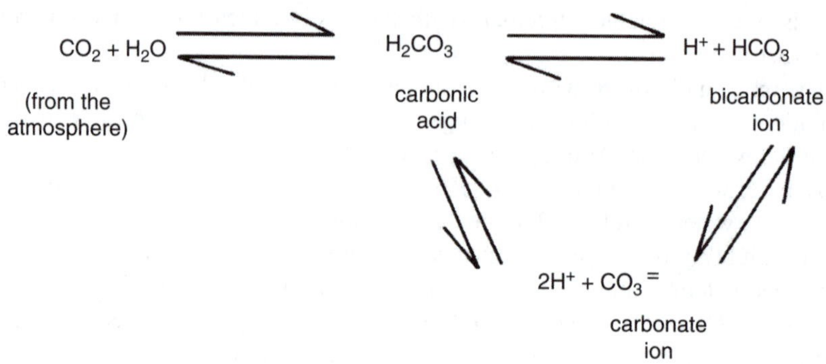

FIGURE 13.3 Ionization patterns of carbon dioxide dissolved in water.

13.3 ORGANIC COMPONENTS IN NATURAL WATERS

There are a wide variety of organic compounds in natural waters, which can range in type from mountain lakes to the oceans. Most are the degradation products of animals and plants following death. There can be low concentrations of fatty acids, fats, hydrocarbons, and many other substances. Generally, low-molecular-weight compounds, such as amino acids and monosaccharides, do not occur in any significant quantities. This is due to the ready degradation of these substances upon formation from more complex substances. This is facilitated by a large and varied population of microorganisms together with many nutrient substances present in natural water bodies. For example, glucose has a half-life of about 10 hours in natural waters at temperatures of about 20°C.

Polysaccharides, such as cellulose, are relatively persistent in natural waters and therefore are present for long periods in relatively large amounts resulting from the death of plants in particular. Cellulose is usually present in plant detritus in water bodies. Similarly, there can be relatively high amounts of proteins and polypeptides that are resistant to degradation.

In many coastal and freshwater areas, there are large zones of swamps and marshlands that are periodically or permanently inundated with water. These areas generally have a high plant biomass, which leads to large amounts of dead plant matter being formed within these zones. This dead plant material is subject to degradation, particularly by microorganisms. After the organic matter has been attacked by microorganisms and other processes, residues remain that are resistant to further degradation. These residues are mainly derived from plant lignins, which are high-molecular-weight polyphenolic substances in plants. The same processes can occur within soil, producing similar substances resistant to biodegradation. As a group, these substances are referred to as **humic substances** and they have a brown or dark color, with molecular weights ranging from the hundreds to the thousands.

This complex group of substances can be fractionated into crude components by acid and base solubility, as shown in Figure 13.4. This allows the isolation of substances that are described as **fulvic and humic acids** and **humin**. In fact, these fractions are complex mixtures with properties covering a range of characteristics,

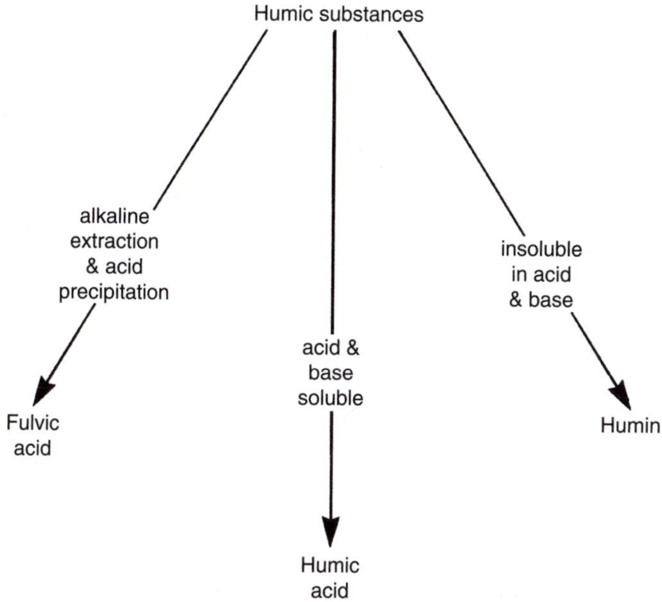

FIGURE 13.4 Fractionation of humic substances into broad component classes.

rather than being specific. The fractions all share somewhat similar properties and chemical structures. Fulvic acid is the lowest-molecular-weight substance and has the highest oxygen content. The chemical behavior of the humic substances is largely governed by the many phenolic and related groups present. as indicated in Figure 13.5. These groups are linked into the larger higher-molecular-weight molecules by a variety of bonding patterns. A possible structure for a fulvic acid component is shown in Figure 13.6.

While the humic substances contain many polar groups, their high-molecular-weight leads to low solubility in water. This is due to the strong intermolecular forces that exist between the molecules of the humic substances related to the high molecular weight. The strong color of these substances is due to extended conjugation with the phenolic ring, which occurs in many cases, leading to strong ultraviolet and visible absorption, as shown in Figure 13.7.

An important environmental property of the humic substances is their ability to chelate, thereby strongly binding metal ions into their structure. This can occur with the phenolic hydroxyl group and the acid groups, as indicated in Figure 13.8. The ions of iron and aluminum are strongly bonded in this system, whereas magnesium is relatively weak, with nickel, lead, and calcium being intermediate in their bonding strength. This may make essential trace metals unavailable to biota in aquatic areas with high concentrations of humic substances.

The humic substances and other organic matter, as well as inorganic matter, may not dissolve in water, but form small particulates that do not settle out over time. These are usually referred to as **colloids**. The presence of colloids in an aquatic area alters the behavior of metals in solution in a somewhat similar manner to the humic

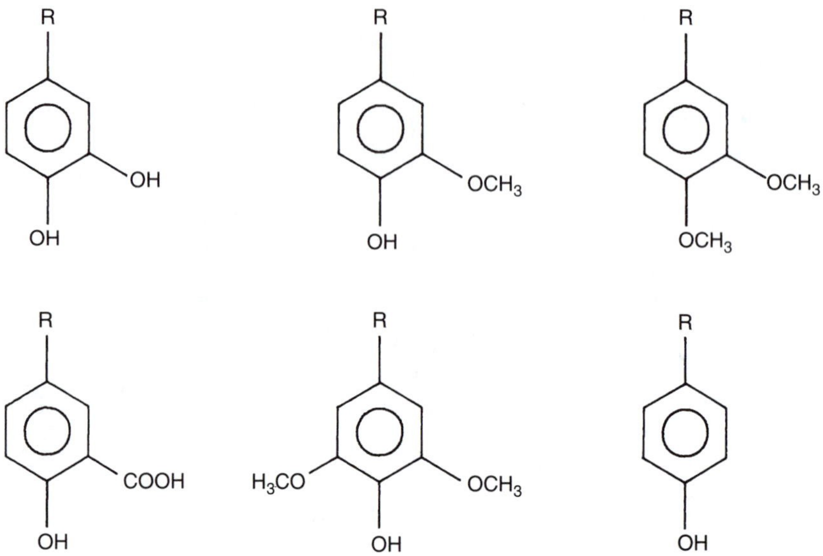

FIGURE 13.5 Phenolic groups present in the structure of humic substances.

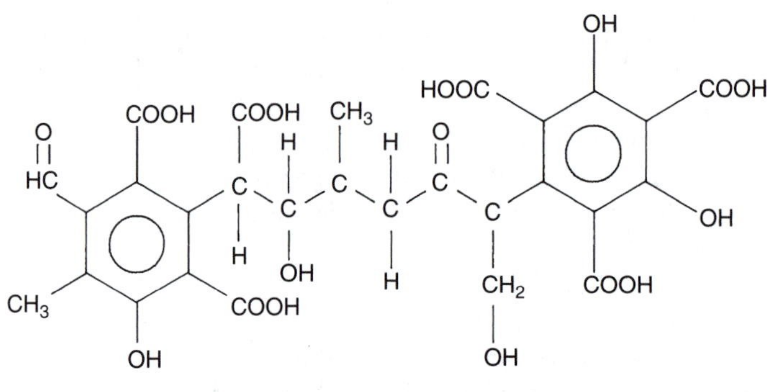

FIGURE 13.6 A possible structure for a fulvic acid component.

substances due to charged surfaces that may exchange and bind metals and other ions. In addition, colloids with lipoid characteristics may alter the behavior of lipophilic substances in water bodies.

13.4 CARBON AND NITROGEN TRANSFORMATIONS IN NATURAL WATER BODIES

13.4.1 The Carbon Cycle

The fundamental process that feeds carbon and energy into aquatic systems is photosynthesis. In this process, carbon dioxide from the atmosphere is dissolved and dissociates into the various forms previously discussed in water bodies. Plants

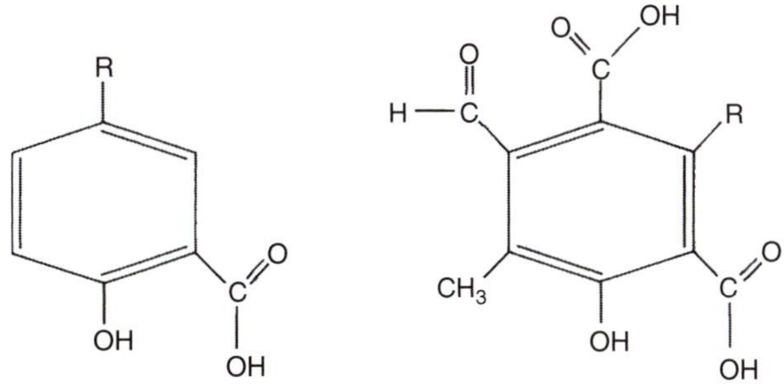

FIGURE 13.7 Examples of extended conjugation in groups present in humic substances.

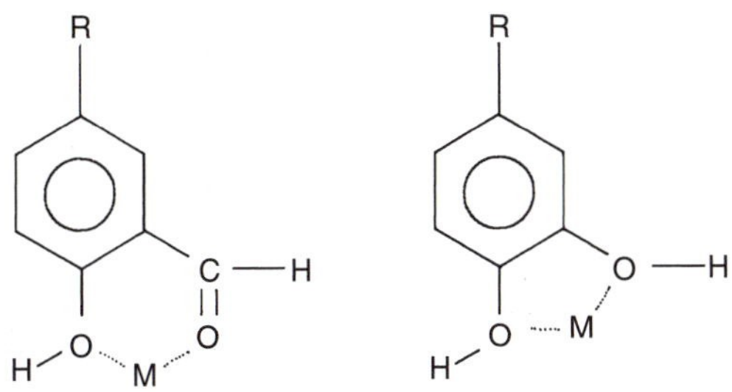

FIGURE 13.8 Chelation of metal ions by phenolic groups present in humic substances.

perform photosynthesis by converting carbon dioxide, water, energy, and various trace components into plant matter. In addition to photosynthesis, the plants consume some of the energy produced by using it in respiration and so release some carbon dioxide back to the water mass. Some plant material is consumed by aquatic animals that utilize the energy in the food by respiration, which results in the release of carbon dioxide to the water mass, as shown in Figure 13.9.

On death, both animals and plants produce biotic fragments referred to as **detritus**. Detritus is rich in carbon and energy in the form of carbohydrate, protein, and fat. Detritus enriches sediments with organic matter where it is acted on by the population of microorganisms present. Some of the resultant respiration occurs by **aerobic** processes and produces carbon dioxide, but often sediments are depleted in oxygen and **anaerobic** processes occur, resulting in the production of carbon dioxide and methane. Both methane and carbon dioxide produced from all respiration processes are released into the water mass and can escape to the atmosphere. These processes can be seen as a cycle of carbon from the atmosphere through the aquatic ecosystem and back to the atmosphere again (Figure 13.9).

13.4.2 The Nitrogen Cycle

Nitrogen exists in aquatic systems in a variety of chemical forms, including both inorganic and organic. The principal inorganic forms all occur in different proportions, depending on the conditions in the water mass. Nitrate (NO^-_3) is the principal

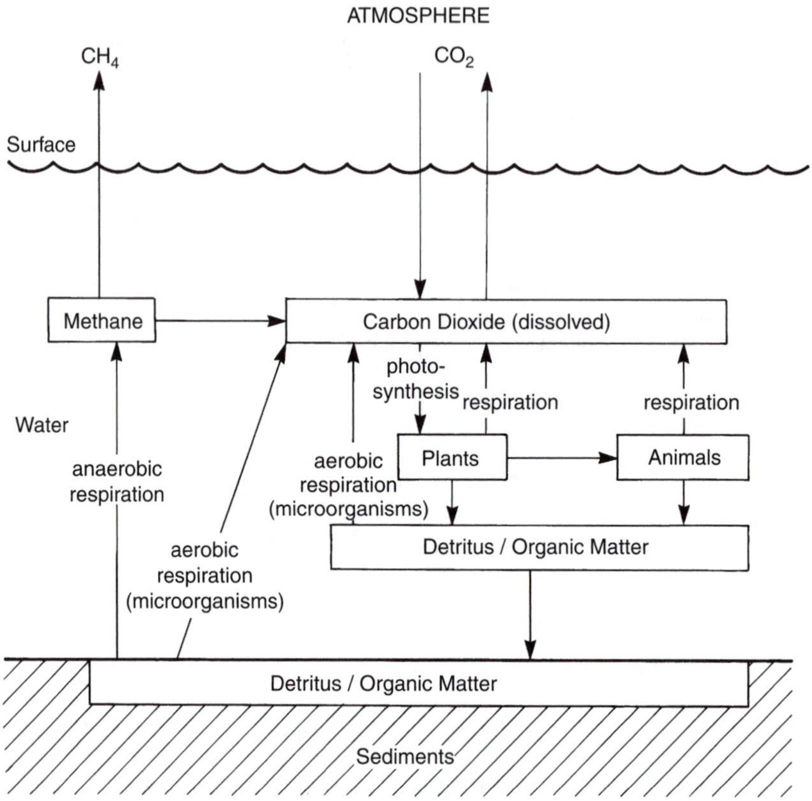

FIGURE 13.9 Transformations of carbon in aquatic systems.

inorganic nitrogen salt in well-aerated water bodies where oxygen is in plentiful supply. Nitrite (NO_2^-) occurs under certain specific conditions, but ammonia (NH_3) is the principal inorganic form of nitrogen under anaerobic conditions. Ammonia dissolves in water to form ammonium hydroxide (NH_4OH), which dissociates into the ammonium (NH_4^+) ion and the hydroxyl ion (HO^-). Oxidation can convert all the inorganic nitrogen forms into the nitrate ion, while reduction converts them all into ammonia.

The oxidation of inorganic nitrogen to nitrate is referred to as **nitrification**, as shown in Figure 13.10. **Denitrification** is the process whereby nitrogen is converted into N_2 through the nitrite ion and then subsequently released to the atmosphere as N_2 or N_2O (nitrous oxide). On the other hand, **nitrogen fixation** is a process whereby atmospheric nitrogen is fixed through ammonia into biological systems. This process requires a substantial amount of energy to convert atmospheric nitrogen into ammonia, and for this reason, it is less favored than other processes by biota. Proteins in detritus, from animals and plants, can be hydrolyzed to the component amino acids and then further degraded to ammonia and the other inorganic nitrogen forms, as

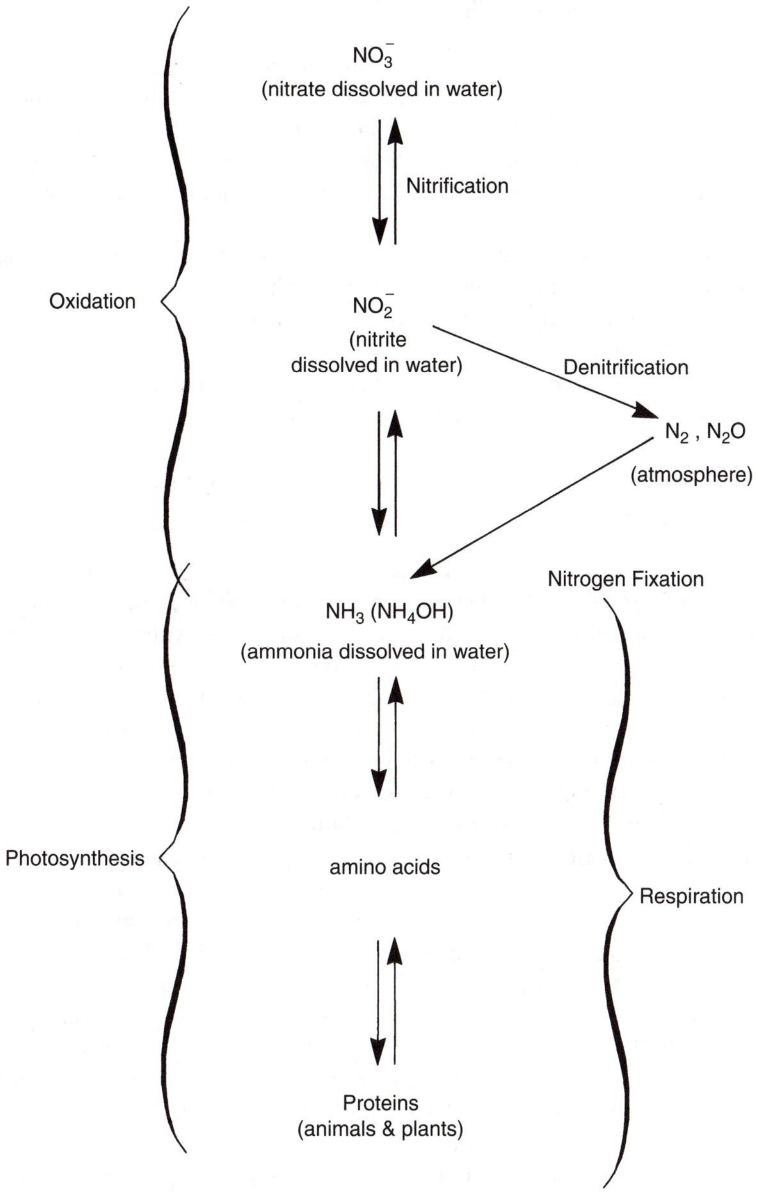

FIGURE 13.10 Transformation of nitrogen in an aquatic system.

shown in Figure 13.10. These processes can also be seen as a cycle of nitrogen, from the inorganic forms in water through the biological ecosystem, finally resulting in the reformation of the inorganic nitrogen forms again.

13.5 ESTUARINE SYSTEMS

Many of the great cities of the world are situated on estuaries. This is because the estuaries form the junction between the oceans and freshwater and often form natural harbors suitable for the development of ports. Estuaries are difficult to define, and there are a range of different water bodies that would be included in this general category. It is generally accepted that the following definition of estuaries forms a reasonable description.

Estuaries are semienclosed coastal bodies of water that have a free connection with the open sea and have waters measurably diluted with freshwater from land drainage. This means that the salinity of the water in estuaries is between that of the ocean and freshwater. Oceanic salinity is usually somewhere around about 35 ppt (3.5% or 35,000 ppm) and freshwater is usually less than 3000 ppm. Estuaries are dynamic areas where sedimentation and erosion are occurring.

13.5.1 Physical Interaction of Fresh- and Seawater in Estuaries

Estuaries are mixing zones for freshwater runoff with seawater. This can occur in a number of different ways. The density of freshwater is about 1.00, while the density of seawater is about 1.03. Since freshwater is slightly less dense than seawater, freshwater tends to float on top of seawater when it flows down from the land and enters an estuary. If the river flows are strong and dominant over the tidal flows, there will be little mixing caused by the tides or by turbulent flow resulting from an irregular bottom profile. In this situation, the estuary is normally classified as **highly stratified** (Figure 13.11). If a vertical profile is taken at any point in the estuary, it exhibits a distinctive pattern of a homogeneous upper layer, a sharp rise in salinity where the fresh- and seawater meet, and a homogenous bottom layer, as shown in the vertical profile in Figure 13.11.

In another type of situation, the river flow can be small compared with the tidal flow, and the bottom profile of the estuary can be irregular. This leads to active mixing of freshwater from the land as it enters the estuary, so that when a vertical profile is taken at any point in the estuary, there is no difference between upper and lower layers in the salinity profile. This type of estuary is referred to as a **mixed** estuary and is illustrated in Figure 13.11. In many situations estuaries exhibit characteristics that change over time depending on rainfall patterns. Thus, an estuary can vary from being **highly stratified** during periods or seasons of heavy rainfall to being **mixed** during periods or seasons of low rainfall. There are also situations where estuaries exhibit characteristics that are somewhere between these two extremes. These are referred to as **semimixed** estuaries and exhibit a vertical profile of salinity, which is somewhere between the **highly stratified** and **mixed**, as illustrated in Figure 13.11.

The chemical composition of estuarine waters exhibits characteristics that reflect both freshwater runoff and seawater. The variations in salinity that can occur in estuaries were previously described, but as a general rule, river waters are relatively high in nitrogen, phosphorus, and some trace metals compared to seawater. On the

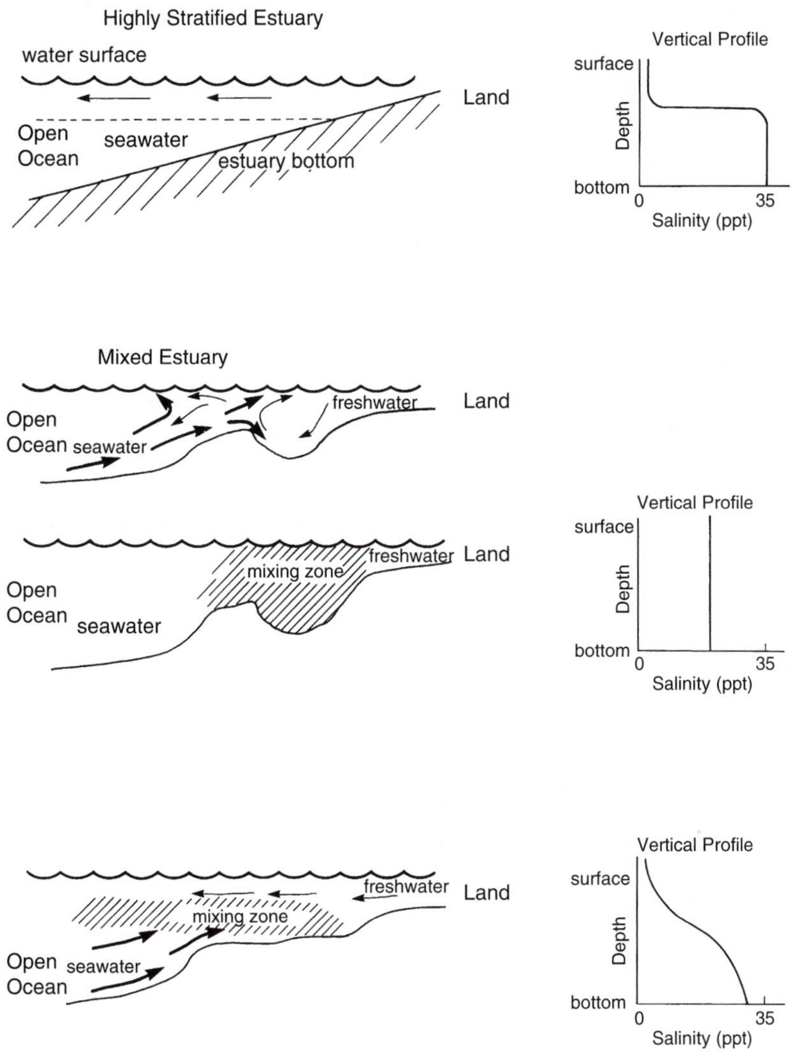

FIGURE 13.11 Different types of estuaries based on physical characteristics.

other hand, seawater often contains an array of trace metals present in only very small concentrations in freshwater. Not unexpectedly, one relationship that is common in estuaries is that the nitrate concentration is negatively related to salinity. This often happens with phosphorus also.

The conditions in estuaries can favor the development of a high biomass of plants and animals due to the presence of relatively high concentrations of nitrogen and phosphorus, together with a wide array of other essential trace elements. At the same time, the conditions do not favor a high number of species. This is due to the unstable nature of conditions in most estuaries. The species present have to be tolerant to tidal submergence and exposure, variable depth and coverage of water, variable salinity, and variations in other chemical factors, such as pH and trace metal concentrations. So most estuaries in fact exhibit a lower number of species than either the comparable adjacent open ocean area or the comparable freshwater area.

13.6 OCEANIC SYSTEMS

The study of the oceans, referred to as **oceanography**, began in a scientific sense in the early 1800s. The first international maritime conference was held in Brussels in 1853. In this period there were wide-ranging studies of the total salt content of the oceans, with the first book on ocean chemistry being published in 1865. In addition, there was considerable interest at this time in ocean biology and ocean currents. While the voyages of Captain Cook could be considered to be among the first scientific voyages in oceanography, the first expedition devoted specifically to oceanography was organized by the Royal Society of London and the ship *Challenger* traveled the oceans of the world from 1872 to 1876 carrying out scientific research. The reports from this voyage totaled 50 large volumes and are still a source of reference for scientists today. Since that time there have been many oceanographic expeditions. The amount of scientific knowledge available on the oceans is quite substantial.

13.6.1 Salinity of Seawater

The salinity of the oceans at the surface varies from about 30 to 40 ppt total soluble solids. There are variations above and below these figures, depending on local conditions. The salinity of the oceans is affected by such factors as rainfall, evaporation, and removal of pure water by the freezing of seawater at the Poles. As seawater freezes, crystals of pure ice are formed and the remaining liquid has a correspondingly higher salinity value.

The average ocean salinity at the surface varies with latitude, as shown in Figure 13.12. The relatively low salinities around the equator result from high precipitation with correspondingly large runoff from the land and relatively light winds giving low evaporation. The salinity rises, reaching a maximum in about the latitudes between 20° and 30°, which are the zones of the **trade winds** giving high evaporation, and also in this region there is less rainfall. At higher latitudes the evaporation falls due to the lower seawater temperatures and precipitation has a tendency to increase, resulting in a fall in the salinity. Increases occur toward the Poles due to the freezing effect mentioned previously.

13.6.2 Vertical Depth Profiles of Seawater Composition

The salinity, in all oceans, has a small range from 34.6 to 34.9 ppt at depths greater than 4000 m. Somewhat similarly, temperature in all the oceans at depths greater than 2000 m is about 4°C, irrespective of location. This suggests that when waters in the ocean reach the maximum density, they accumulate in the bottom waters and remain there for an indefinite period.

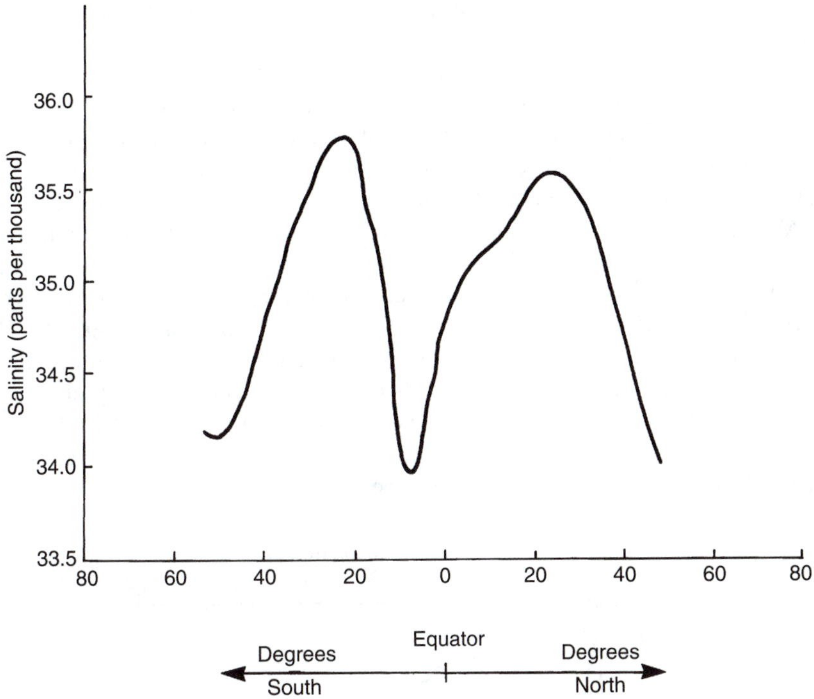

FIGURE 13.12 Variation in the salinity of surface waters in the oceans with altitude.

Seawater contains the following positive ions in order of concentration, Na^+, Mg^{++}, Ca^{++}, K^+, and Sr, and the following negative ions in order of concentration, Cl^-, SO_4^{2-}, HCO_3^-, Br^-, and F^-, as well as boron, as $H_3 BO_3$. These substances account for about 99.9% of the dissolved solids present in seawater. The remarkable thing about these 11 constituents of seawater is that although the total concentration may vary, the ratio of one to another is very close to constant in all situations. In oceanography, a common measure of salinity is the **chlorinity**. This is a measure of the chloride ion content of seawater and can be carried out accurately by chemical titration techniques. **Chlorinity** is related to salinity by the following equation:

$$\text{Salinity} = 1.805\ \text{Chlorinity} + 0.03$$

There are a wide range of minor elements present in seawater that can be detected in concentrations in the parts-per-million range. These include zinc, copper, iron, chromium, molybdenum, silica, aluminum, and many others. The biologically important minor elements, nitrogen and phosphorus, are present in seawater in relatively low concentrations. Nitrogen can be present in the inorganic forms of nitrate, nitrite, and ammonium ions, as well as organic compounds and amounts to a total of about 0.5 mg/l in all ocean waters. Phosphorus is present as orthophosphate (PO_4^{3-}) and related ions HPO_4^{2-}, $H_2 PO_4^{1-}$, and $H_3 PO_4$. The concentration of total phosphorus amounts to 0.07 mg/l on average within all ocean water. Importantly, this concentration is the average concentration throughout the water mass within the oceans, but at the surface, the concentrations are much lower. In large areas of the oceans, the concentrations of total nitrogen are somewhere in the region 0.1 mg/l, and phosphorus is about 0.001 mg/l. However, concentration increases with depth with both of these nutrients, so that at depths greater than about 1000 m, the concentrations are in the range of 0.2 to 0.5 mg/l of total nitrogen and about 0.04 mg/l of total phosphorus. This is illustrated by the diagram in Figure 13.13.

As a general rule, nutrient-rich waters have total nitrogen concentrations in excess of 1.0 mg/l and total phosphorus concentrations in excess of 0.004 mg/l; thus, the surface waters of the oceans are deficient in nitrogen and phosphorus, while the waters at depths greater than 1000 m are relatively rich in these nutrients. This means that large areas of the oceans will be deficient in life since photosynthesis occurs in the upper layers of the water where light is abundant but nutrients are in low concentrations. So productivity in the oceans is usually around land masses where nutrient-rich runoff waters enter the surface waters, allowing the growth of marine plants and subsequent growth of other marine organisms.

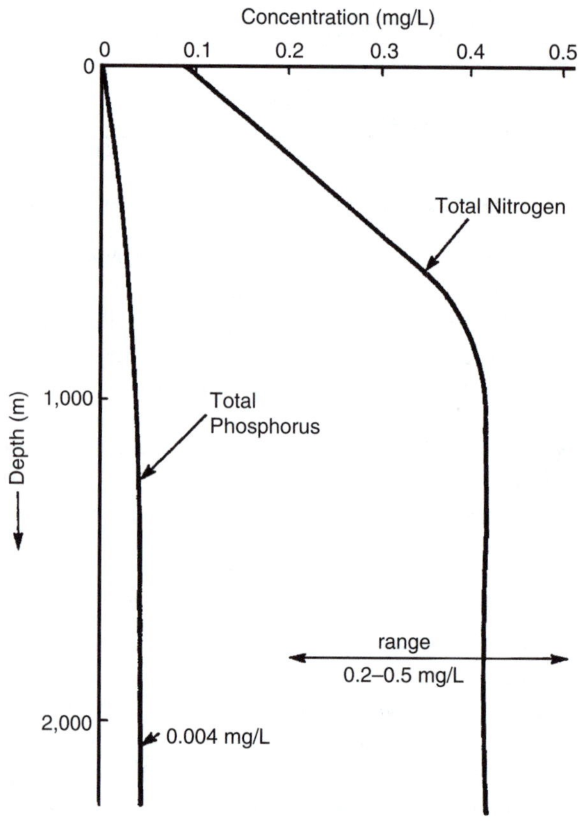

FIGURE 13.13 Change in concentration of total nitrogen and total phosphorus with depth generally observed in the oceans.

Another phenomena increasing the concentration of nutrients in the surface waters of the oceans is called **upwelling.** Upwelling is the movement of the nutrient-rich bottom waters to the surface, where photosynthesis and primary production can occur. Upwelling can be caused by persistent winds blowing offshore, and so moving the surface waters offshore and causing the bottom waters to come to the surface near the land mass. This process occurs off the west coasts of the Americas and Africa. Upwelling can also be caused by deep ocean currents moving to the surface due to the bottom topography. This occurs in the North Atlantic due to the movement of the Gulf Stream and the North Pacific due to a similar current in that ocean. Upwelling can also be caused by the mixing and turbulence where deep ocean currents meet. Areas where this occurs are in the Central Pacific, Atlantic, and Indian Oceans. Its interesting to note that the regions of upwelling throughout the world are generally regions where extensive fisheries have been established, for example, the North Pacific, North Atlantic, West African, and West American zones. There is extensive upwelling in the Antarctic, but in that case, this results in the growth of krill, which is not yet extensively harvested in a fishery.

13.7 DEOXYGENATING SUBSTANCES IN WATER BODIES

13.7.1 Background

Surprisingly, the major water pollutants produced by human society are substances that have comparatively little direct harmful effect. These are relatively nontoxic animal and vegetable wastes, which are often discharged into streams and inshore marine areas. In certain limited quantities, these materials can be beneficial in some bodies of water by releasing nutrients in the form of nitrogen and phosphorus salts, which stimulate plant and animal growth. However, in excessive quantities their secondary effects are usually extremely harmful. These wastes are rich in organic carbon and are commonly discharged into waterways in the form of sewage, food processing wastes from the fruit, meat, and dairy industries, as well as wastes from paper manufacturing and a variety of other industries.

The organic carbon in these wastes usually exists in the form of carbohydrates, proteins, fats, humic substances, surfactants, and a range of derived and other substances. Organic wastes are produced in large quantities in all urban population centers, but primary, secondary, and tertiary wastewater treatment can reduce the effects of discharges. Organic pollution can also result from urban storm water runoff, which produces nonpoint discharges to waterways. In some situations, non-urban catchments can yield high organic inputs where certain types of activities, such as logging, are undertaken.

On degradation by microorganisms in waterways, organic wastes can stimulate microorganism respiration, which consumes oxygen and reduces the dissolved oxygen in the ambient water. In addition, the degradation process releases nitrogen and phosphorus compounds, which stimulate plant growth leading to nutrient enrichment and eutrophication. These two broad areas of environmental impact — dissolved oxygen reduction and nutrient enrichment — are discussed below.

13.7.2 Influence of Chemical Processes on Dissolved Oxygen

Photosynthesis and respiration are the fundamental chemical processes supporting life on Earth; they are discussed in detail in Chapter 12. The chemical processes involved can be simply expressed as

Photosynthesis

$$6CO_2 + 6H_2O \rightleftharpoons 6(CH_2O) + 6O_2$$

Carbon dioxide from the atmosphere	**Respiration**	Plant matter	Oxygen released to the atmosphere

These processes involve carbon dioxide and oxygen in the Earth's atmosphere, which in aquatic ecosystems are dissolved in the water mass. The basic transformations of carbon in natural aquatic ecosystems are shown in Figure 13.9. Carbon from dissolved carbon dioxide is incorporated by photosynthesis into the plant biomass of the ecosystem. Consumption of this plant biomass by animals and finally the decomposition of body tissues after death by microorganisms result in respiration. Both of these processes result in the formation of carbon dioxide, which is returned to the water and ultimately the atmosphere.

If substances rich in organic carbon are added to the system, some of the pathways of carbon, shown in Figure 13.9, are increased in magnitude, and also some of the pools of organic carbon are increased in size. This results in an increase in respiration mainly through the respiration of microorganisms, giving rise to increased amounts of carbon dioxide and methane, which is produced by anaerobic respiration. Importantly for aquatic ecosystems, the transformations of carbon have a strong influence on the related transformations of oxygen, as shown in Figure 13.14. The oxygen needed for respiration is obtained from the dissolved oxygen in the water mass. This oxygen is usually substantially derived from solubilization from the atmosphere and converted into carbon dioxide by respiration, which is then discharged into the water mass and ultimately to the atmosphere. Increased respiration due to an increased amount of organic carbon in the water results in changes in the magnitude of the pathways and changes in the pools of oxygen involved, as shown in Figure 13.14. Most importantly, there is a demand on the reservoir of dissolved oxygen in the water mass, which is comparatively small since oxygen has limited solubility in water. The solubility of oxygen in water usually ranges from about 6 to 14 mg/l. Therefore, substantial reductions in dissolved oxygen can occur that have significant implications for aquatic organisms.

In waters that are high in dissolved oxygen, i.e., having dissolved oxygen levels approaching 100% saturation, **aerobic respiration** occurs mediated by aerobic microorganisms. The chemical reaction involved can be simply expressed as

$$6(CH_2O) + 6O_2 \rightarrow 6CO_2 + 6H_2O$$

Carbohydrate

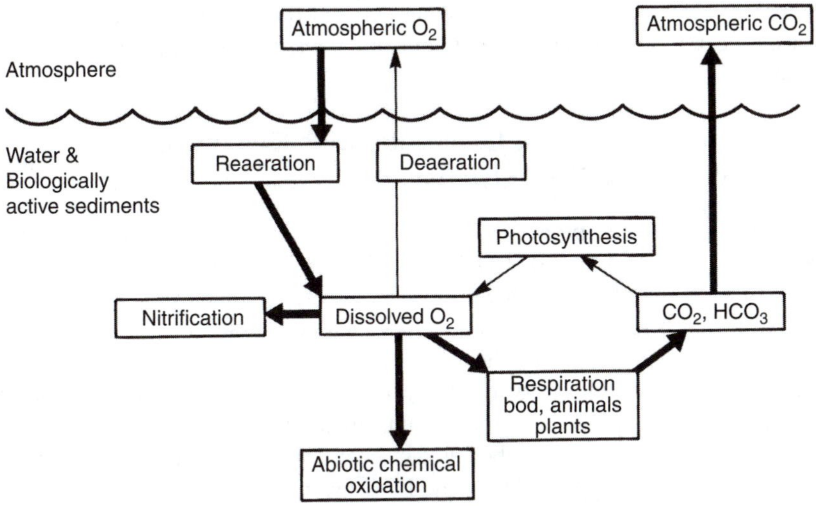

FIGURE 13.14 Oxygen transformations in aquatic systems, with processes increased or accelerated by organic discharges indicated by heavy lines.

In most aquatic bodies, communities of microorganisms exist that have a wide range of capabilities to degrade different types of organic matter under different conditions. In the absence of oxygen, anoxic or hypoxic conditions develop, which means that the dissolved oxygen has been completely removed. In this situation, anaerobic microorganisms take over the degradation and decomposition of organic matter. The anaerobic respiration reaction can be simply expressed as

$$6(CH_2O) \rightarrow 3CH_4 + 3CO_2$$

Carbohydrate Methane

This process makes no demand on the dissolved oxygen present in the water mass, but it is very important for the natural removal of oxygen-demanding substances in waterways. The methane and carbon dioxide produced are released into the water mass, then to the atmosphere, resulting in the removal of organic carbon and oxygen demand from the system. This process occurs in swamps, bottom muds enriched with organic matter, and bodies suffering from pollution by organic wastes.

Anaerobic respiration with sulfur- and nitrogen-containing organic substances gives rise to hydrogen sulfide and ammonia, respectively. Ammonia in the presence of oxygen is readily oxidized to nitrate ion. The nitrite ion is formed under suitable oxidation and reduction conditions, but is less commonly the end product of nitrogen metabolism. The formation of the nitrate ion from organic matter is described as the **nitrification** reaction and is mediated by a variety of microorganisms. A series of reactions occur that can be simply expressed as

$$NH_4^+ + 2O_2 \rightarrow NO_3^- + 2H^+ + H_2O$$

Ammonium ion

This process can also make an important demand on dissolved oxygen in natural bodies, especially where sewage contamination containing suitable microorganisms and ammonium ion occur.

13.7.3 Measurement of the Oxygen-Reducing Capacity of a Wastewater: The BOD Test

The oxygen-demanding processes affecting a water body can be measured by the **biochemical oxygen demand (BOD) test**. This test measures the loss of dissolved oxygen in water samples incubated over a period of 5 days under standard conditions, often with seed microorganisms added. The BOD of wastewaters can range up to hundreds of milligrams per liter. In addition, the BOD test can be used as a measure of the quality of water in waterways. In general, clean water has a BOD of less than 1 mg/l and seriously polluted water contains greater than 10 mg/l. The nitrification reaction occurs in what is described as the second stage of the BOD test at periods greater than 5 days. The precision of the BOD test is about + or –17% and may not satisfactorily reflect the actual conditions existing in a natural body. The test conditions may differ from those existing in the environment in factors such as temperature, microorganisms present, and so on.

13.7.4 Kinetics of BOD Reduction

The kinetics of BOD changes in waterways has been subject to intense investigation. One of the principles governing BOD in natural waterways was first formulated by Phelps as "the rate of biochemical oxidation of organic matter is proportional to the remaining concentrations of unoxidized substance measured in terms of oxidizability." Thus, the oxidation of organic matter follows first-order decay reaction kinetics. From this, the following series of mathematical expressions can be derived:

$$-\frac{dL}{dt} \propto L$$

Thus,

$$-\frac{dL}{dt} = K_1 L$$

which can be integrated to

$$\ln(L/L_o) = -K_1 t$$

or

$$\log(L/L_o) = -0.434\ K_1 t = k_1 t$$

and

$$L/L_o = 10^{-k_l t}$$

where L_0 is the total BOD debt at time zero, L the BOD debt at time t, t the period since time zero, and K_l and k_l the empirical decay rate constants using base e and base 10, respectively ($0.434\ K_l = k_l$).

By actual measurements of BOD in waterways it has been found that in general, at 20°C, $k_l = 0.1$ day^{-1} if the time period is measured in days. Since first-order kinetics apply, it can be shown that $t_{1/2} = 0.301/k_l$, and thus the half-life normally exhibited by BOD-demanding substances in natural waterways is equal to about 3 days. Also, from the mathematical treatment outlined above, a plot of log (percentage remaining BOD) vs. time will give a straight line (Figure 13.15).

The BOD decay rate coefficient (k_l) varies with temperature. It can be shown that the deviation of this constant from the 20°C value can be derived from the equation

$$k_T = k_{20°C} \times 1.047^{(T-20)}$$

13.7.5 The Dissolved Oxygen Sag

One of the most importance sources of dissolved oxygen in waterways is oxygen in the atmosphere that dissolves in the water mass at the water surface. The oxygen content of the atmosphere is about 20%. Reductions in dissolved oxygen present in water can be caused by the BOD present in the water, leading to a deficit below the

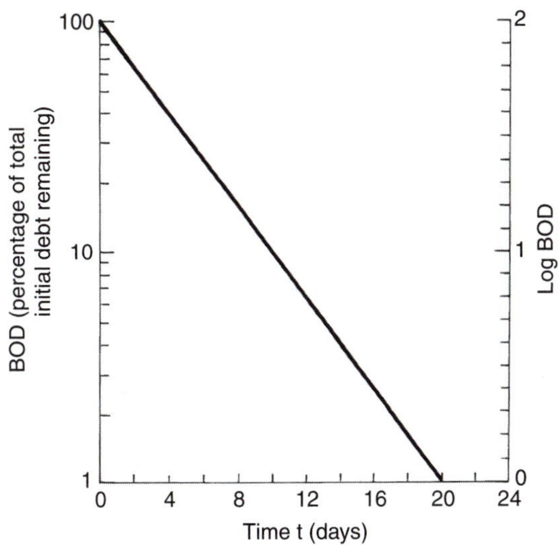

FIGURE 13.15 Normal depletion of organic BOD debt at 20°C ($k_l = 0.1$ day^{-1}) in an aquatic area.

saturation level. The rate of reaeration is proportional to the oxygen deficit in the water mass. Thus,

$$\frac{dC}{dt} = k_2 D$$

where C is the concentration of oxygen, k_2 the reaeration coefficient, and D the oxygen deficit, i.e., $C_{saturation} - C_{actual}$.

The rate of loss of BOD follows first-order kinetics; thus, the rate of loss of oxygen is proportional to the BOD present at the time involved:

$$\frac{dD}{dt} = k_1 L$$

where D is the oxygen deficit, L the BOD debt or concentration of BOD, and k_1 the BOD decay rate coefficient. By combining the two equations above, the actual oxygen deficit can be obtained by subtracting the uptake from the deficit due to BOD. Thus,

$$\frac{dD}{dt} = k_1 L - k_2 D$$

These two processes are illustrated diagrammatically in Figure 13.16. Curve e is the deoxygenation curve, which would represent dissolved oxygen values if no reaeration were to occur. However, as soon as an oxygen deficit occurs, which is soon after time zero, reaeration commences, yielding curve d. The reaeration

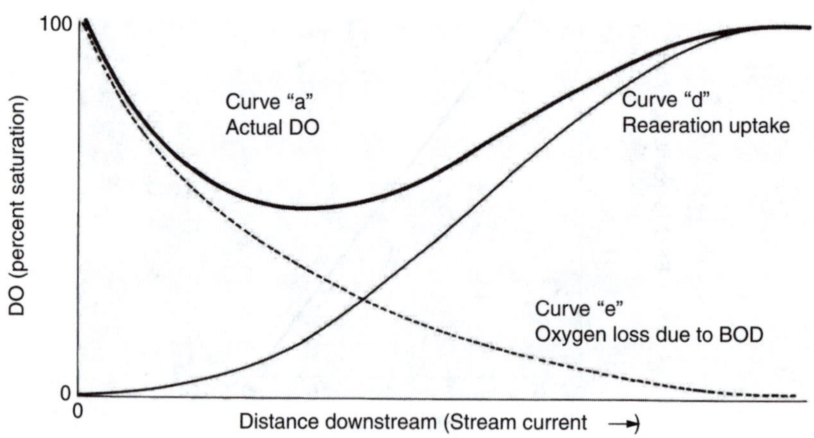

FIGURE 13.16 Deoxygenation, reaeration, and oxygen sag curves in a stream as a result of an organic discharge at point zero, giving a large increase in BOD.

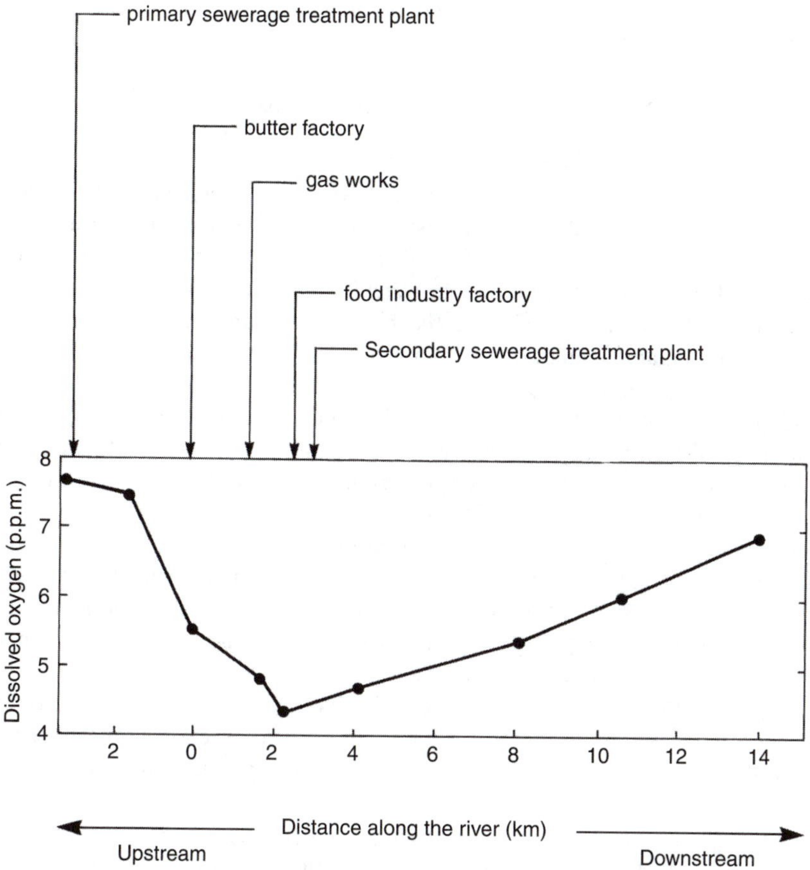

FIGURE 13.17 Dissolved oxygen profile of a river receiving wastewater discharges rich in organic matter.

increases as the oxygen deficit increases. Curve d represents the oxygen input to the water as a result of reoxygenation. The summation of these curves gives the actual oxygen deficit profile, the dissolved oxygen sag (curve a). An expression for this curve can be mathematically obtained by integrating the equation above.

The production of an oxygen sag in a river is illustrated by the data and plot in Figure 13.17. The dissolved oxygen concentrations at various points along a river receiving wastewaters rich in organic matter were measured and plotted. The oxygen sag is related to the position of the discharges. Downstream of the discharges, natural oxygen replenishment occurs by exchange with air at the surface, leading to a rise in the dissolved oxygen levels present in the water.

In the discussion above, only oxygen demand due to BOD is considered, which is due to organic matter present as suspended or dissolved matter in the water mass. However, organic matter can be sedimented out of the water mass to form bottom sediments that are rich in organic matter. These organic-rich layers may produce anaerobic degradation in the bottom waters, but above this, aerobic respiration can occur. This can lead to an oxygen demand not measured by BOD, which can exceed the BOD in some situations.

13.7.6 Seasonal Variations and Vertical Profiles of Dissolved Oxygen

Turbulent water conditions give well-mixed waters in which vertical profiles of dissolved oxygen, and other characteristics, are constant from the surface to the bottom waters. However, thermal stratification, due to solar heating of the surface waters, leads to isolation of the bottom waters. This was described in Section 13.2 and is illustrated in Figure 13.2. If the hypolimnion and bottom sediments are enriched with organic matter, the bottom waters may become depleted in dissolved oxygen, while the surface waters remain unaffected. In this manner, a vertical profile showing considerable variation in dissolved oxygen concentration can occur. Many environmental factors vary seasonally. For example, the incidence of sunlight has an important effect on photosynthesis and resultant dissolved oxygen concentrations in many aquatic areas.

In temperate areas, vertical stratification usually occurs in a seasonal pattern, with stratification in the summer and mixing, or overturn, in the winter. Corresponding to this, there is usually a seasonal pattern of dissolved oxygen variation in bottom waters, which is common in lakes and relatively small bodies of water, but can also occur on the continental shelves of the oceans.

Figure 13.18 shows the seasonal effect of stratification on dissolved oxygen concentration in waters in the New York Bight, located on the continental shelf of the northeastern U.S. The bottom waters and sediments have been enriched by

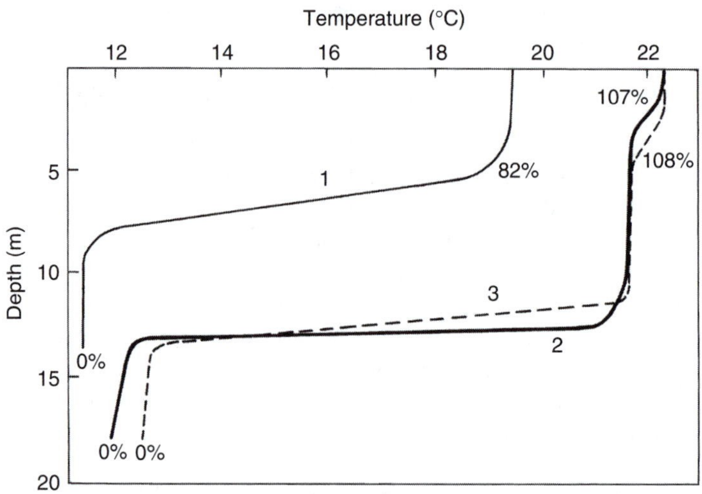

FIGURE 13.18 Temperature profiles and associated dissolved oxygen values (as percentage of the 100% saturation value in water) at three ocean stations in the vicinity of Little Egg Inlet, NJ, in July 1976.

phytoplankton decay, river discharge containing oxygen-demanding substances and nutrients, and dumping of sewage sludge in the area. In 1976 there was an early arrival of the relatively warm conditions of spring, leading to elevated surface temperatures that coincided with high discharge of water from the adjacent Hudson and Delaware Rivers. This led to a prolonged period of stratification in conjunction with the presence of relatively large quantities of organic matter, causing severe depletion of dissolved oxygen in the bottom waters and a widespread mortality of marine organisms.

13.7.7 Anaerobic Processes: Production of Poisonous Gases from Animal and Vegetable Wastes

A more advanced state of pollution occurs when the dissolved oxygen content of natural waters drops to zero and microorganisms that do not require free oxygen become active. This is an **anaerobic** system. When oxygen is present, carbon dioxide and water are the major substances produced by microorganisms, but the major products of the anaerobic process are methane and carbon dioxide, with lesser amounts of ammonia, hydrogen sulfide, and a variety of other organic and inorganic compounds. Some of these compounds, particularly hydrogen sulfide, are poisonous and have a strong nauseous odor. Streams exhibiting such strong odors are almost invariably heavily contaminated.

13.7.8 The Effects of Organic Wastes

The effects of pollution by organic wastes will depend mainly on the volume and strength of the discharge and the volume and flow rate of the receiving water. In a flowing river, a number of different aquatic environments and communities of aquatic organisms are usually created that are essentially related to dissolved oxygen content and presence of organic matter enrichment. These can be classified into zones, and a variety of different types of classification systems have been developed. However, it is convenient to consider a simple zone system related to the discharge point. First, upstream of the discharge point is the **unaffected** zone. At the discharge point itself and for a distance below this point, there is a zone of **degradation**, which may grade into a zone of **active decomposition**, where dissolved oxygen falls to zero and anaerobic processes occur. This zone, which may extend for some kilometers, grades into a **recovery** zone, finally ending in clean water resembling that upstream of the discharge point.

In the unpolluted zone, there will generally exist a large number of species of animals, plants, and insects. Some of these will be affected by the introduction of small quantities or organic wastes. For example, turbidity will reduce light penetration, and the number of bottom-dwelling photosynthetic plants will in turn diminish. Further additions of wastes may cause depletion of dissolved oxygen and result in the disappearance of most species of fish. Fish are sensitive to reduced oxygen, as is illustrated by the data in Table 13.3. Only a limited number of species of animals and plants will survive in the zone of active decomposition. However, some of these may be well adapted to this new kind of environment. Sludge worms, such as *Tubifex*,

TABLE 13.3
Examples of Limiting Dissolved Oxygen Concentrations for Aquatic Organisms[a]

Organism	Temperature	Dissolved Oxygen Concentration (mg/l)
Brown trout (*Salmo trutta*)	6.4–24	1.28–2.9
Coho salmon (*Oncorhynchus kisutch*)	16–24	1.3–2.0
Rainbow trout (*Salmo gairdnerii*)	11.1–20	1.05–3.7

[a] Limiting values for continued existence.

can survive on trace quantities of dissolved oxygen and actively feed on organic sludge sinking to the bottom. In the recovery zone, as the dissolved oxygen rises, conditions improve and the number of species increases. When recovery is complete, there is a return to the array of species present upstream of the discharge.

13.8 NUTRIENT ENRICHMENT AND EUTROPHICATION

13.8.1 BACKGROUND

The enrichment of aquatic areas with plant nutrients is an important process in aquatic pollution, and a significant aspect of this is **eutrophication.** Eutrophication was described by Weber in 1907 when he introduced the terms *oligotrophic*, *mesotrophic*, and *eutrophic*. These terms describe the eutrophication process as a sequence from a clear lake to a bog by enrichment with plant nutrients and increased plant growth. The Organization for Economic Cooperation and Development (OECD) has defined eutrophication as "the nutrient enrichment of waters which results in of an array of symptomatic changes among which increased production of algae and macrophytes, deterioration of fisheries, deterioration of water quality and other symptomatic changes are found to be undesirable and interfere with water uses."

In the natural eutrophication process, plant detritus, nutrient salts, silt, and so on, from a catchment are entrained in runoff water and deposited in the water body over geological time. This leads to nutrient enrichment, sedimentation, infilling, and increased biomass. Figure 13.19 illustrates, in general terms, how eutrophication is related to aging. As well as lakes, dams, and enclosed bodies of water, nutrient enrichment also occurs in streams, estuaries, the continental shelf, and the open seas. These water bodies may show many of the characteristics of eutrophication and are often referred to in eutrophication terms.

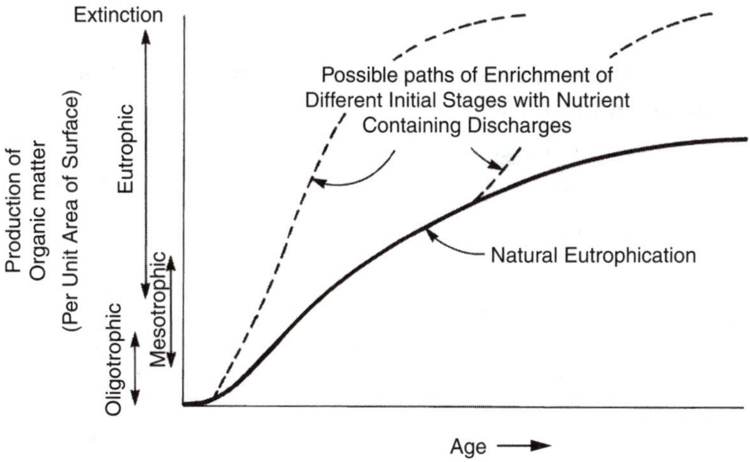

FIGURE 13.19 Hypothetical curves for the course of eutrophication in a water body. The broken lines show possible courses of accelerated eutrophication when enrichment from pollution discharges occurs.

Nutrient enrichment and eutrophication can be greatly accelerated by human activities. In fact, many lakes have been shown to have been rapidly enriched with nutrients over the last 100 years due to pollution. Discharges, such as domestic sewage, septic tank runoff, some industry wastes, urban runoff, runoff from agriculture and managed forests, and animal wastes contain plant nutrients that often lead to nutrient enrichment and accelerated eutrophication.

Eutrophication can cause quite a number of important problems in water use. An increase in the populations of plants can lead to a decrease in the dissolved oxygen content of the water on plant death and decomposition of the plant detritus by microorganisms. This decreases the suitability of the area as a habitat for many species of fish and other organisms. The increase in turbidity and color that occurs during eutrophication renders the water unsuitable for domestic use or difficult to treat to a suitable standard for this purpose. Odors are also produced by many of the algal growths, which create problems in domestic use. **Blooms**, **pulses**, and so on, of aquatic plants become more frequent and, if toxic, lead to the death of a variety of aquatic organisms and also of terrestrial organisms using the water. Floating macrophytes and algal scums can render a water body unsuitable for recreation and water sports and also cause navigation problems.

13.8.2 Nutrients and Plant Growth

If the growth of algal cells is not limited by any environmental or nutrient factor, then population growth occurs according to an exponential function. Exponential growth of the kind outlined above cannot be maintained for very long by algae due to limitations of various kinds. For example, elements, such as carbon, nitrogen, hydrogen, and oxygen, are needed to construct plant tissue. In addition to these

TABLE 13.4
Relative Quantities of Essential Elements in Plant Tissue (Demand) and Their Supply in River Water

Element	Demand by Plants (%)	Supply by Water (%)	Demand by Plants/Supply Water (approximate)
Oxygen	80.5	89	1
Hydrogen	9.7	11	1
Carbon	6.5	0.0012	5000
Silicon	1.3	0.00065	2000
Nitrogen	0.7	0.000023	30,000
Calcium	0.4	0.0015	<1000
Potassium	0.3	0.00023	1300
Phosphorus	0.08	0.000001	80,000
Magnesium	0.07	0.0004	<1000
Sulfur	0.06	0.0004	<1000

elements, sulfur, calcium, silicon, and sodium and a variety of other elements are needed to construct vital components that are present in plants in smaller amounts. Carbon, hydrogen, and oxygen are the major elements in carbohydrates, proteins, and fat, which are the principal chemical components of biota. Nitrogen is required in proteins particularly for the formation of the peptide bond (–CO–NH–). Phosphorus is a component of adenosine di- and triphosphate (ADP and ATP) and a variety of other substances needed for the transfer of energy in biota. Magnesium and iron are required to biosynthesize chlorophyll and hemoglobin, which are fundamental to the existence of life.

Table 13.4 lists the relative quantities of essential elements that occur in plant tissue. The function of all of these elements in plant processes is not clear at the present time, but they appear to be essential for growth. In most aquatic areas, carbon and oxygen are readily available from carbon dioxide in the atmosphere, and hydrogen and oxygen can also be readily obtained from water. On the other hand, the other elements mentioned above are usually obtained from dissolved salts in the water or sediments. However, these substances are not always available in the quantities required to maintain maximum growth. For example, Table 13.4 shows a comparison of the relative quantities of the elements required for plant growth with their occurrence in river water. This indicates that phosphorus and nitrogen are in comparatively short supply compared with all the other elements and that phosphorus is likely to be less available than nitrogen. This situation is generally applicable to aquatic areas. Thus, these elements are often growth limiting, and addition of them to a water body will stimulate plant growth. Of course, these data are generalized, and in individual cases other elements or combinations of elements may be limiting.

13.8.3 Sources and Losses of Nutrients

A convenient way to divide discharges of nutrient-containing wastewaters is to categorize them as **point** source and **diffuse** source discharges. Point source discharges are those arising from a specific location, whereas diffuse discharges arise from many dispersed sources, e.g., through eroded soil and sediment derived from cultivated land. Some of the major sources and sinks of nutrients are shown in Table 13.5. The discharge of untreated or primary treated sewage into an aquatic area causes deoxygenation, and during this process nutrient salts are released. Secondary treated sewage has had BOD substantially removed during treatment, but contains the same nutrient salts as would be released by untreated sewage. So untreated and primary and secondary treated sewage all contain nutrients that are active in aquatic areas.

The most important nutrients are in the form of ammonia (NH^{+}_{4}), nitrate (NO_3–), and orthophosphate (PO^{-3}_{4}), which occur substantially as a result of the degradation of proteins and detergents. The builders used in detergents to enhance the activity of the surfactant are usually polyphosphates. Phosphates occur in crude sewage and survive treatment in a secondary sewage treatment plant. Phosphates in detergents are a major source of phosphate in sewage and storm water runoff in agricultural and urban areas. Overflow from septic tanks and leachate from fertilized gardens are major contributors to the nutrient content. In addition, Table 13.5 lists a number of natural sources of nutrients, including leachate from leaves and miscellaneous vegetation debris, animal

TABLE 13.5
Sources and Sinks for the Nitrogen Budget of a Lake

Sources	Sinks
Leachate from leaves and miscellaneous debris	Water outflow
Agricultural (cropland) and drainage	Groundwater recharge
Animal waste runoff	Fish harvesting
Marsh drainage	Weed harvesting
Runoff from uncultivated and forest land	Volatilization (of NH_3)
Urban storm water runoff	Denitrification
Domestic waste effluent	Irreversible sediment deposition of detritus
Industrial waste effluent	Sorption of ammonia onto sediments
Natural groundwater	
Subsurface agricultural and urban drainage	
Nitrogen fixation	
Sediment leaching	

waste runoff, and marsh drainage. Once these nutrients enter a water body, there are a number of losses or sinks that can be identified. These include denitrification, which is the conversion of nitrogen into nitrogen gas, which disperses into the atmosphere, and irreversible sediment deposition of detritus-containing nutrients.

13.8.4 Nutrient Transformations in a Water Body

The carbon and nitrogen transformations generally occurring in aquatic bodies irrespective of their nutrient status have been described in Chapter 14. The same processes occur in enriched water bodies. However, stratification is probably more significant in nutrient-enriched bodies than in unenriched bodies. The influence of this process on the nitrogen transformations in water bodies is shown in Figure 13.20. Inputs of nutrients can occur from various sources, as shown in Table 13.5. The inorganic nitrogen salts of nitrate and ammonia are readily taken up by plants in photosynthesis. The death of the plants followed by sedimentation transfers this organic matter to the **hypolimnion**, where microbiological degradation occurs, resulting in the formation of carbon dioxide, orthophosphate, ammonia, nitrite, and nitrate (under different conditions) with the consumption of dissolved oxygen.

In an oligotrophic situation, these processes may not result in any marked differences between the hypolimnion and the epilimnion since all processes occur at a low rate. However, in eutrophic conditions, the hypolimnion is usually depleted in dissolved oxygen since direct reaeration from the atmosphere cannot occur. In addition, there is usually an accumulation of the inorganic materials formed during anaerobic respiration, for example, ammonia, hydrogen sulfide, and so on. These processes are illustrated in Figure 13.20. Oxygen depletion of the hypolimnion occurs in many water bodies throughout the world due to the introduction of plant nutrients, causing excess plant growth. Somewhat similar processes have been occurring in the oceans over geological time. In many areas, the bottom waters of the oceans have dissolved oxygen levels of zero due to the accumulation of organic matter from primary production in the photic zone in the past.

If dissolved oxygen is zero in the bottom waters, then the nitrification reaction leading to the formation of nitrate cannot proceed and organic nitrogen yields ammonia as the principal end product. This substance is not formed in surface waters since these are well aerated by atmospheric and photosynthetic oxygen. Here the principal nitrogen form is nitrate. Oligotrophic and eutrophic conditions produce vertical profiles of nitrate and ammonia, which correspond with the above observations and are characteristic of these conditions. In Figure 13.20, the conditions are outlined and the related vertical profiles are illustrated.

13.8.5 Photosynthesis and Diurnal Variations in Dissolved Oxygen

Many nutrient-enriched areas have high plant biomass, dominated by rooted aquatic plants and attached algae. Plants in this situation can have an important influence on the dissolved oxygen content of water through photosynthesis and respiration. Photosynthesis occurs during the daylight hours, but respiration by plants occurs

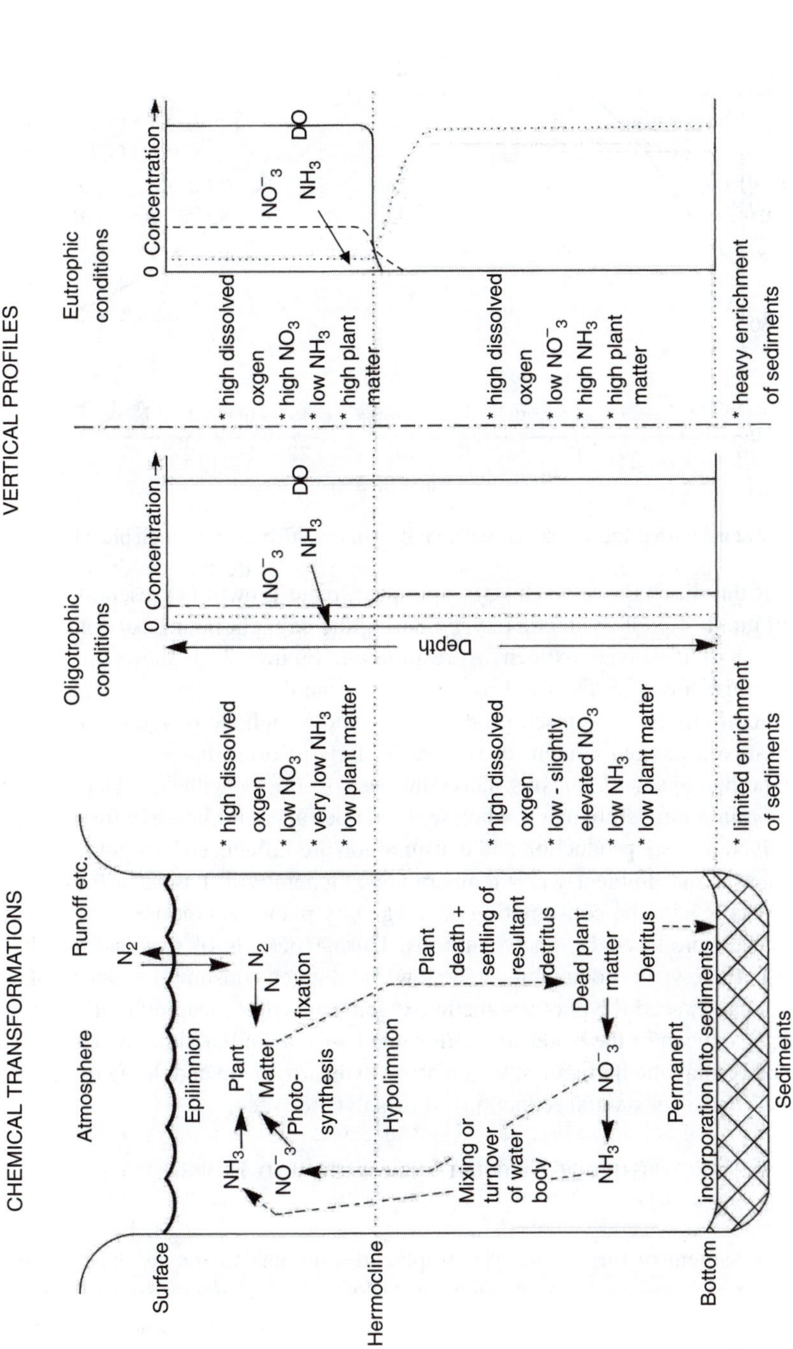

FIGURE 13.20 Typical (generalized) chemical transformations and vertical profiles of dissolved oxygen and nitrogen forms in stratified water bodies.

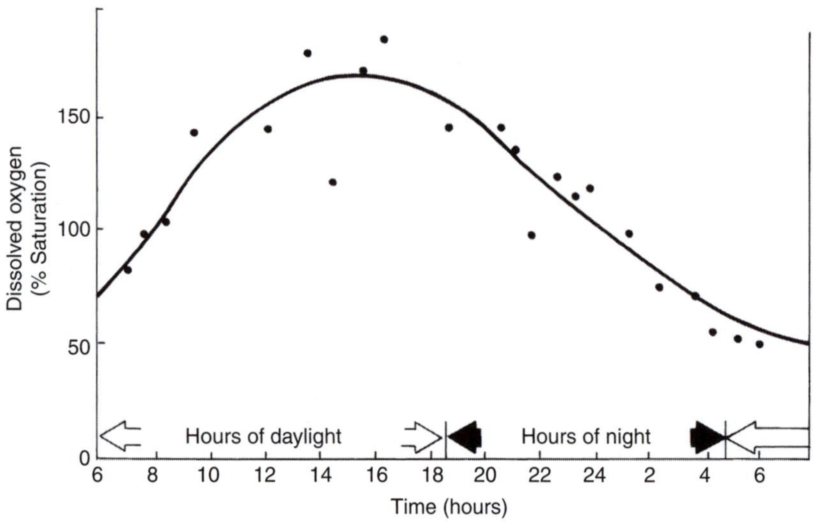

FIGURE 13.21 Diurnal variations in dissolved oxygen in a stream with a high plant biomass.

throughout the **diurnal** cycle. Thus, if significant plant growth is present, this will lead to an input of photosynthetic oxygen during the daylight hours, but continuous consumption of dissolved oxygen by respiration. Figure 13.21 shows that these processes cause a rise in dissolved oxygen during the daylight hours, with a maximum in the afternoon. At sunset, production of photosynthetic oxygen ceases, and so the dissolved oxygen content of the water starts to drop due to respiration by plants and other aquatic organisms. This continues overnight, with dissolved oxygen reaching a minimum before dawn, when sunlight once again initiates photosynthesis.

Dissolved oxygen production and consumption are influenced by light intensity, plant biomass, and ambient water temperatures. Generally, it has been found that over a diurnal cycle, the consumption of oxygen by plant respiration is about 0.75 of the oxygen produced by photosynthesis. During periods of overcast weather, photosynthetic oxygen production can be low and oxygen consumption due to plant respiration can exceed the photosynthetic oxygen production, leading to a decrease in dissolved oxygen in the water mass. In some cases, a similar situation can occur after the use of aquatic herbicides, which cause extensive death and decay of aquatic plants, leading to substantial reductions in dissolved oxygen.

13.8.6 Characteristics of Nutrient Enrichment and Trophic Status

A clear assessment of the degree of eutrophication is needed for the development of procedures for water quality management. Most of the characteristics of trophic status presently available have been based on freshwater lakes, as outlined in Table 13.6. As a general rule, the total phosphorus is measured in winter when stratification is not present and the water body is likely to be more homogeneous. The chlorophyll-a content is a measure of the amount of green pigment present and

TABLE 13.6
Some Quantitative Limit for Several Characteristics That Define the Trophic State

	Oligotrophic[a] ≤	Eutrophic[a] ≥
μg l^{-1} total P (winter)	10–15	20–30
μg l^{-1} chlorophyll-a (summer)	2–4	6–10
m Secchi desk (summer)	3–5	2–1.5
Total annual rate of primary productivity g C/cm²/year	7–25	75–700

[a] The intermediate class, mesotrophic, has intermediate characteristics.

considered to be indicative of the population of plants present. The Secchi disc is a circular disc attached to a rope that is lowered into the water until the disc cannot be seen. This value gives an indication of the turbidity of the water. The measure of total annual rate of primary production gives an evaluation of the amount of plant matter that is produced throughout the year.

Different phytoplankton also have different dynamics and requirements for nitrogen, phosphorus, carbon dioxide, and other factors that produce changes in the community composition with increasing eutrophication. The most prominent change with increasing nutrient enrichment is that the blue-green algae or cyanobacteria (*Cyanophyceae*) become increasingly dominant, as shown in Figure 13.22.

13.8.7 The Blue-Green Algae (*Cyanobacteria*)

The blue-green algae are best classified scientifically as cyanobacteria. Prolific growth of algae are usually referred to a **blooms**, and blooms of blue-green algae

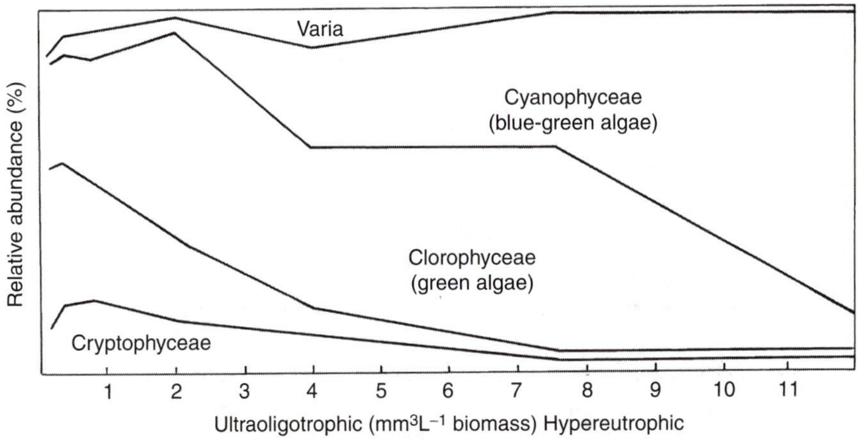

FIGURE 13.22 Composition of phytoplankton and algal blooms expressed by volume in relation to increasing lake fertility.

seem to have increased on a worldwide scale. These algae contain a range of toxins that can be released into the surrounding water. Blooms of blue-green algae have caused the deaths of many animals, both domestic and natural. The existence of toxic blue-green algae in water supplies has been known for some time, but it has only been in recent times that the potential for these toxins to affect human health through use of contaminated water supplies has been fully recognized. The most serious mammalian effects are due to two classes of toxins: the liver toxins (hepatotoxins) and those affecting the nervous system (neurotoxins). The toxicity of the various types is variable, some toxins being only moderately toxic compared to botulism and tetanus, but highly toxic compared to substances such as strychnine or sodium cyanide.

13.8.8 The Future

Nutrient problems are likely to continue in waterways into the foreseeable future. Even if all nutrient discharges could be halted now, the large reservoir of nutrients in aquatic sediments ensures that nutrient levels in overlying waters will continue to follow current trends. Over time, the nutrient levels in sediments would be expected to decline due to processes such as denitrification, dispersal, and covering by fresh sediments. High nutrient levels and other environmental conditions in aquatic areas can be expected to continue to stimulate major growths of algae and other aquatic plants.

13.9 KEY POINTS

1. Due to its relatively high polarity, water has boiling point, melting point, and vapor pressures higher than comparable compounds of similar molecular weight. The polarity of water also results in this substance being a good solvent for ionic salts, such as sodium chloride, and a poor solvent for lipids.
2. Atmospheric gases, such as oxygen, nitrogen, methane, and carbon dioxide, are all dissolved in natural water bodies and have an important influence on chemical and biological processes. The behavior of gases in natural water bodies is conveniently described by the Henry's law constant (H), which is defined as

   $$H = P/S$$

   The Henry's law constant is useful for calculating concentrations of gases in water and the atmosphere.
3. An important group of organic substances in natural waters are the humic substances, which are the residues left after the degradation of plant matter, particularly lignin. These substances have strong brown to black coloration and are of high molecular weight, ranging from hundreds to thousands, and contain a range of phenolic groups. Many metals, including

the essential trace metals, are strongly bonded by chelation into the humic substances.

4. The carbon cycle in natural water bodies starts with carbon dioxide in the atmosphere, which is converted by photosynthesis into plant material, which is consumed by other organisms in the system. Respiration by plants, animals, and microorganisms results in the release of carbon dioxide back into the water mass, and subsequently to the atmosphere, thereby completing the cycle.
5. The nitrogen cycle in natural water bodies commences with inorganic nitrogen forms, principally the nitrate ion within the water mass. Nitrate ion is taken up by plants during photosynthesis and converted into ammonia and amino acids, and finally into plant matter. The consumption of plant matter by animals and microorganisms results in the conversion of the proteinaceous food material into amino acids, ammonia, and subsequently nitrite and nitrate, thereby completing the cycle.
6. Estuaries can be classified on the basis of the interaction between fresh- and seawater into three types: the **highly stratified**, **mixed**, and **semimixed** estuaries. The semimixed estuary lies somewhere between the highly stratified and fully mixed estuaries.
7. The salinity of the oceans varies with latitude. At the equator the salinity is relatively low, at an average value of about 34 ppt, due to light winds and high rainfall. At higher latitudes the salinity rises due to the onset of the trade winds and lower rainfall, reaching a maximum at latitudes of about 25°. The salinity then falls due to rainfall exceeding evaporation.
8. The composition of seawater shows quite distinctive vertical patterns, particularly the nutrients nitrogen and phosphorus, which are of major biological importance. In large areas of surface waters, nitrogen exhibits total concentrations of less than 0.1 mg/l, while phosphorus exhibits concentrations of about 0.001 mg/l. With depth, the concentration of both P and N increases to a maximum below 1000 m depth, where nitrogen is between 0.2 and 0.5 mg/l and phosphorus is at about 0.04 mg/l. These nutrient-rich lower waters can reach the surface by **upwelling**.
9. **Photosynthesis** and **respiration** are the two fundamental chemical processes supporting life on Earth. These processes can be simply expressed as

$$\text{Photosynthesis}$$
$$6CO_2 + 6H_2O \leftrightarrow 6(CH_2O) + O_2$$
$$\text{Respiration}$$

10. The addition of wastewaters rich in organic matter can stimulate the respiration process through the stimulation of growth of microorganisms. This draws on the dissolved oxygen content in the ambient water and can result in substantial reductions. The oxygen-reducing capacity of a wastewater is measured by the **biochemical oxygen demand** (BOD) test.

Wastewater discharges rich in biochemical oxygen demand can cause a reduction in the dissolved oxygen content of receiving waters, which can rise again as reaeration from the atmosphere occurs. This is often referred to as the **dissolved oxygen sag** when it occurs in rivers and streams.

11. Anaerobic processes can occur when the dissolved oxygen content declines to zero as a result of the biochemical oxygen demand depleting the dissolved oxygen contents of the water. When these conditions occur, there can be the production of poisonous gases such as hydrogen sulfide and ammonia. These anaerobic processes can be simply represented by the equation set out below:

$$\underset{\text{Carbohydrate}}{6(CH_2O)} \rightarrow 3CO_2 + 3CH_4$$

12. The discharge of wastewaters rich in biochemical oxygen demand can result in substantial changes to the ecosystem in a stream. In the zone where maximum effects are observed, there can be very large populations of some organisms that are adapted to the low oxygen and high organic matter present. As the dissolved oxygen content returns to normal, a series of different communities occur downstream.
13. Many aquatic areas are deficient in nitrogen and phosphorus in terms of the elements present that will support the growth of aquatic plants. Thus, the addition of nitrogen and phosphorus salts to many aquatic areas will cause an increase in the growth of plants. The increase in nutrients present in an aquatic area can result from both natural and pollution processes. The different states of enrichment with nutrients are described as **oligotrophic** when the nutrients are very low, **eutrophic** when the nutrients, and resultant plant growth, are very high, and **mesotrophic** for an intermediate situation.
14. The nutrients in waterways can result from a variety of discharges. Untreated, primary treated, and secondary treated sewage all contain significant quantities of nutrients that can stimulate the growth of aquatic plants. In addition, storm water runoff from agricultural and urban areas can also contain significant concentrations of nutrients.
15. In all water bodies a common set of nitrogen and phosphorus transformations occur. But these transformations occur in different parts of the water body and to different extents, depending on such factors as depth, light penetration, water flows, and so on. Also, stratification is an important physical factor influencing the chemical transformations that occur in a water body. Oligotrophic and eutrophic water bodies produce different vertical profiles of chemical constituents when the bodies are stratified.
16. Water bodies with a high mass of plant matter can exhibit distinctive diurnal variations in dissolved oxygen. With these water bodies the dissolved oxygen content during the day can rise to very high levels due to

photosynthetic oxygen production, but drop to very low levels overnight when photosynthesis does not occur.

17. The occurrence of the toxic blue-green algae (*Cyanobacteria*) becomes more common under highly eutrophic conditions. There seems to be a worldwide trend toward the increasing occurrence of these algae.

REFERENCES

Connell, D.W. and Miller, G.J., *Chemistry and Ecotoxicology of Pollution*, John Wiley & Sons, New York, 1984.

Manahan, S.E., *Environmental Chemistry*, 7th ed., CRC Press, Boca Raton, FL, 1999.

Meadows, P.S. and Campbell, J.I., *An Introduction to Marine Science*, Blackie and Son Ltd., Glasgow, 1978.

QUESTIONS

1. A pond with dimensions 100 m square and 1 m deep is used to contain wastewaters. It is now filled with water having a BOD of 0 and a DO reading of 8 mg/l. If 10 m^3 of a wastewater containing 5000 mg/l of glucose was discharged to the pond, what would you expect the DO reading to be 24 hours later, when the glucose would be completely decomposed? Assume that there is no uptake of atmospheric oxygen.
2. The BOD is a measure of the oxygen-reducing capacity of a wastewater containing organic matter. If k_1 is 0.1 day^{-1}, what fraction of the total or ultimate BOD is measured by the BOD_5 (BOD measured over 5 days)?
3. Sulfur is an important micronutrient in the growth of aquatic plants. Explain, in chemical terms, why this element is required.
4. In a stratified lake there are distinct vertical profiles from surface to bottom for the various chemical forms of plant nutrients. Draw out the vertical profile that would be expected for the element phosphorus (PO_4^{-3}) in a stratified eutrophic lake and explain why this profile would be expected.
5. Explain why the surface waters of the oceans are depleted in N and P and will remain indefinitely in this state.
6. Nitrogen gas in the atmosphere will dissolve in natural water bodies. Using the Henry's law constant in Table 13.1 and the composition of the atmosphere at 78% nitrogen, calculate the concentration that would be expected in pure water at 25°C.
7. Carbon dioxide dissolves in water to form carbonic acid. Calculate the pH of pure water that is in equilibrium with carbon dioxide in the atmosphere at 25°C. Dry air contains 0.0314% CO_2 by volume, the vapor pressure of water at 25°C is 0.03 atm, the Henry's law constant for CO_2 is 29 atm/mol^{-1}/l, and the dissociation constant in the carbonic acid–bicarbonate system is 4.45×10^{-7}.

ANSWERS

1. The complete degradation of glucose would be expected to follow this equation:

$$6(CH_2O) + 6O_2 \rightarrow 6CO_2 + 6H_2O$$

Thus, 180 g of glucose requires 192 g of O_2 for complete oxidation. The volume of water is 10,000 m^3, and at 8 mg/l it contains 80,000 g of O_2. The wastewater contains 50,000 g of glucose, which will consume 53,000 g of oxygen. The water in the pool will thus contain 27,000 g of oxygen when degradation is complete, giving a concentration of 2.7 mg/l.

2. From the kinetic equation for loss of BOD, then

$$\log (L/L_o) = k_1 t$$

If k_1 is 0.1 day^{-1} and t is 5 days, then

$$\log (L/L_o) = -0.5$$

$$\therefore\ L/L_o = 0.0316$$

$$\therefore\ L = 0.316\ L_o$$

This means that after 5 days there is 0.316 of the ultimate, or total, BOD remaining. Thus, 68.4% of the ultimate BOD is measured by the BOD_5.

3. Sulfur is needed in plant growth for the following reasons:
   - To form biologically important chemical groups such as –SH (sulfhydryl) and–S–CH_3.
   - Sulfur-containing groups such as –SH and –SCH_3 are contained in essential amino acids such as cysteine

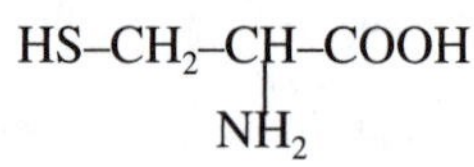

   and methionine

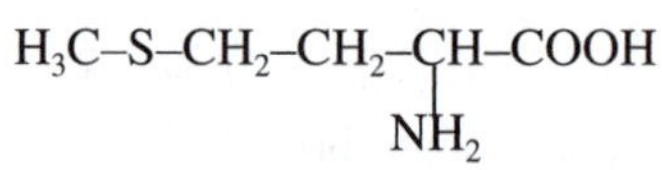

   - The sulfur-containing group, the disulfide bridge –S–S–, plays a major role in the secondary and tertiary structures of proteins.
   - Sulfur is needed in the formation of enzymes containing all the groups and compounds mentioned above.

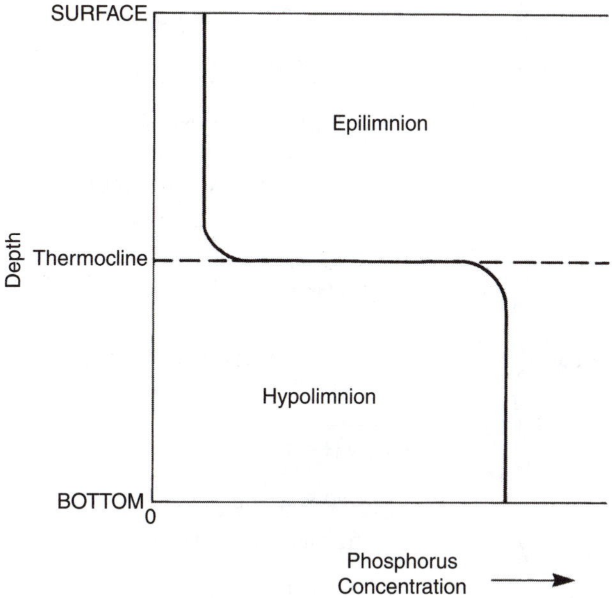

FIGURE 13.23 The depth concentration profile that could be experienced for phosphorus in a stratified eutrophic lake (answer to question 4).

4. The phosphorus would be incorporated into plant matter during photosynthesis, and orthophosphate ($PO_4^{\equiv}$) would be released from plant matter and detritus on degradation. Degradation usually occurs after the plant has died and fallen to the bottom. This would be expected to be significant in a eutrophic lake. Thus, phosphorus would be taken in at the surface (epilimnion), giving reduced concentrations and produced in the hypolimnion, leading to elevated concentrations. This would lead to the profile shown in Figure 13.23.
5. Photosynthesis results in the removal of N and P from the surface waters in the photic zone and the formation of cells of phytoplankton. On death, these cells fall downward into the zone below the photic zone, taking the N and P with them, and there is no corresponding way in which surface waters are renewed with nutrient. Thus, should any nutrients appear in the photic zone, they will be removed by this process. This means the oceans surface waters will remain in a continually depleted state.
6. The atmosphere contains 78% nitrogen by volume and allows for 0.03 atm of water vapor at 25°; the partial pressure of nitrogen is 0.804 atm.

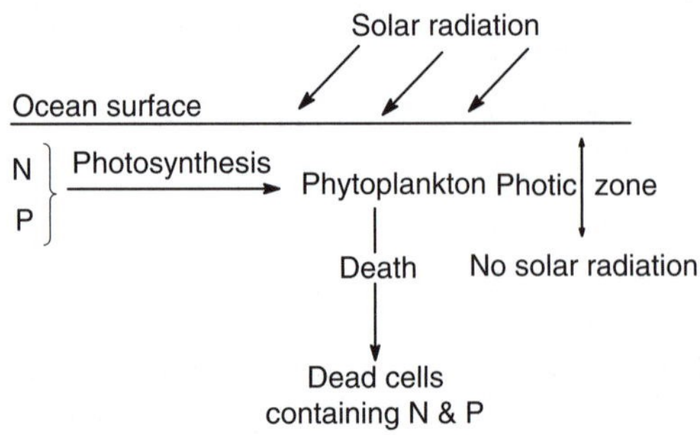

From Henry's law,

$$S = P / H$$

$$= \frac{0.804 \text{ atm}}{1610 \text{ atm / mol / l}}$$

$$= 4.99 \times 10^{-4} \text{ mol / l}$$

Thus, concentration in pure water equals 14.0 mg/l. This is in reasonable agreement with the figure reported from direct measurement in Table 13.1 of 13.4 mg/l.

7. Dry air contains 0.0314% CO_2 by volume, the vapor pressure of water at 25° is 0.0313 atm, the Henry's law constant for CO_2 is 29 atm/mol/l, and the dissociation constant for the carbonic acid–bicarbonate system is 4.45 $\times 10^{-7}$.

Partial pressure of air	=	(1.000 – 0.03) atm
	=	0.97 atm
Partial pressure of CO_2	=	$(0.97 \times 0.0314 \times 10^{-2})$ atm
	=	0.0305×10^{-2} atm

Since S = P/H, the CO_2 concentration in water equals

$$= \frac{0.0305 \times 10^{-2} \text{ atm}}{29 \text{ atm / mol/ l}}$$

$$= 1.05 \times 10^{-5} \text{ mol/l}$$

$$H_2\,CO_3 = H^+ + H\,CO^-_3 \text{ and } k = 4.45 \times 10^{-7}$$

Thus,

$$4.45 \times 10^{-7} = \frac{[H^+][HCO_3^-]}{[H_2CO_3]}$$

$$= \frac{[H^+]^2}{[H_2CO_3]}$$

Thus,

$$[H^+]^2 = (4.45 \times 10^{-7})\ [H_2\ CO_3]$$

Since the molar concentration of CO_2 in the water is the same as the molar concentration of H_2CO_3, then

$$[H^+]^2 = (4.45 \times 10^{-7}) \times (1.05 \times 10^{-5})$$

$$= 4.67 \times 10^{-12}$$

Thus,

$$[H^+] = 2.16 \times 10^{-6}$$

$$\therefore\ pH = 5.66$$

This means that atmospheric carbon dioxide in equilibrium with pure water will have a pH of 5.66.

14 Soil Contamination

14.1 INTRODUCTION

Increasing human populations have led to the rapid expansion of existing urban areas and the development of new ones. Areas used by industry in the past have often become valuable land for the development of domestic housing in many parts of the world. As a result, many former industrial areas are the location of extensive housing developments containing large numbers of people.

In the 1970s, with the expansion of investigations into the occurrence of contamination in the environment, attention was directed to areas of soil contamination. Throughout the world it became apparent that many substances, which are well-known toxicants or carcinogens, were present in relatively high concentrations in the soil in some urban areas (Amdur et al., 1991). Many of these areas became the focus of public controversy and have become famous as contaminated sites. So, heavily contaminated sites such as Love Canal, the Picello Farm, and the Valley of Drums have become household names associated with soil contamination in many countries.

Essentially, the problems of soil contamination can be considered to consist of two fundamental types: (1) those that result in the exposure of the human population to contaminants and (2) those that result in the exposure of natural ecosystems. Human exposure results from development of schools, kindergartens, domestic housing, and offices and workplaces on contaminated sites. Exposure of natural systems can occur in several ways. Natural terrestrial biota in contaminated areas are directly exposed by contact with soilborne contaminants. In addition, water runoff and groundwater seepage from contaminated soils to natural aquatic systems result in exposure of adjacent aquatic ecosystems. Dust and vapors can also distribute contaminants into wider areas.

The management of contaminated soils is now a major governmental activity (Shineldecker, 1992). In many countries the problem is of a considerable magnitude. For example, in the Netherlands, over 100,000 sites have been identified as being potentially contaminated, with 10,000 sites confirmed as being contaminated. Likewise, in Germany, over 50,000 potentially contaminated sites have been identified. In the U.S., there are an estimated 100,000 sites that have been nominated as contaminated, with some 10,000 of these designated as priority areas.

14.2 SOURCES OF SOIL CONTAMINATION

The contamination of soil can originate essentially from the following activities:

- Industrial operations
- Agricultural activities
- Domestic and urban activities

TABLE 14.1
Examples of Activities Resulting in Soil Contamination

Industrial Operations
Chemical industries, gas and electricity supply, wood preserving, oil refining, service stations, smelters, mining, tanning, dockyards, waste dumps
Agricultural Activities
Treatment of crops, handling and storage of agricultural chemicals, use of cattle dips
Domestic and Urban Activities
Solid waste disposal, sewage sludge disposal, sewage works and farms, motor vehicle discharges, usage of chemicals

Some examples of activities within these categories that result in soil contamination are shown in Table 14.1.

Deliberate disposal of industrial waste to land has been a common disposal method. Generally, this has not been carried out in disregard for the environment or government regulation, but through a lack of regulation by government and a lack of understanding of potential adverse consequences. In fact, many of the land disposal operations were approved by governments as the most appropriate disposal operation for hazardous chemicals. Trenches and pits have been used in which wastes from such industrial operations as tanneries and coal gas plants were disposed. Accidental spills also are a major cause of soil contamination. Accidental spillages have occurred frequently in such operations as wood preserving, petrol stations, fuel depots, and similar activities. The operation of smelters giving atmospheric discharges of contaminated particulates that subsequently deposit in soils is another source of contamination that occurs in many areas. Mining wastes have often been disposed into special dams and other land-based operations, resulting in soil contamination. The broad use of pesticides on crops has resulted in widespread contamination of soil in some areas. More intense contamination has often occurred in specific rural areas where pesticides are stored, distributed, and loaded onto vehicles. In addition, the use of dips for treatment of cattle for pests has often resulted in contamination of relatively small areas.

Activities in normal domestic and urban situations also result in soil contamination. Perhaps the major source of contamination in this area is the disposal of solid waste to land areas. Sewerage sludge disposal can contain high levels of contaminants and also be disposed of to soil. The use of motor vehicles results in discharges of lead and other contaminants in particulate form, which accumulate in soils in the vicinity of busy roads. A range of chemicals are used in domestic situations. For example, pesticides and other chemicals are used in the maintenance of gardens and lawns. The usage of chemicals and the disposal of waste following the usage can result in contamination of soils in urban areas.

It should be kept in mind that natural soils are not necessarily free of hazardous compounds, which may have deleterious biological effects. Of course, compounds that have only originated as a result of synthetic chemical processes would not be present in natural soils. Thus, the synthetic pesticides, such as DDT and dieldrin, as well as industrial chemicals such as the polychlorinated biphenyls (PCBs), would not be present in natural soils. The occurrence of these substances would be as a result of human contamination. On the other hand, polycyclic aromatic hydrocarbons (PAHs) (described in Chapter 9), polychlorodibenzo dioxins (PCDDs) (described in Chapter 6), and polychlorodibenzo furans (PCDFs) (described in Chapter 6) are all produced in low concentrations by combustion of organic matter, including natural organic matter such as wood and paper. These substances have been found in low concentrations in natural soils throughout the world. However, there is a tendency for these substances to occur in the vicinity of urban and industrial areas in higher concentrations than the natural background levels. Metals such as lead, mercury, and arsenic occur naturally in high concentrations in various geological strata. Where these formations reach the surface, the associated soils can contain relatively high concentrations. Also, there can be the migration of these substances to other areas, which can lead to the contamination of other soils in the vicinity. Thus, natural soils may contain levels of substances that can be considered hazardous to human health and natural ecosystems.

14.3 CHEMICAL NATURE OF SOIL CONTAMINANTS

Some common soil contaminants and their associated sources are shown in Table 14.2. Clearly, soil contaminants do not fall into a single, or several simple, classes of chemicals, but are very diverse in chemical nature. In fact, the only common property of soil contaminants is that these substances can be harmful to the natural environment or human health.

TABLE 14.2
Chemical Identity of Some Soil Contaminants

Chemical	Source
Arsenic, copper, chromium	Tanneries, wood preserving, mining wastes, cattle dips, smelters
Lead	Smelters, motor vehicles
Petroleum hydrocarbons	Petrol stations, fuel depots, accidental spills, disposal of waste chemicals
Polyaromatic hydrocarbons (PAHs), polychlorodibenzo dioxins (PCDDs), polychlorodibenzo furans (PCDFs)	As for petroleum hydrocarbons, coal gas plants
Pesticides (DDT, dieldrin, etc.)	Agricultural areas, disposal of waste chemicals, cattle dips
Industrial chemicals (solvents, PCBs, acids, alkalis, etc.)	Disposal of waste chemicals, industrial operations

The **heavy metals** (see Chapter 11) are common soil contaminants. For example, **lead** originating from motor vehicles is a very common soil contaminant in urban areas, particularly in the vicinity of roadways. There are also many examples of lead contamination where the lead originates from smelters and mining operations. Although lead in paints has been banned in many countries, there can be buildings that were painted prior to the ban. Building renovation in recent times can result in soil contamination or adverse effects in the environment. **Copper** and **chromium** are also common heavy metal soil contaminants and primarily originate from tanneries and wood-preserving plants. Another common soil contaminant is **arsenic**, which has found wide use in the past as a preservative of hides in tanning, pesticide with cattle in dips, and wood preservative. It is also a natural soil component in many areas. Strictly speaking, arsenic is not a metal, but it is convenient to consider this substance in the heavy metal group.

A wide range of organic substances also occur as contaminants in soil. Almost all of the organic substances used in industry and society in general can occur as soil contaminants. **Petroleum hydrocarbons** are particularly common due their wide distribution and use in petroleum products. Similarly, PCDDs and PCDFs are also widespread as a result of the use of combustion for many different purposes. The concentrations of both these groups of chemicals are generally low, but can be relatively high in particular localities. Of particular note here is the occurrence of former coal gas plant sites in many cities. With many of these plants, the waste coal tar produced was disposed into wells and pits. Coal tar consists of a highly complex mixture of aromatic hydrocarbons, phenols, and PAHs. As a result, many cities contain sites with very high levels of these soil contaminants.

Organic liquids are used as solvents in many industrial processes and products such as paints. Petroleum products, including gasoline, naptha, toluene, and xylene, are common solvents used in industry. In addition, chlorohydrocarbons such as dichloroethylene and carbon tetrachloride are commonly used as solvents. It is important to note that chlorinated solvents are being phased out of industrial uses due to their adverse effects on the ozone layer (see Chapter 12).

In agricultural areas, pesticides and other agricultural chemicals are often present in soil. Even pesticides having discontinued use, such as DDT and the other chlorohydrocarbon pesticides, can be present in agricultural soils due to previous usage. Of course, the concentrations of all of these substances would be expected to be declining due to environmental transformation and degradation processes (see Chapters 3 and 8).

14.4 IMPORTANT ENVIRONMENTAL PROPERTIES OF SOILS

Soils are a complex mixture of substances that vary in composition from area to area. In dry areas and beaches, the soil consists essentially of silica sand with some calcium carbonate components but very little else. In most agricultural and urban areas, the soil components that affect the environmental properties of contaminants are principally clay and organic matter. Clay consists of various hydrous silicates and oxides that can be characterized by such measures as cation exchange capacity and the specific surface area (m^2/g). These properties give measures of how clay affects the behavior of polar organic molecules and metal ions. The cationic exchange capacity is a measure of the capacity of the soil to sorb cations with which it comes in contact. Thus, strongly **cationic pesticides** such as diquat (Figure 14.1) are strongly sorbed by the clay component in soil. It is interesting to note that diquat is also very soluble in water, and so is highly hydrophilic, as illustrated by the data in Table 14.3. But the sorption to clay is sufficiently strong to overcome the highly hydrophilic properties of this compound, which would favor its occurrence in water. Glyphosate exhibits somewhat similar properties (Table 14.3). The structure of glyphosate is shown in Figure 14.1. It is a molecule with several sites for cationic and anionic effects and in fact exists as a zwitterion. Thus, a hydrogen ion can move internally between ionic groups depending on the ambient pH. This compound is highly hydrophilic and sorbs strongly to clay minerals in soil.

The organic compounds DDT, dieldrin, and benzo(a)pyrene are also strongly sorbed by soil, as indicated by the values of log K_{OW} shown in Table 14.3. The K_{OM} value is the concentration in soil organic matter/concentration in water at equilibrium, and the K_{OM} values of DDT are 63,000; dieldrin, 10,000; and benzo(a)pyrene, 32,000. So all of these compounds are strongly sorbed to soil, but in this case the compounds are all very insoluble in water, with solubilities less than 0.16 mg/l, and are highly nonpolar and lipophilic (Figure 14.1). These compounds do not sorb to the clay fraction of the soil to any significant extent.

Soil organic matter usually ranges in content from 0.1 to 7% of the total soil. It provides the most important component for the sorption of nonionic and nonpolar chemicals. The organic matter is derived principally from plant detritus, which enters soil as plant litter and as a result of the death of plants. The activities of microorganisms on this plant detritus produce the organic matter that is a major component of soil. Chemically, the main component in the plants that contributes to soil organic matter is plant **lignin**. This is a complex phenolic compound of high molecular weight and is relatively resistant to degradation in the environment. This plant material is acted on by microorganisms and other processes, which results in the formation of **humic substances**, often referred to as **humic** and **fulvic acids**. Humic and fulvic acids are similar and related chemically to lignin, and are separated principally by the properties of solubility in acid and alkaline solutions.

The molecules of these substances are relatively large and probably vary depending on the source material and degradation conditions. A hypothetical structure is shown in Figure 14.2. An important aspect of the structure of humic and fulvic acids is that these substances contain a range of phenolic groups, as shown in Figure 14.3. These groups have the capacity to complex metal ions, as shown in Figure 14.4. Many of the groups have the capability of forming chelation complexes with attachment of metal ions to more than one site in the organic molecule. These chelation complexes usually exhibit considerable stability and resistance to breakdown in the environment.

Importantly, the humic and fulvic acids also have lipophilic properties. There are many polar and ionic sites in these molecules, but their large size probably gives

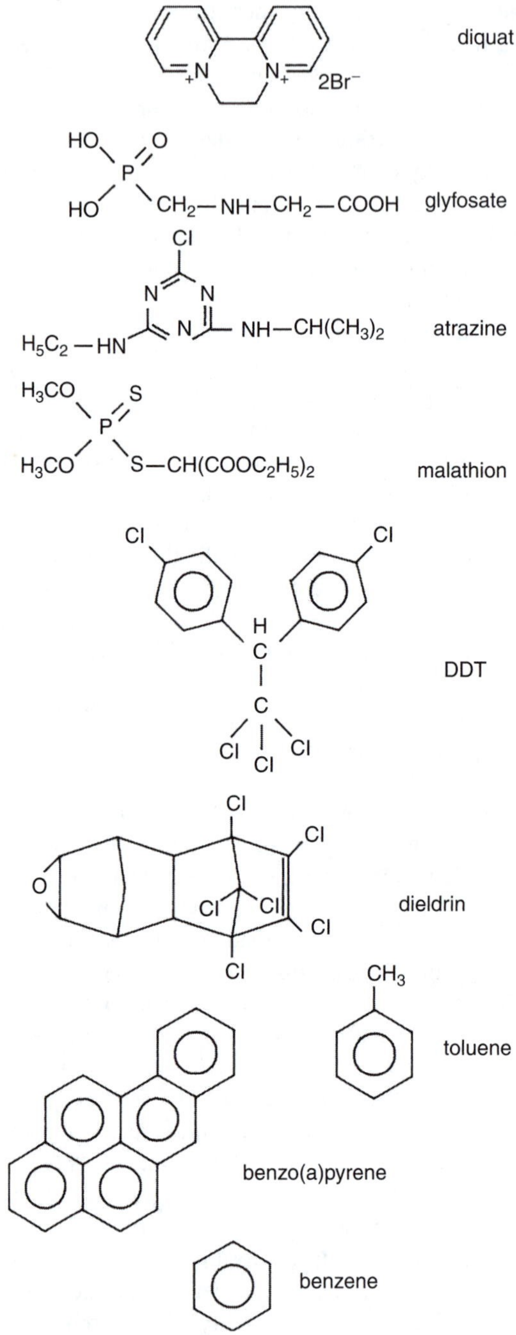

FIGURE 14.1 Chemical structure of some soil contaminants.

TABLE 14.3
Measured Properties of Some Compounds Related to Their Behavior on Soils

Compound	Log K_{OW}	Aqueous Solubility (mg/l)	Log K_{OM}	$T^1/_2$ (days)	VP[a] (mm)	MI[b]
Diquat	—	700,000	Highly sorbed to clay	—	—	—
Glyphosate	–1.7	1200	Highly sorbed to clay	50–70	—	—
Atrazine	2.75	30	2.0	1–8.0	3×10^{-7}	–7.06
Malathion	2.36	143	3.3	3–7	4×10^{-5}	–5.54
DDT	6.2	0.0032	4.8	700–6000	2×10^{-7}	–13.99
Dieldrin	4.3	0.17	4.0	175–1100	1×10^{-7}	–11.76
Toluene	2.69	515	3.5	4–22	10	+0.21
Benzo(a)pyrene	6.0	0.004	5.5	57–490	—	—
Benzene	2.24	1.780	3.3	5–16	76	+1.83

[a] Vapor pressure.
[b] Mobility index.

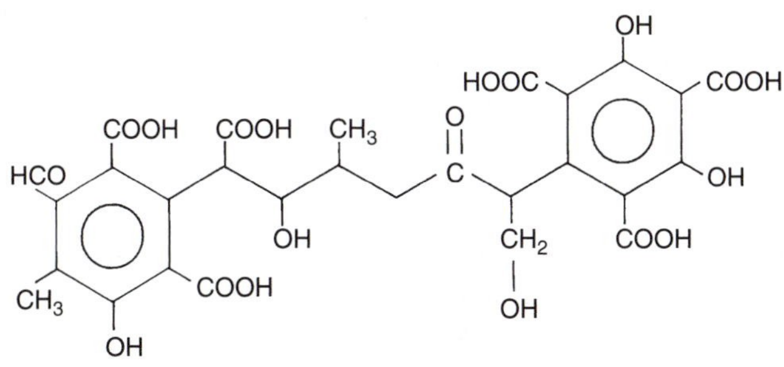

FIGURE 14.2 A proposed hypothetical chemical structure for fulvic acid.

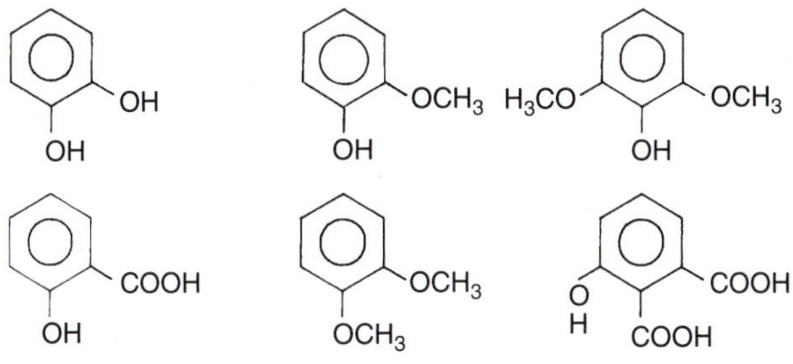

FIGURE 14.3 Some of the structural groups present in humic and fulvic acids.

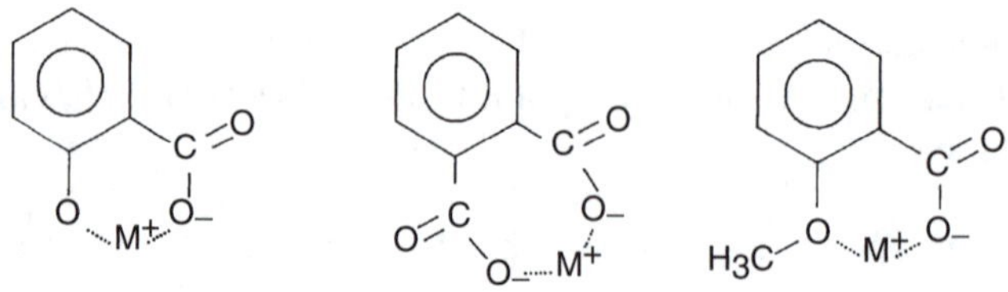

FIGURE 14.4 Examples of possible complexes formed by humic and fulvic acid active groups with metal ions.

them the lipophilicity property since lipophilicity generally increases with molecular size. So lipophilic compounds exhibit strong sorption to organic matter in soils.

14.5 IMPORTANT ENVIRONMENTAL PROPERTIES OF SOIL CONTAMINANTS

One of the most important properties of contaminants in soil is their persistence. Organic compounds will be degraded by microorganisms as well as abiotic and other processes. Also, they will be volatilized from the soil and removed by water leaching processes. Of course, metals and organometallic compounds are not susceptible to degradation beyond the elemental state. However, these substances can be removed from soils by transformation to volatile organic forms or organic complexes, which can result in evaporation into the atmosphere, leaching into groundwaters, or loss through the action of storm water runoff. Thus, generally metals would be expected to be more persistent in soils than organic compounds.

Substances that are removed from soil by environmental processes usually follow first-order kinetics as described in Chapter 3. This chapter should be referred to to obtain background on the approaches to considering the kinetics of these processes. The half-life ($T_{1/2}$) is the characteristic usually used to measure the persistence of a compound in soil. Environmental persistence, in general, is often considered in this phase of the natural environment because soil is a major repository for contaminants in the environment.

A characteristic of particular importance with soils is the moisture content, since this exercises a control over potential for growth of microorganisms. Dry soils do not support an actively growing population of microorganisms, whereas moist soils can support large populations of microorganisms. The type of microorganism present can also influence the degradation processes, as can environmental variables such as temperature and the availability of oxygen. Because of the variability in these factors, it would be expected that compounds in soil would exhibit a corresponding variability in persistence measured as the $T_{1/2}$. Some ranges of $T_{1/2}$ values found in soil are shown in Table 14.3. The longest $T_{1/2}$ values are evident with the chlorohydrocarbons, DDT, and dieldrin, which have $T_{1/2}$ values from 175 to 6000 days. These substances contain a limited range of bond types that are not susceptible to oxidation or hydrolysis, which are common degradation and transformation processes. The hydrocarbons benzene and benzo(a)pyrene exhibit $T_{1/2}$ values from 5 to 490 days, which is less than that of the chlorohydrocarbons. The remaining compounds in Table 14.3 all exhibit shorter $T_{1/2}$ values due to their relatively high aqueous solubility, making them more readily available to microorganisms. Also, the presence of active chemical groups within the molecule renders the compounds susceptible to attack.

Volatilization is a major process for the removal of contaminants from soil. A diagrammatic illustration of the process of evaporation from soil is indicated in Figure 14.5. Several processes are illustrated in this figure. First, a compound can partition between the soil particle and the pore water present between the particles (illustrated as ($C_s \Leftrightarrow C_w$)). Diffusion in the pore water can then occur and some chemical molecules eventually reach the pore water surface and evaporate together with water molecules as well. Volatilization depends on:

- Inherent properties of a chemical
- Properties of soil
- Environmental conditions

The inherent properties of the molecule are properties such as molecular weight, polarity, and other characteristics that govern its vapor pressure and Henry's law constant (water–air distribution coefficient). The properties of the soil that influence the soil–water partition process can be seen as part of the volatilization process, as shown in Figure 14.5. Thus, the organic carbon content of the soil influences the

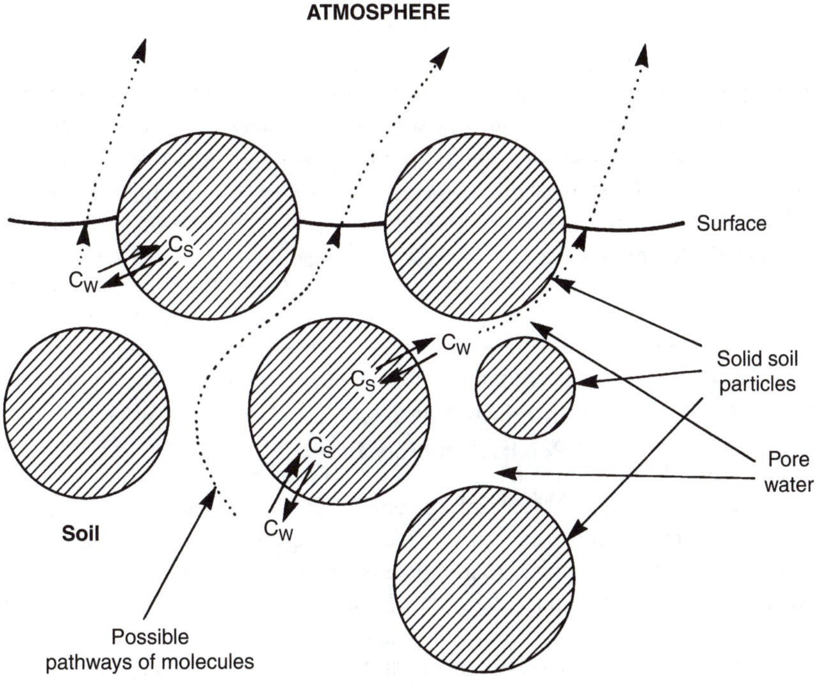

FIGURE 14.5 Processes involved in the volatilization of a contaminant from soil, where C_S is the contaminant in soil solids and C_W the contaminant in water.

volatilization rate. The higher the organic carbon content, the more the lipophilic organic compounds are retained and the lower the evaporation rates. The moisture content of the soil is also a key characteristic. High soil moisture contents give higher contaminant evaporation rates, which may be due to greater water loss from the soil, and with it a greater amount of the contaminant. This effect is often referred to as the **Wick effect**. These processes are influenced by environmental conditions such temperature and surface air speeds, with increases in both these factors leading to increased rates of volatilization.

A variety of expressions have been derived to calculate the loss of a chemical by volatilization. In general the models indicate that as K_{OM} and sorption of the chemical to soil particles increase, the rate of loss declines and $T_{1/2}$ increases. On the other hand, as the vapor pressure (P) increases, the volatilization increases and $T_{1/2}$ declines.

Often a chemical can also be removed by leaching. In this process the pore water is displaced by water movement, and in doing so, the chemical in the pore water is removed from the soil. Thus, contaminated water from a soil can move to other areas, possibly groundwater or surface water, and lead to contamination in those areas. A simple measure of the leaching capacity of a chemical, R, can be calculated using the following equation:

$$R = \frac{1}{K_D(1-\varphi^{2/3})d_s}$$

where K_D is the soil–water partition coefficient, ϕ pore water fraction of the soil, and d_s the density of the soil solids. Thus, leachability of a chemical declines as K_D increases and increases as the pore water fraction increases.

Often, an overall measure of mobility of organic compounds in soil is useful. This can be used as a measure of the likely decline in concentration of a soil contaminant due to losses from volatilization and leaching. A soil mobility index (MI) can be calculated using the following equation:

TABLE 14.4
Interpretation of the Relative Mobility Index

Mobility Index	Description
>5.00	Extremely mobile
5.00 to 0.00	Very mobile
0.00 to –5.00	Slightly mobile
–5.00 to –10.00	Immobile
< –10.00	Very immobile

$$MI = \log\left[\frac{S.V}{K_{OM}}\right]$$

where S is water solubility (mg/l), V is vapor pressure at ambient temperature (mm), and K_{OM} is the soil sorption partition coefficient in terms of organic matter.

Some MI values for different compounds are shown in Table 14.3. The meaning of the MI values as measures of mobility is shown in Table 14.4. This indicates that DDT and dieldrin are clearly immobile in soil and have little potential to decline as a result of movement to other environmental phases. Of course, particles containing sorbed substances may move to contaminate other areas in addition to the water itself. Atrazine and malathion are more mobile and move slightly in soils and have some ability to contaminate other environmental phases. Toluene and benzene are very mobile and can readily contaminate surface and groundwaters adjacent to the contaminated area.

14.6 DISTRIBUTION OF SOIL CONTAMINANTS

Mobile chemicals, and even slightly mobile chemicals, in soil can redistribute from soil into various other environmental phases. The major phases influencing the redistribution of soil contaminants are the abiotic phases: the atmosphere, soil solids, and pore water. The partition processes involved are illustrated in Figure 14.6. Chemicals distribute between these phases by sets of two-phase processes. Some of these are illustrated below, where C_A, C_W, and C_S are the chemical concentrations in air, water, and soil solids, respectively:

Atmosphere (C_A)

↓↑

Soil solids (C_S)

↓↑

Pore water (C_W)

Movement to groundwater is a transfer of chemical by actual movement of the water phase. Other mass transfer processes occur when storm water runs off a contaminated area containing dissolved and particulate sorbed chemicals.

14.6.1 THE SOIL–WATER PARTITION PROCESS

A lipophilic chemical, in concentrations below its solubility in water, when placed in a soil–water system will reach equilibrium by movement of molecules between the two phases, as described in Chapter 15. In this condition, the concentrations in the soil solids (C_S) and water (C_W) are constant. Thus, C_S/C_W is constant and is referred to as the soil–water partition coefficient, K_D. K_D can be measured in laboratory experiments by repeated experiments at different concentrations to obtain more reliable results. When this is done, the concentration in soil is plotted against

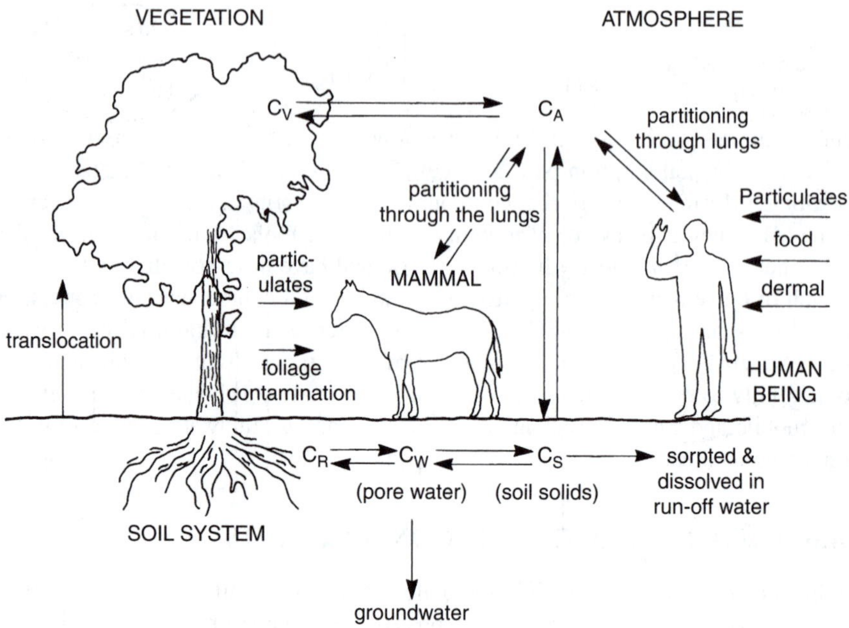

FIGURE 14.6 Distribution patterns of soil contaminants in soil ecosystems.

the concentration in water, as shown in Figure 14.7a, and the slope of the line is the K_D value. Thus,

$$K_D = C_S/C_W$$

As previously mentioned, it has been found that the organic matter in soil is the prime component of the soil that sorbs lipophilic chemicals. Organic matter is often measured as the **organic matter content** of the soil. The partition coefficient obtained by expressing the concentration in the soil in terms of organic matter gives much more consistent results between different soil types. This means that

$$K_{OM} = C_{SOM}/C_W$$

where C_{SOM} is the concentration in the soil organic matter. In this case, C_{SOM} can be plotted against C_W at equilibrium, as shown in Figure 14.7b. The K_{OM} value is the slope of the regression line produced in this plot. Generally, the plot of C_{SOM} against C_W has a better correlation coefficient and a higher slope than the C_S against C_W line, as shown in Figure 14.7a. In these relationships the following can be derived:

$$C_{SOM} = C_S/f_{OM}$$

so

$$K_{OM} = C_S/f_{OM}C_W$$

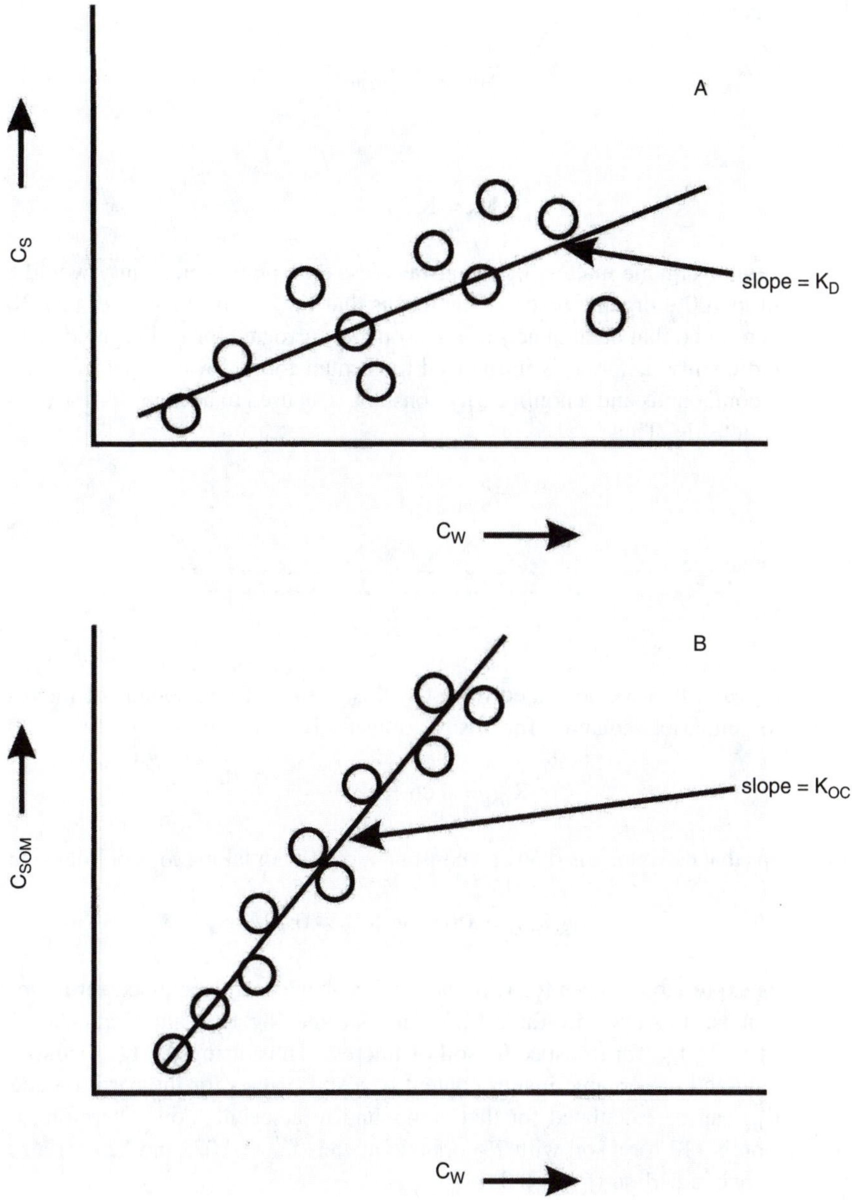

FIGURE 14.7 Experimental partition experiment results in a soil–water system with a lipophilic chemical. Plots of the concentration in soil (C_S) against the concentration in water (C_W) at equilibrium for a range of soil types (plot a), and the same set of data using concentration in soil organic matter (C_{SOC}) (plot b).

This means that

$$K_{OM} = K_D/f_{OM}$$

and

$$K_D = K_{OM}f_{OM}$$

The fraction of organic matter (f_{OC}) is always less than unity, since unity would be equivalent to 100% organic matter. This means that K_{OM} is always greater than K_D.

Its been found that octanol acts as a reasonable surrogate for soil organic matter if a proportionality factor, x, is introduced to account for its lower capacity to sorb lipophilic compounds and a nonlinearity constant, a, is used to account for curvature of the relationship. Thus,

$$K_{OM} = x\ K_{OW}^{a}$$

and this means that

$$K_D = x.f_{OM}.K_{OW}^{a}$$

Thus, K_{OM} can often be calculated from the K_{OW} values of compounds of interest. A common empirical equation for this relationship is

$$K_{OM} = 0.66\ K_{OW}^{1.03}$$

This means that constant x is 0.66 and constant a is 1.03, or taking logs of both sides,

$$\log K_{OM} = 1.03 \log K_{OW} - 0.18$$

Using this expression, values for K_{OC} and log K_{OC} have been calculated for a range of compounds, as shown in Table 14.5. The K_D and log K_D values can then be calculated using f_{OM} for the specific soil of interest. Thus, using the K_{OW} value for a compound and the organic matter content of a soil, values for the partition coefficient (K_D) can be calculated for that compound in a specific soil. Therefore, K_D for diazinon is 112 for a soil with 2% organic matter ($f_{OM} = 0.02$) and 224 for a soil with 4% organic carbon ($f_{OM} = 0.04$).

14.6.2 The Soil–Organism Process

There are a variety of animal and plant organisms in soil. Roots are very common soil biotic components, and a variety of invertebrate organisms, including earthworms, can reside in soil as well. A simple three-phase model can be used to estimate the transfer of contaminants from soils to organisms, as illustrated in Figure 14.8.

TABLE 14.5
The K_{OM} Values for Some Compounds Calculated from the K_{OW} Value

Compound	K_{OW}	Log K_{OW}	K_{OM}	Log K_{OM}
Diazinon	8,000	3.81	5,600	3.74
Dieldrin	21,000	4.32	19,000	4.27
Tetrachlorophenol	174,000	5.24	165,000	5.22
Hexachlorobenzene	204,000	5.31	195,000	5.29

The soil–water partition process, which represents part of this three-phase process, can be described as outlined in the previous section. The next part of the process can be addressed as a pore water–biota partition process, in which biota could be plant tissue, such as roots, or animals, such as earthworms. This is also considered

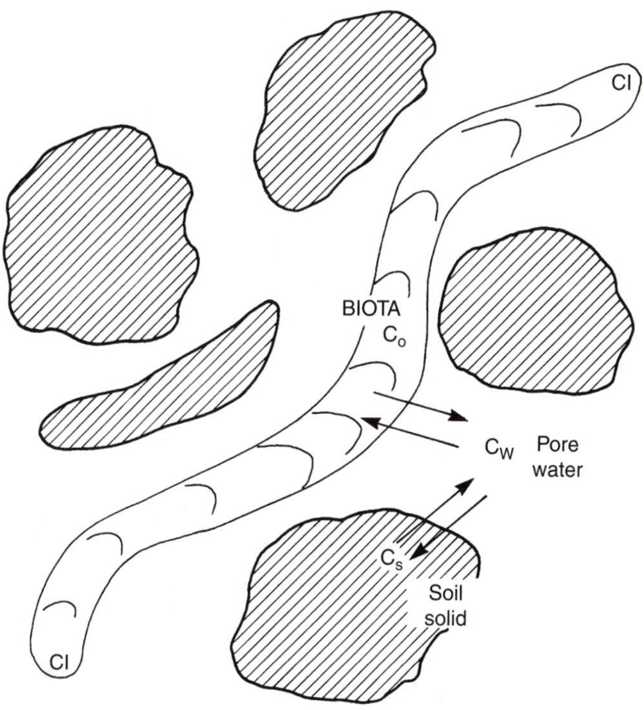

FIGURE 14.8 The three-phase partitioning process resulting in a transfer of contaminants from soil solids to soil biota, e.g., plant roots and earthworms.

in Chapter 15. If we assume that biota lipids take up the lipophilic organic compounds, then the biota–water partition coefficient (K_B) can be represented as follows:

$$K_B = C_B/C_W \text{ at equilibrium}$$

where C_B and C_W are the concentrations in the biota and water, respectively. This means that a plot of C_B against C_W would be expected to be linear at low concentrations. If octanol is a good surrogate for biota lipid, then

$$K_B = f_{lipid}\, K_{OW}{}^{b}$$

where f_{lipid} is the lipid fraction in the biota and constant b is an empirical nonlinearity constant. It is commonly found that constant b is usually about 0.95. Now the relationship between soil solid and biota can be considered. The bioaccumulation factor (BF) is the ratio of the concentration in the biota (C_B) and the concentration in the soil (C_S); so, BF = C_B/C_S. This can be expanded by inserting the water phase; thus,

$$BF = \frac{C_B}{C_W} \cdot \frac{C_W}{C_S} = \frac{K_B}{K_D}$$

This can be expressed in terms of the K_{OW} values by inserting into the expression the previous expressions derived for K_B and K_{OC} in terms of K_{OW}. Thus,

$$BF = f_{lipid}\, K_{OW}{}^{b}/x\ f_{OM}.K_{OW}{}^{a}$$

and

$$BF = (f_{lipid}/x\ f_{OM}).K_{OW}{}^{b-a}$$

Often the nonlinearity constants in both relationships (a and b) are close to unity. In this case, constant a is 1.03 and constant b is 0.95, and so (a – b) is 0.08, which is close to zero and means that $K_{OW}{}^{0.08}$ is close to unity. Substituting these values into the previous equation, we get

$$BF = (f_{lipid}/0.66.f_{OM}).K_{OW}{}^{0.08}$$

and, to simplify this expression, we can assume

$$BF \approx f_{lipid}/0.66\ f_{OM}$$

This means that the BF values should show little dependence on K_{OW} and other properties of the chemical. It primarily depends on the properties of the soil and the biota, particularly the ratio of lipid in biota to the organic carbon content of the soil. This equation can be used to give an approximate idea of the bioaccumulation of lipophilic chemicals by soil biota.

14.7 ECOLOGICAL AND HEALTH EFFECTS OF SOIL CONTAMINATION

Soil contaminants can lead to obvious adverse effects on natural ecosystems, particularly in agricultural areas. Occasionally, in some agricultural areas **fish kills** are observed after storms. In many cases, these are due to soil contaminants being swept from the land into streams. In some situations, these contaminants may exceed the lethal level to fish and a fish kill occurs. In many situations, lethal levels may not be reached, but the presence of contaminants in aquatic systems can lead to deleterious sublethal effects on reproduction, growth, and other important biological characteristics.

Human exposure to soil contaminants can also have important adverse effects. The general pattern of distribution of soil contaminants leading to uptake by human beings and animals is shown in Figure 14.6. Several pathways from soil to human beings can be identified as follows:

- Partitioning of vapors of contaminants originating from soil through the lung walls into blood
- Lodging of particulates containing contaminants in the respiratory system
- Diffusion through the skin from dermal contact with soil
- Ingestion in food
- Direct consumption of soil

This final pathway identified above is important with children since significant amounts of soil are directly taken up by children. These pathways are of different importance with different chemicals in different localities.

Lead is a common contaminant in urban areas and is probably one of the most important soil contaminants from the human health point of view. Urban lead often originates from its use in fuel for motor vehicles to enhance the octane rating. Even though this use is being discontinued, in many areas lead contamination in soil can be expected to remain for a considerable period. Lead is also a common soil contaminant in many mining or former mining areas. Lead in soil in contaminated communities has been found in a number of areas to have several adverse effects on human well-being. Possibly the most important of these is the deleterious effect on intellectual development that occurs with children.

The levels of occurrence of toxic substances in soil is an important aspect of the management of soil contamination. There can be some difficulties in assessing the meaning of concentrations of toxic substances that occur in soil. This difficulty relates to the normal levels of a substance that can be expected in the soil without adversely effecting human health. The normal levels of toxic metals in soil relate to such factors as:

- Geochemistry of soil formation from rocks; some rocks are relatively rich in many toxic metals, e.g., lead, chromium, nickel, and so on.
- Soil-forming processes, e.g., weathering and leaching.
- Erosion and deposition processes leading to the actual physical movement of soil.
- The normal levels of toxic metals in an area can be considered to include levels that result from historical activities leading to soil contamination.

When management authorities set levels, and perhaps define normal levels, it is possible that these levels may exceed guidelines for control of adverse effects on human health and the environment. The management response to soil contamination may not be simple in terms of the need for any actions to protect public health and natural ecosystems. There are a variety of factors to be considered, including:

- The nature of the contaminant, e.g., its potential to contaminate adjacent areas
- The current or proposed use of the land, including industrial activities, recreational use, domestic housing, and so on
- The sector of the human population exposed to the contaminated soil, e.g., children, adults, adults undertaking specific activities, and so on

Actions for remediation need to take into account all of these factors before remedial activities can be planned.

14.8 KEY POINTS

1. Soil can become contaminated due to a wide range of human activities, including agriculture, industry, and urban development. Human populations and natural ecosystems can be exposed to contaminants through emissions of vapors, dispersal of dust, contamination of food and water, and direct uptake through the skin.
2. Important soil contaminants include toxic metals and related substances, such as lead, copper, chromium, and arsenic; organic substances, such as pesticides, petroleum hydrocarbons, polycyclic aromatic hydrocarbons (PAHs), and polychlorinated biphenyls (PCBs); and industrial chemicals, such as various solvents, acids, and alkalis.
3. The properties of soil have a strong influence on any contaminants present. Clay consists of various hydrous silicates and oxides that have a capacity to sorb cations such as metal ions, polar organic molecules, and organic cations. For example, diquat and glyphosate are strongly sorbed. The organic matter in soil contains plant lignin and related degradation products, described as humic substances or humic and fulvic acids. These

substances are high molecular weight and contain many phenolic groups, but are lipophilic in nature. Highly hydrophobic compounds, such as DDT, dieldrin, and benzo(a)pyrene, are strongly sorbed to the organic matter in soil.

4. The persistence of a compound in soil is of major environmental importance. The loss of a compound usually follows first-order kinetics.
5. Leaching of contaminants is an important process for the loss of contaminants from soil. A simple measure of the leachability of a chemical from soil, R, is

$$R = \frac{1}{K_D(1-\varphi^{2/3})d_s}$$

6. Mobility of a chemical in soil is a major aspect of environmental behavior and can be estimated using the mobility index (MI). Thus,

$$MI = \log\left[\frac{S.V}{K_{OM}}\right]$$

7. Soil – water partitioning of lipophilic chemicals is expressed as the K_D value; thus,

$$K_D = \frac{C_S}{C_W}$$

This partition coefficient can also be expressed more consistently in terms of organic carbon; thus,

$$K_D = K_{OM}f_{OM}$$

8. The K_{OM} value of a lipophilic compound in the soil–water system can be estimated from the octanol–water partition coefficient K_{OW}. Thus, $K_{OM} = x\ K_{OW}^{a}$, where constant x is usually about 0.66 and constant a is 1.03. By taking logarithms,

$$\log K_{OC} = 1.03 \log K_{OW} - 0.18$$

9. The uptake of lipophilic compounds by soil biota, such as earthworms and roots of plants, can be expressed as follows:

$$BF = (f_{lipid}/0.66.f_{OM})K_{OW}^{0.08}$$

$$BF \approx f_{lipid}/0.66\ f_{OM}$$

10. Human exposure to soil contaminants can be calculated by estimating the uptake through the possible exposure routes, i.e., vapor sorption, lodging of particulates in the respiratory system, dermal contact, ingestion in food, and direct consumption. This last route is particularly important with young children.
11. Contaminants in soil can have adverse effects on human health and the natural environment. For example, lead in soil can lead to impaired intellectual development of children, and pesticides in agriculture soil can have deleterious effects on aquatic ecosystems.
12. The normal levels of substances in soil, such as toxic metals, relate to rock geochemistry, soil-forming processes, erosion and deposition processes, and historical pollution-causing activities. These may exceed guideline levels for protection of human health and the natural environment.

REFERENCES

Amdur, M.O., Poull, J., and Klassen, C.D., Water and soil pollutants, in *Casarett and Doull's Toxicology: The Basic Science of Poisons*, 4th ed., McGraw-Hill, New York, 1991, chap. 26, p. 872.

Shineldecker, C.L., *Handbook of Environmental Contaminants: A Guide for Site Assessment*, Lewis Publishers, Boca Raton, FL, 1992.

QUESTIONS

1. A spill of a petroleum product onto soil in an urban area has occurred. Analysis of the soil (2.0% organic carbon) has revealed the following average concentrations of the major contaminants:

Benzene	150 mg/kg
Toluene	100 mg/kg
Benzo(a)pyrene	5 mg/kg

The environmental management authority has asked the following question: After what period will the soils reach acceptable concentrations? The authority has set the acceptable concentrations at benzene, 0.04 mg/kg; toluene, 0.05 mg/kg; and benzo(a)pyrene, 0.004 mg/kg.
Provide an answer to these questions using the data in Table 14.3.

2. A crop of carrots (0.5% lipid) has been grown in soil (1% organic carbon) contaminated by dieldrin and atrazine (3.5 and 10.0 mg/kg, respectively). Estimate the approximate levels of these contaminants that could be expected in the carrots?
3. You are a member of a team evaluating an industrial site with contaminated soil. Your task is to evaluate the potential for soil contaminants to contaminate the groundwater. You have been able to collate the property data on the contaminants as follows:

Compound	Vapor Pressure (mm Hg)	Aqueous Solubility (mg/l^{-1})	Soil Sorption Coefficient (K_{OM})
Phenol	0.2	67,000	2
Styrene	9.5	280	120
Tetrachloroethane	5	2900	480
Chloropyrifos	1.9×10^{-5}	2	13,000

Evaluate the possibility of groundwater contamination by these compounds.

ANSWERS

1. To enable calculations to be made for the period to reach acceptable levels, the first-order decay equation will be used since it usually provides a reasonably accurate representation of the behavior of contaminants in the environment:

$$C_t = C_o^{-kt}$$

Thus,

$$\ln C_t = \ln C_o - kt$$

To use this equation, values for k are needed. These can be calculated using the most conservative $T_{1/2}$ values (longest) from Table 14.3 and the equation $k = 0.693/T_{1/2}$. Thus, the following k values can be obtained: benzene, 0.04 day^{-1}; toluene, 0.03 day^{-1}; and benzo(a)pyrene, 0.0014 day^{-1}. Substitute in the logarithmic kinetic equation above to obtain the periods to reach an acceptable level. Then,

Benzene:

$$\ln 0.04 = \ln 150 - 0.04t$$

Thus, t = 206 days

Toluene:

$$\ln 0.05 = \ln 100 - 0.03t$$

Thus, t = 253 days

Benzo(a)pyrene:

$$\ln 0.004 = \ln 5 - 0.0014t$$

$$\text{Thus, } t = 5092 \text{ days (14 years)}$$

Thus, the periods to attain acceptable levels are 206 days, benzene; 253 days, toluene; and 5092 days (14 years), benzo(a)pyrene.

2. The first step is to calculate the approximate bioaccumulation factor of the chemicals in the soil for carrots using the following equation:

$$BF = f_{lipid}/0.66\ f_{OM}$$

Dieldrin:

$$BF = 0.76$$

$$\text{Thus, } C_B/C_S = 0.76$$

$$\textit{Thus, } C_B = 2.66 \textit{ mg/kg wet weight}$$

Atrazine:

$$BF = 0.76$$

$$\text{Thus, } C_B/C_S = 0.76$$

$$\textit{Thus, } C_B = 7.6 \textit{ mg/kg wet weight}$$

3. The mobility index (MI) will give an estimation of the capacity of a chemical to move and contaminate groundwater,

$$MI = \log\left(\frac{S.V}{K_{OM}}\right)$$

The following MIs were calculated: phenol, +3.83; styrene, +1.35; tetrachloroethane, + 1.48; and chloropyrifos, –8.53.

By consulting Table 14.4, these data indicate that while chloropyrifos is immobile and has little potential to contaminate groundwater, all of the other compounds are very mobile and so have considerable potential for groundwater contamination.

15 Distribution of Chemicals in the Environment

15.1 INTRODUCTION

Prediction of the possible effects of chemicals in the environment in terms of human health or natural ecosystems is now a major environmental management activity. Some aspects of this are described in Chapter 18. A vital part of these procedures is to evaluate the potential exposure of biota to contaminants. This requires methods to understand and estimate the patterns of distribution in the environment (Mackay, 1991). When a chemical is discharged to the environment, it is necessary to know how much will deposit in soil and sediments and so on.

The first step in understanding the distribution of chemicals in the environment is to simplify the environment. The environment can be seen to consist of phases that are physically distinctive sections that are relatively homogeneous and within which a chemical behaves in a uniform manner. Commonly, the environment is considered to consist of the atmosphere, water (lakes, dams, streams, oceans, etc.), soils, sediments, suspended sediments, and biota. When a chemical is discharged to the environment, it will move and distribute between the phases according to the properties of the chemical and the properties of the phases. The distribution between the phases can be considered as a result of a set of two-phase partition processes, as illustrated in Figure 15.1. Here a number of two-phase partition processes are presented, including atmosphere–water, vegetation–atmosphere, soil–atmosphere, biota–water, sediment–water, and suspended sediment–water. Phases that are not in contact, e.g., atmosphere and sediment, cannot exhibit a two-phase partition interaction. Thus, the key to understanding the distribution of chemicals in the environment is to develop an understanding of the two-phase partition processes that occur. The fundamental nature of partitioning and properties influencing this process are described in Chapter 2.

15.2 TWO-PHASE PARTITION PROCESSES IN THE ENVIRONMENT

Two-phase partition processes can be described by the **Freundlich equation**. Thus, at equilibrium

$$C_1 = K(C_2)^n$$

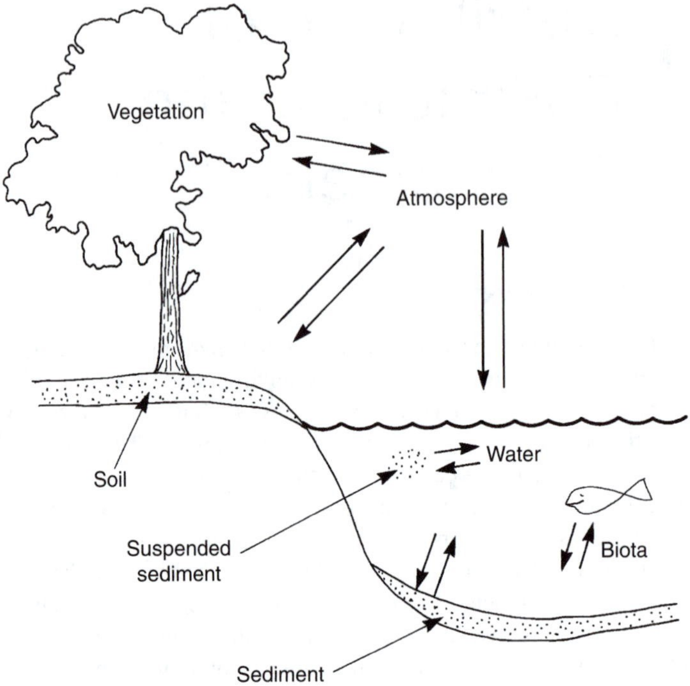

FIGURE 15.1 Distribution of a chemical between phases in the environment, with various two-phase partition processes involved, e.g., atmosphere–water, sediment–water, biota–water.

where C_1 and C_2 are the concentrations in phases 1 and 2, respectively, K the partition coefficient, and constant n a nonlinearity constant. At equilibrium, the concentrations in the phases (C_1 and C_2) are constant. In this treatment the solubility of the substance should not be exceeded. If this occurs, then another phase is formed and this theory is not applicable. The relationship between C_1 and C_2 for a series of concentrations can be plotted as shown in Figure 15.2a. With many environmental processes, the concentrations of chemical are low and the value of constant n can be considered to be unity. Also, the relationship can be converted into a more useful form by taking logarithms. Thus,

$$\log C_1 = \log K + n \log C_2$$

This can be plotted as shown in Figure 15.2b. This gives a linear relationship from which log K can be obtained as the intercept at a log C_2 value of zero and n is the slope.

Some environmental partition processes have been extensively investigated. The air–water process was described as **Henry's law** and investigations started in the early 1900s. Other processes have been subject to limited evaluation, for example, the vegetation–air process.

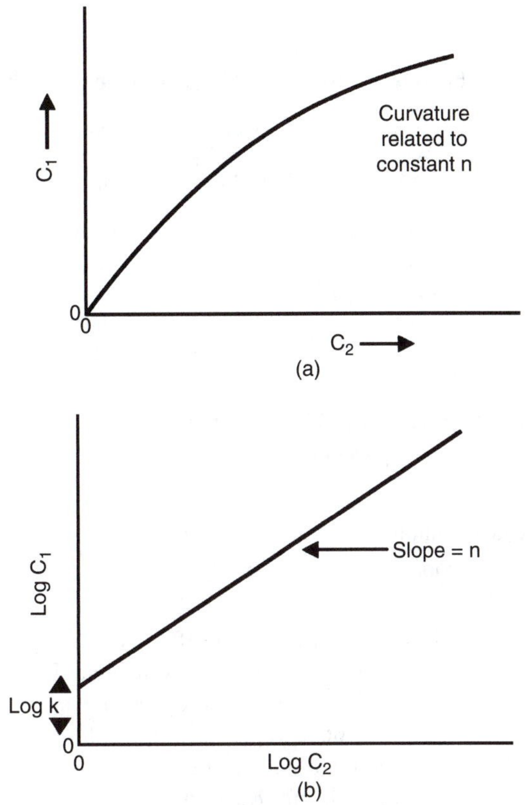

FIGURE 15.2 a: Plot of the corresponding concentrations in a two-phase system at equilibrium. b: Plot of the same data in a logarithmic form.

Table 15.1 lists some properties of organic chemicals that influence the chemicals' environmental distribution. Two key properties are aqueous solubility and vapor pressure. The relationship between these properties is described as the Henry's law constant, H, which is essentially the partition coefficient of the compound between air and water. When equilibrium is established, the partial pressure of the compound in air and the corresponding concentration in water are constant. Then,

$$H = \frac{\text{partial pressure of compound in air}}{\text{concentration in water}} \quad \text{(at equilibrium)} \qquad (15.1)$$

In this case, the partial pressure is the method of expressing the concentration of the gas. The value of H can often be calculated from the basic properties of compounds as the vapor pressure divided by the water solubility.

A valuable environmental characteristic, which is measured in the laboratory, is the **octanol–water partition coefficient, K_{OW},** which is the ratio of the concentration

TABLE 15.1
Physicochemical Properties of Some Organic Chemicals at 25°

Chemical	Molar Mass (G)	Bp (°C)	Water Solubility (mol/m³)	Vapor Pressure (Pa)	H (Pa·m³/mol)	K_{OW}
Chloroform	119.4	81	69	23,080	336	$10^{1.97}$
Benzene	78	80	23	12,700	557	$10^{2.13}$
Naphthalene	128	218	0.25	10.4	42	$10^{3.35}$
p,p-DDT	354.5		0.0000087	0.0002	2.3	$10^{6.19}$
2,3,7,8-TCDD (dioxin)	322		0.0006	1×10^{-7}	0.0017	$10^{6.80}$

in *n*-octanol to that in water, at equilibrium. This can be measured in the laboratory in an octanol–water two-phase system. Thus,

$$K_{OW} = \frac{\text{concentration of chemical in octanol } (C_O)}{\text{concentration in water } (C_W)} \quad \text{(at equilibrium)}$$

The significance of octanol is that it is a useful surrogate for the weakly polar organic matter present in soils and the lipid tissues of biota. High K_{OW} compounds tend to partition strongly into these organic-rich environmental phases. This means that they sorb onto soil or sediments and are accumulated by biota. The numerical value of K_{OW} can be used to estimate the partition coefficients for these processes.

The biota–water partition coefficient, K_B, is the bioconcentration factor at equilibrium. It is measured by placing fish, or other aquatic biota, in aquaria and measuring the concentration of the chemical in the biota after exposure to a fixed concentration in the water when an equilibrium is established. Thus,

$$K_B = \frac{\text{concentration of chemical in biota } (C_B)}{\text{concentration in water } (C_W)} \tag{15.2}$$

Experimental values of K_B are known for many chemicals and species of aquatic biota. However, an approximate theoretical relationship between K_B, K_{OW}, and the lipid fraction of the biota can be derived to estimate K_B from K_{OW} as set out below.

With lipophilic compounds, the nonlipoid phases in the aquatic organism can be considered not to play any role in the partition process and only the biota lipids participate. After partitioning between biota lipid and water has occurred in the aquaria through the gills of the organism and equilibrium is established, then

$$\frac{C_L}{C_W} = \text{constant}$$

where C_L is the concentration of the chemical in the biota lipid. If f_{lipid} is the fraction of lipid present in the biota, then

$$C_B \text{ (whole biota weight)} = f_{lipid} C_L$$

and

$$K_B = \frac{C_B}{C_W} = \frac{f_{lipid} C_L}{C_W}$$

If octanol is a perfect surrogate for biota lipid, then C_L is equal to C_O and

$$K_B = \frac{f_{lipid} C_L}{C_W} = \frac{f_{lipid} C_O}{C_W} = f_{lipid} K_{OW}$$

$$\text{so} \quad K_B = f_{lipid} K_{OW} \tag{15.3}$$

$$\text{and} \quad \log K_B = \log K_{OW} + \log f_{lipid}$$

But often octanol is not a perfect surrogate for lipid and a power coefficient, b, is used to better express the relationship. The value of coefficient b would be unity if octanol was a perfect surrogate for the biota lipid, and it often approaches unity since octanol is quite a good surrogate for lipid in most situations:

$$\text{so} \quad K_B = f_{lipid} . K^b_{OW}$$

$$\text{and} \quad \log K_B = b.\log K_{OW} + \log f_{lipid}$$

Often the lipid fraction, f_{lipid}, for aquatic biota is about 0.05 or 5% and the coefficient b is unity. Then, taking logarithms,

$$\log K_B = 1.\log K_{OW} - 1.30$$

The partition coefficient between the abiotic solid phases (soil, sediment, or suspended particulates) and water is often referred to as K_D (see Section 14.6.1); here it is defined as follows:

$$K_D = \frac{\text{chemical concentration in whole solid phase } (C_S)}{\text{concentration in water } (C_W)} \quad \text{(at equilibrium)} \tag{15.4}$$

Partitioning occurs between the organic matter in the solid phase and water. The other solid phase components, such as silica sand, clay, and calcium carbonate, are not considered to play a significant role in this process. Thus, at equilibrium

$$K_{OM} = \frac{C_{SOM}}{C_W} = \text{constant}$$

where C_{SOM} is the concentration in the solid organic matter in terms of organic matter. If f_{OM} is the fraction of organic matter in the solid, then

$$C_S \text{ (whole solid phase)} = f_{OM}.C_{SOM}$$

and

$$K_D = \frac{C_S}{C_W} = \frac{f_{OM}C_{SOM}}{C_W}$$

If octanol is a good surrogate for organic matter, but not as good as it is with the biota lipids, then C_{SOM} is equal to C_O (concentration in octanol in the octanol–water system) multiplied by constant x, a proportionality factor related to the efficiency of octanol as a surrogate for the organic matter in the solid phase. Thus,

$$K_D = \frac{xf_{OM}C_O}{C_W} = xf_{OM}.K_{OW}$$

and

$$K_{OM} = x.K_{OW}$$

The constant x would be unity if octanol was a perfect surrogate, but it generally has a value of about 0.66 with soils, although other values are used with specific sets of compounds and specific solid phases. Also, the assumption that octanol is a perfect surrogate for organic matter when the proportionality factor x is used is not fully accurate, and a power coefficient, a, is used with K_{OW} to allow for this. Thus, with soils

$$K_D = 0.66.f_{OM}.K^a{}_{OW} \tag{15.5}$$

The value of f_{OM} is usually about 0.02 to 0.04 (2 to 4%) for soils, sediments, and suspended solids, and often the value of coefficient a is about unity. It would be unity if octanol was a perfect surrogate for the organic matter.

These two-phase partition processes are valuable for understanding each of the processes involved, but do not indicate how the chemical will behave when all the systems are operating together. An overall process for establishing an equilibrium distribution when all of these systems are operating together is now needed.

15.3 THE FUGACITY CONCEPT

Imagine two balloons, one inflated and the other slack, connected by a closed tube. If the closure is opened, then clearly gas escapes into the slack balloon until the pressures equalize. The gas molecules in both balloons have a tendency to escape their confinement. The molecules in the inflated balloon are more crowded and have a greater escaping tendency, so that when the balloons are connected, there is a net flow of molecules to the slack balloon. The net flow of molecules stops when their escaping tendencies become equal.

This **escaping tendency** of molecules is called **fugacity**, f, after the Latin *fugere*, to flee. The fugacity of gases is closely related to their pressures, and has the same units (e.g., Pascal, Pa). Indeed, for an ideal gas, the fugacity *is* the pressure (P):

$$f = \text{P (ideal gases)} \tag{15.6}$$

For nonideal or real gases, the fugacity is the effective pressure, as if the gas were behaving ideally. The fugacity of real gases may be more or less than the observed pressure, depending on how molecular interactions and sizes cause the gas to deviate from ideal behavior. However, organic contaminants usually occur at low concentrations or pressures in the air around us. As gases, they often behave roughly ideally, and we will assume that their fugacity and partial pressure are the same, as indicated in Equation 15.6.

The concept that a substance exerts an escaping tendency or fugacity can also be applied to where the substance is contained in a condensed phase, such as a solid, liquid, or solution. Consider for a moment a flask into which we pour some benzene, then stopper the flask as indicated in Figure 15.3. Immediately after addition of the

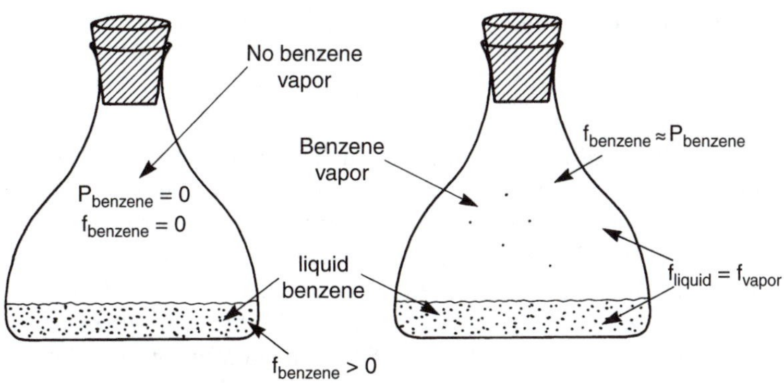

FIGURE 15.3 The changes in pressure and fugacity when benzene is added to a flask and it is stoppered. Left: The situation immediately after adding benzene to the flask; the system is not at equilibrium. Right: The situation when the system reaches equilibrium.

benzene, there is no benzene vapor in the air space. But benzene is a volatile liquid, and its molecules exert an appreciable fugacity. Some benzene molecules escape across the liquid–air boundary to form a vapor. This evaporation continues, gradually building up the pressure of benzene vapor until it reaches saturation, when net evaporation stops. An equilibrium is reached, where the escaping tendencies of the benzene molecules from the liquid and gas phases are equal. At this point the number of benzene molecules escaping from the liquid is the same as the number from the vapor redissolving in the liquid, and the concentrations in vapor and liquid are constant. Thus,

$$f_{\text{liquid}} = f_{\text{vapor}} \text{ (at equilibrium)}$$

Just as with the two-balloon example, the escaping pressures of the compound in the two phases have equalized. If the flask and its contents are at, say, 25°C, then we can place a value on this fugacity for liquid benzene; it is about 13 kPa, the value of the saturation vapor pressure for benzene (Table 15.1).

We can extend our model to a chemical discharged into the environment. The chemical will disperse to reduce its fugacity. It may move into several phases, such as air, water, or biota, or sorb onto sediments or soil. A simple outcome is an equilibrium, where the concentrations of the chemical in each phase become steady and the fugacities become equal:

$$f_{\text{air}} = f_{\text{water}} = f_{\text{biota}} = f_{\text{sorb}} \text{ (at equilibrium)}$$

It is the concentration (C) of a chemical that usually concerns us, so we need to relate fugacity to concentration. At low chemical concentrations, such as those usually found in the environment, the two are roughly proportional; thus,

$$C \propto f$$

By inserting a proportionality constant, Z, then

$$C = Zf \qquad (15.7)$$

The proportionality constant Z is called the **fugacity capacity** factor. It has units of concentration and reciprocal pressure, e.g., mol/m³·Pa. We must be careful to be consistent with the units we use in these calculations; i.e., concentrations are in terms of mol m³ and pressure in terms of Pa. The value of Z usually depends on the temperature, the properties of the chemical, and the nature of the environmental phase into which the chemical is dispersed. At equilibrium, when there is a common fugacity value for a chemical in all phases f, the ratio of a chemical's concentrations in any two phases is the same as the ratio of Z values. For example,

$$\frac{C_{\text{air}}}{C_{\text{water}}} = \frac{Z_{\text{air}} f}{Z_{\text{water}} f} = \frac{Z_{\text{air}}}{Z_{\text{water}}}$$

Similarly,

$$\frac{C_{biota}}{C_{water}} = \frac{Z_{biota}}{Z_{water}}$$

$$\frac{C_{air}}{C_{sediment}} = \frac{Z_{air}}{Z_{sediment}}$$

and so on.

The Z values are the key to calculating the distribution of a chemical in the environment. These values are a property of the chemical and how it behaves in the phase. Chemicals will tend to accumulate in those phases that have high Z values (fugacity capacity factors) and avoid those phases with low Z values. The solubility of the chemical in any phase should not be exceeded, as another phase will form and this theory will no longer apply.

15.4 FUGACITY AND CHEMICAL DISTRIBUTION

Shortly, we will show how Z values can be determined, but for the moment, we will illustrate how to exploit fugacity to evaluate a chemical's distribution among several phases. Imagine an aquarium of total volume 1.0 m^3 that is two thirds filled with water with a tight-fitting lid that prevents the escape of the benzene and water (Figure 15.4). A small amount of benzene (0.02 mol) is added and allowed to equilibrate between the air space and the water. We wish to know the final masses of benzene in the air and the water, given that for benzene Z_{air} and Z_{water} are 4.0×10^{-4} and 1.8×10^{-3} mol/m^3·Pa, respectively.

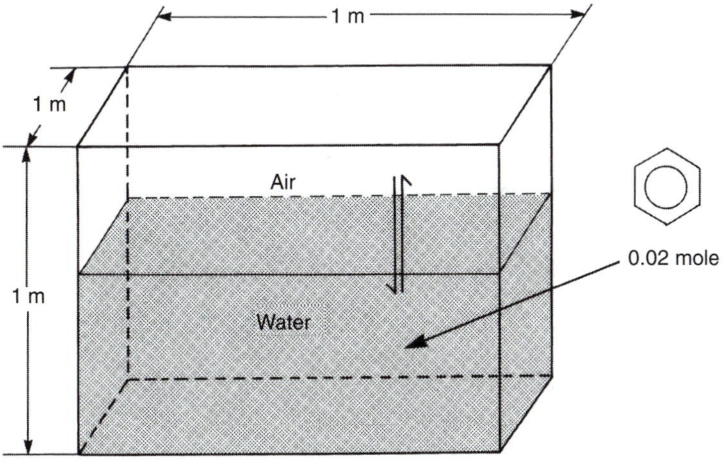

FIGURE 15.4 The distribution of benzene in an aquarium containing air and water.

Since the mass of benzene, M, is related to its concentration, C, and the phase volume, V, from Equation 15.7

$$M = CV = f\,ZV \quad (15.8)$$

for both phases. Thus,

$$M_{air} = f\,Z_{air}\,V_{air} \text{ and } M_{water} = f\,Z_{water}\,V_{water}$$

The benzene fugacity is the same in both phases at equilibrium and the total benzene mass is unchanged, so that

$$M_{total} = M_{air} + M_{water} = f\,Z_{air}\,V_{air} + f\,Z_{water}\,V_{water}$$

or

$$M_{total} = f\,(Z_{air}\,V_{air} + Z_{water}\,V_{water}) \quad (15.9)$$

Now we just need to substitute in the values to find the prevailing fugacity:

$$M_{total} = 0.02 \text{ mol} = f(4.0 \times 10^{-4} \times 1/3 + 1.8 \times 10^{-3} \times 2/3)$$

Thus,

$$f = 15 \text{ Pa}$$

This fugacity value is inserted back into Equation 15.8 to generate the benzene mass in each phase. Thus, for air the mass of benzene is

$$M_{air} = f\,Z_{air}\,V_{air} = 15 \times 0.0004 \times 1/3 = 0.0020 \text{ mol}$$

and for water,

$$M_{water} = f\,Z_{water}\,V_{water} = 15 \times 0.0018 \times 2/3 = 0.018 \text{ mol}$$

The benzene total of 0.02 mol is obtained from the sum of both phase masses as a check on the calculations. The concentrations can be obtained by dividing the mass obtained by the phase volumes, that is, air is 0.002/0.33 = 0.006 and water is 0.018/0.66 = 0.027 mol/m^3.

We can modify the aquarium by placing several small fish in it. We will assume again that an equilibrium redistribution of benzene occurs (and that the fish tolerate the toxicity of the benzene). The fish have a volume of 10^{-4} m^3 (100 ml), and the Z value for benzene in fish is 0.012 mol/m^3·Pa.

To accommodate the new phase, we only need to insert an extra term in Equation 15.9, so that

$$M_{total} = f\,(Z_{air}\,V_{air} + Z_{water}\,V_{water} + Z_{fish}\,V_{fish})$$

The volumes of air and water are virtually unchanged. The new equilibrium benzene fugacity is $f = 14.99 \approx 15$ Pa, also little changed, mainly because of the small volume of the fish. The benzene mass in the fish is

$$M_{fish} = f\, Z_{fish}\, V_{fish} = 15 \times 0.012 \times 10^{-4} = 1.8 \times 10^{-5} \text{ mol}$$

and

$$C_{fish} = 1.8 \times 10^{-5}/10^{-4} = 0.18 \text{ mol m}^{3}$$

We can generalize the changes to Equations 15.8 and 15.9 that were made in this second example. For any number of phases, denoted i, into which the chemical distributes until the fugacities become equal and equilibrium is established, the mass in each phase is given by

$$M_i = f\, Z_i\, V_i \qquad (15.10)$$

and the total mass by

$$M_{total} = \Sigma M_i = \Sigma\, (f\, Z_i\, V_i) = f\, \Sigma Z_i\, V_i \qquad (15.11)$$

15.5 THE FUGACITY CAPACITY FACTORS (Z VALUES)

Expressions for the **fugacity capacity factors**, Z, for a chemical dispersed in the major environmental phases — air, water, biota, soil, sediments, and suspended particulates — can be obtained as outlined below. For a chemical dispersed as a gas in air, we need only to relate the partial pressure of the gas to its concentration. The usual form of the ideal gas equation is

$$PV = nRT$$

where P is now the chemical's partial pressure in air. Rearranged, this equation gives

$$P = \frac{n}{V}\, RT$$

The ratio n/V is the number of moles per unit volume, or concentration, C_{air}. Thus,

$$P = C_{air} RT$$

or

$$C_{air} = \frac{P}{RT}$$

Since $f = P$ (Equation 15.6), and recalling that $C = f\,Z$, then

$$C_{air} = f\,Z_{air} = \frac{f}{RT}$$

Thus,

$$Z_{air} = \frac{1}{RT} \tag{15.12}$$

where R is the universal gas constant and T the temperature in Kelvin. Thus, at a given temperature, Z_{air} is the same for *any* gas. At 25°C (298 K), Z_{air} is equal to 4.04×10^{-4} mol/mPa.

In water, the concentration of dissolved chemical is related to its **equilibrium** vapor pressure in air by Henry's law (Equation 15.1), or

$$P = HC_{water} \text{ (at equilibrium)}$$

Since $f = P$ and $C = f\,Z$, then

$$f = HC_{water} = H\,f\,Z_{water}$$

and

$$Z_{water} = \frac{1}{H} \tag{15.13}$$

The Henry's law constant, H, varies from substance to substance (Table 15.1). Thus, each substance has its own unique Z_{water} value in water.

Knowing Z_{water} provides a link in evaluating the Z values for the chemical sorbed onto solid phases, such as soils, sediments, or suspended particulates. If a mixture of the chemical and (the sorbing) solid phase and water is allowed to equilibrate, then the concentration of sorbed chemical is often simply related to its concentration in water (Equation 15.14); thus,

$$C_{sorb} = K_{sorb}\,C_{water} \tag{15.14}$$

Since at equilibrium

$$C_{sorb} = f\,Z_{sorb} \quad \text{and} \quad C_{water} = f\,Z_{water}$$

then from Equation 15.14

$$f\,Z_{sorb} = K_{sorb}\,C_{water} = K_{sorb}\,f\,Z_{water}$$

or

$$Z_{sorb} = K_{sorb}\,Z_{water} = \frac{K_{sorb}}{H} \tag{15.15}$$

and thus for soil,

$$Z_{soil} = K_D/H$$

The Z_{sorb} values vary from chemical to chemical in a specific solid phase. Because of its dependence on K_{sorb}, the Z_{sorb} value also depends on the properties of the sorbing soil, sediment, or particulate phase.

For a chemical that distributes into aquatic biota, the equilibrium concentration of chemical is proportional to its concentration in water (Equation 15.2), or

$$C_{biota} = K_B\,C_{water} \tag{15.16}$$

As with K_{sorb}, K_B values are known from experiments or can be estimated from K_{OW}, as shown in Section 15.2. Since Equations 15.14 and 15.16 are analogous, by making the appropriate substitutions, Equation 15.16 becomes

$$Z_{biota} = \frac{K_B}{H} \tag{15.17}$$

15.6 CHEMICAL DISTRIBUTION IN A MODEL ENVIRONMENT

Now that we are armed with methods to calculate Z values, estimations can be made of the distribution of some chemicals in the environment. To do this, we need further important information: the **volumes** of the environmental phases, just as we did in the benzene aquarium example. Estimations of these volumes could be made by actually going out and measuring the sizes of the biota, soil, and air that are likely to equilibrate with our chemical. Another approach is to adopt an imaginary or **model** environment, where the sizes of the phases are an educated guess. We are then not so concerned with the accuracy of our model environment, but can focus instead on using the properties of the chemical to evaluate the way it distributes.

A model environment is described in Figure 15.5. The model has the six major phases: air, water, soil, sediments, suspended solids, and biota. This model environment has an area of 1 km^2 and has an atmosphere 10 km high. Soil to a depth of 3 cm covers 30% of the surface, while the remainder is covered with water to an average depth of 10 m. The water has a 3-cm layer of sediment, contains 5 ml of suspended solids per cubic meter, and contains 0.5 ml/m^3 biota. Each of the six phases is assumed to be homogeneous. We will also assume that a chemical discharged into this model environment disperses until eventually steady concentrations in each phase are reached.

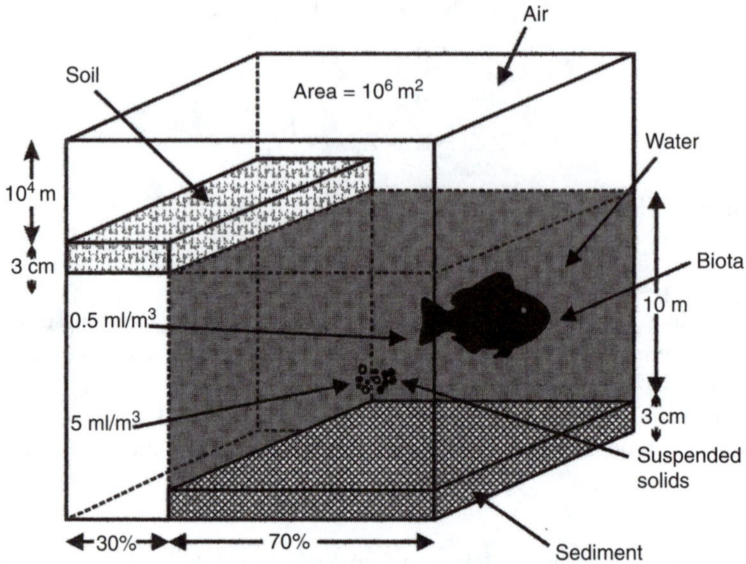

FIGURE 15.5 A six-phase model environment used to evaluate the distribution of a chemical (dimensions not to scale). (Adopted from Mackay, D., Multimedia Environmental Models, Lewis Publishers, Chelsea, Michigan, 1991. With permission.)

Example Distribution Calculation

An arbitrary amount of benzene, 100 mol (78 kg), is added to the model environment in Figure 15.5. The benzene is assumed to distribute to equilibrium and the temperature is taken as 25°. Six steps are followed to complete the calculation of distribution, as shown below and in Table 15.2.

TABLE 15.2
Summary Distribution Calculation for Benzene

Phase	Air	Water	Soil	Suspended Solid	Sediment	Fish	Total
V (m^3)	10^{10}	7×10^6	9×10^3	35	2.1×10^4	3.5	—
Z (mol/m^3·Pa)	4.04×10^{-4}	1.8×10^{-3}	2.0×10^{-3}	4.0×10^{-3}	4.0×10^{-3}	1.2×10^{-2}	—
ZV (mol/Pa)	4.04×10^6	12.6×10^3	18	0.14	84	4.2×10^{-2}	4.05×10^6

$$\text{Fugacity } f = M_{total}/\Sigma Z_iV_i = \frac{100}{4.05 \times 10^6} = 2.47 \times 10^{-5} \text{ Pa}$$

Phase	Air	Water	Soil	Suspended Solid	Sediment	Fish	Total
M = fZV (mol)	99.69	0.31	4.4×10^{-4}	3.5×10^{-6}	2.07×10^{-3}	1.0×10^{-6}	100
C = M/V (mol/m^3)	1×10^{-8}	4.4×10^{-8}	5×10^{-8}	1×10^{-7}	1×10^{-7}	3×10^{-7}	—

Note: Benzene mass, M_{total} = 100 mol.

Step 1: Assemble the important physical chemical properties for benzene. These are:

H	557 Pa m^3/mol
K_{OW}	135
K_{sorb} (soil, 2% organic carbon)	1.1
K_{sorb} (sediment, suspended solids, 4% organic carbon)	2.2
K_B (fish, 5% lipid)	6.7

The K values for soil, sediment, suspended solids, and biota have been estimated from K_{OW} using Equations 15.3 and 15.5.

Step 2: Estimate the volumes of each phase. These volumes are listed in Table 15.2.

Step 3: Calculate the Z values for each phase as follows using the expressions derived in Section 15.5:

Z_{air}	$= 1/RT$	$= 4.04 \times 10^{-4}$ m^3/mol·Pa
Z_{water}	$= 1/H$	$= 1.8 \times 10^{-3}$ m^3/mol·Pa
Z_{soil}	$= K_{sorb}(\text{soil})/H$ or K_D/H	$= 2.0 \times 10^{-3}$ m^3/mol·Pa
$Z_{sediment, suspended solids}$	$= K_{sorb}(\text{sediment, suspended solids})/H$	$= 4.0 \times 10^{-3}$ m^3/mol·Pa
Z_{biota}	$= K_B/H$	$= 1.2 \times 10^{-2}$ m^3/mol·Pa

Step 4: Calculate ZV for each phase and sum these products as $\Sigma Z_i V_i$ (4.05×10^6 mol/Pa), and calculate the prevailing equilibrium fugacity, recalling that $f = M_{total}/\Sigma Z_i V_i$ (Equation 15.11) as 2.47×10^{-5} Pa (Table 15.2).

Step 5: Use this fugacity value to calculate the benzene mass in each phase, recalling that $M_i = f Z_i V_i$ (Equation 15.10) (Table 15.2).

Step 6: Finally, calculate the benzene concentration in each phase using ($C_i = M_i/V_i$), as shown in Table 15.2.

These calculations indicate that greater than 99% of the benzene is dispersed into the air. This is expected, since benzene is a volatile compound and most of the model environment is taken up by air (>99.9% by volume). But what if the compound under investigation is p,p-DDT? The relevant physical chemical properties for DDT are:

H	2.3 Pa m^3/mol
K_{OW}	1,555,000
K_{sorb} (soil)	12,700
K_{sorb} (sediment, suspended solid)	25,400
K_B	77,400

Compared to benzene, DDT is much less volatile, more hydrophobic, and has a strong potential to bioaccumulate. Pursuing the same sequence of calculations as before, a summary of the distribution of 100 mol DDT in the model environment is as follows:

DDT Equilibrium Fugacity = 3.46×10^{-7} Pa

	Air	Water	Soil	Suspended Solid	Sediment	Fish	Total
Mass (mol)	1.4	1.1	17.2	0.13	80.2	0.04	100
Concentration (mol/m^3)	1.4×10^{-10}	1.4×10^{-7}	1.9×10^{-3}	3.8×10^{-3}	3.8×10^{-3}	1.1×10^{-2}	—

The distribution profile for DDT is quite different from that of benzene. Only a small fraction of the DDT distributes into air. Most DDT partitions into the soil and sediment, since these phases are rich in organic material that has an affinity for DDT. The lipid-rich fish contain the highest concentrations, but still contain the smallest DDT mass due to the relatively tiny volume of this biotic phase.

A caution in these calculations: The amount of organic substance chosen as the total should be within the capacity of the model environment. For example, water solubilities or saturation vapor pressures should not be exceeded or a new phase would have to be introduced — the pure compound. In the DDT example, the estimated fugacity is one thousandth the saturation vapor pressure, and the water concentration about 10% of the maximum solubility. But with about a 0.04% body burden, any real fish would probably have long since succumbed to DDT poisoning.

Finally, a reminder that a simple, static view of chemical distribution has been illustrated: Other factors may influence a pollutant's fate. For example, environmental phases such as air and water move about, carrying pollutants with them. The pollutant discharge may be continuous, and the chemical can decompose. These factors and others need to be covered when distributions in specific environments are considered.

15.7 KEY POINTS

1. Predictions of possible distribution patterns of chemicals in the environment are needed to assess exposure in evaluating potential effects of contaminants on human health and natural ecosystems.
2. The first step in evaluating distribution is to classify the environment into phases; the following phases are commonly used: air, water, soil, suspended solids, sediments, and aquatic biota.
3. The distribution patterns in the environment can be seen as a set of two-phase equilibria, such as air–water, air–soil, biota–water, and so on. These processes are described by the Freundlich equation:

$$C_1 = K(C_2)^n$$

or

$$\log C_1 = \log K + n \log C_2$$

4. The Henry's law constant, H, is an important environmental characteristic defined as

$$H = \frac{\text{partial pressure of compound in air}}{\text{concentration in water}} \quad \text{(at equilibrium)}$$

5. The *n*-octanol–water partition coefficient, K_{OW}, is a valuable characteristic of organic compounds that can be measured in the laboratory; thus,

$$K_{OW} = \frac{\text{concentration in octanol}}{\text{concentration in water}} \quad \text{(at equilibrium)}$$

6. The bioconcentration factor for aquatic organisms, K_B, is the ratio between the chemical concentrations in the biota and water (C_B/C_W) at equilibrium. The solid sorption factor is the ratio between the concentration in the solid phase (soil, sediments, suspended sediments) and water at equilibrium; thus,

$$K_B = \frac{\text{concentration in biota}}{\text{concentration in water}}$$

$$K_{sorb} = \frac{\text{concentration in solid}}{\text{concentration in water}}$$

7. The K_B and K_{sorb} values including the K_D value for soils can be calculated from the K_{OW} value using the following relationships and assuming the power coefficients, a and b, are unity:

$$K_B = f_{lipid} K_{OW}$$

$$K_D = 0.66.f_{OM}.K_{OW}$$

8. The fugacity approach is needed to describe sets to two-phase equilibria and calculate distributions of chemicals. Fugacity (f), the escaping tendency, is measured in units of pressure. It is related to concentration, at concentrations usually encountered in the environment, and is defined as

$$f = C/Z$$

9. The fugacity capacity constants, Z values, can be calculated from a set of expressions for each phase in the environment using partition coefficients. Thus,

Z_{air} $= 1/RT$

Z_{water} $= 1/H$

Z_{soil} $= K_{sorb}(soil)/H$ or K_D/H

$Z_{sediment, suspended solids}$ $= K_{sorb}(sediment, suspended solids)/H$

Z_{biota} $= K_B/H$

10. Calculations can be made of distributions in model environments using the volumes of the phases and the calculated Z values with the following expressions:

$$f = M_{total}/\Sigma Z_i V_i$$

$$M_i = f Z_i V_i$$

$$C_i = M_i/V_i$$

11. Calculations made using the approach outlined above provide a simple static view of the distribution of a chemical. Other factors, such as movement of phases and degradation of the chemical, are needed to provide a more accurate assessment.

REFERENCES

Mackay, D., *Multimedia Environmental Models*, Lewis Publishers, Chelsea, MI, 1991.

QUESTIONS

1. If soil is analyzed indicating a benzene concentration of 0.003 mol/m^3 (about 1 ppm), what is the equilibrium concentration of benzene in the air pore spaces in the soil?
2. Given that naphthalene has the following characteristics at 25°C,

K_{sorb} (soil, 2% organic carbon)	18
K_{sorb} (sediment, suspended solids, 4% organic carbon)	36
K_B (fish, 5% lipid)	112

calculate the missing Z values for naphthalene,

Z_{air} $= 4.04 \times 10^{-4}$ mol/m^3·Pa
Z_{water} =
Z_{soil} =
Z_{fish} =
$Z_{suspended solids}$ = 0.86 mol/m^3·Pa
$Z_{sediment}$ =

3. Given that 100 mol naphthalene is distributed in the model environment shown in Figure 15.5, work out the mass and concentrations in each phase, filling in the empty columns in the table below.

Naphthalene, 100 mol at 25°C

Phase	Air	Water	Soil	Suspended Solids	Sediment	Fish	Total
Volume, m^3	10^{10}	7×10^6	9×10^3	35	2.1×10^4	3.5	—
Z mol/m^3·Pa	4.04×10^{-4}			0.86			—
ZV mol/Pa	4.04×10^6	1.67×10^5		30.1	1.81×10^4		
Fugacity, $f = \dfrac{100}{\Sigma Z_i V_i}$ Pa							
Mass, mol (fZV)		3.95			0.43		100
Concentration mol/m^3 (M/V)		5.64×10^{-7}	1.02×10^{-5}				—

ANSWERS

1. Assume the soil and air have the same Z values as in Section 15.4. From

$$\frac{C_{air}}{C_{water}} = \frac{Z_{air}}{Z_{water}}$$

then

$$C_{air} = \frac{Z_{air}}{Z_{water}} C_{water}$$

So

$$C_{air} = \frac{4.04 \times 10^{-4}}{1.8 \times 10^{-3}} \times 0.003 = 0.00067 \text{ mol m}^{-3}$$

Thus,

$$C_{air} = 0.00067 \text{ mol m}^{-3}$$

2. The missing Z values for naphthalene can be calculated using the expressions in Section 15.4 as:

$Z_{air} = 4.04 \times 10^{-4}$ mol/m^3·Pa

$Z_{water} = 0.0239$ mol/m^3·Pa

$Z_{soil} = 0.43$ mol/m^3·Pa

$Z_{fish} = 2.67$ mol/m^3·Pa

$Z_{suspended\ solids} = 0.86$ mol/m^3·Pa

$Z_{sediment} = 0.86$ mol/m^3·Pa

3.

Naphthalene, 100 mol at 25°C

Phase	Air	Water	Soil	Suspended Solids	Sediment	Fish	Total
Volume, m^3	10^{10}	7×10^6	9×10^3	35	2.1×10^4	3.5	—
Z mol/m^3·Pa	4.04×10^{-4}	0.0239	0.43	0.86	0.86	2.667	—
ZV mol/Pa	4.04×10^6	1.67×10^5	3870	30.1	1.81×10^4	9.335	4.229×10^6

$$\text{Fugacity, } f = \frac{100}{\Sigma Z_i V_i} = 2.365 \times 10^{-5} \text{ Pa}$$

Phase	Air	Water	Soil	Suspended Solids	Sediment	Fish	Total
Mass, mol (fZV)	95.5	3.95	0.0915	7.12×10^{-4}	0.43	2.21×10^{-4}	100
Concentration mol/m^3 (M/V)	9.55×10^{-9}	5.64×10^{-7}	1.02×10^{-5}	2.03×10^{-5}	2.05×10^{-5}	6.31×10^{-5}	—

16 Genotoxicity: The Action of Environmental Chemicals on Genetic Material

16.1 INTRODUCTION

The linking of particular cancers to chemicals first occurred in the 1760s and 1770s. In 1761, John Hill hypothesized that the habit of sniffing snuff caused nasal cancer. Shortly after, in 1775, Percival Potts noted that many of his patients with scrotal cancer were chimney sweeps. Another important discovery was made by a German physician, Rehn, in 1895, who noted an increased incidence of urinary bladder cancer among workers in the dye industry. All of these discoveries were later verified using laboratory animals. Some of the important cancer discoveries are presented in Table 16.1.

The first suggested link between deformities of children and their development was in 1651 by William Harvey. He suggested that hare lips developed by the cessation of growth of the lip during development of the fetus. He based this hypothesis on the fact that a hare lip-like condition normally occurs during development but disappears. The term *teratology* was coined in 1832 by Eteinne Geoffrey

TABLE 16.1
Key Events in the History of Cancer Research

Discoverer	Test Species	Affected Tissue	Carcinogen
Pott (1775)	Human	Scrotum	Soot
Rehn (1895)	Human	Bladder	Dye intermediates
Van Trieben (1902)	Human	Skin	X-rays
Müller (1939)	Human	Lung	Tobacco smoking
Molesworth (1937)	Human	Skin	Sunlight
Wagner (1960)	Human	Lung	Asbestos
Yamagiwa and Itchikawa (1915)	Rabbit	Ear skin	Coal tar
Findley (1928)	Mouse	Skin	UV light
Edwards (1941)	C3H mouse	Liver	Carbon tetrachloride
Innes (1969)	Mouse	Liver	DDT
Arnold (1979)	Rat	Bladder	Saccharin

St. Hilaire and his son. It was derived from the Greek word *teras* and literally means the study of monsters. Until the 1940s, it was a commonly held view that the placenta protected the developing fetus from most, if not all, chemical and physical threats. This was dispelled in 1941 when Gregg showed that pregnant mothers exposed to rubella (German measles) led to deformed children. The independent discovery of the teratogenic properties of thalidomide in 1961, by McBride and Lenz, made the world realize the sensitivity in pregnancy to chemical exposure. It has only been since the 1950s that it has been widely accepted that genotoxic effects could be prevented if genotoxins could be identified and exposure prevented. This idea is now central to all efforts in genotoxicity research and prevention.

Strictly speaking, genotoxicity is the study of the adverse effects of compounds on the genetic material of cells (DNA) and the subsequent expression of these changes. As such, it deals only with mutagenesis (the formation of inheritable mutations), carcinogenesis (the formation of cancers), and some forms of teratogenesis that involve damage to the DNA. However, in this chapter we have also included teratogenic deformities that are not caused by interaction with DNA.

All the forms of genotoxicity that are discussed in this chapter occur naturally. They occurred even when humans existed in the most natural state, as there are numerous natural genotoxins (Ames et al., 1987). Many people have a couple of misconceptions regarding genotoxicity: (1) only substances made by humans could be genotixic, and (2) natural products are inherently safe, nontoxic, and neither carcinogenic, mutagenic, or teratogenic. This has absolutely no basis in fact; there are numerous chemicals isolated from nature that are genotoxic, e.g., radon gas (Ames et al., 1990; Marshall, 1990).

Almost all knowledge of and research into genotoxicity is concerned with the effects on humans rather than the potential effects on animals and the environment. The animals used to determine the genotoxicity of chemicals are simply a means of obtaining the necessary data, as human experimentation is not permitted. Our knowledge of the effect on animals is simply a by-product of this quest; it is not applied or used to protect the animals or the environment.

The importance of genotoxicity is evident from the fact that cancer is one of the three main causes of death in most countries. The others main causes are heart attack and stroke. Cancer kills approximately one in every four citizens of the developed countries. Cancer has caught the public imagination and is the form of genotoxicity of most concern and on which we spend the most time and effort. Even though mutations and deformities of fetuses are much less publicized, it has been estimated that they cause approximately 40% of all infant deaths and a similar percentage of all spontaneous abortions.

16.2 THE GENETIC CODE

Deoxyribonucleic acid (DNA) is the genetic material that codes for all characteristics of life, from the synthesis of proteins and enzymes to the color of the eye. DNA generally consists of two long strands of nucleotides that spiral around each other, forming a helix (Figure 16.1).

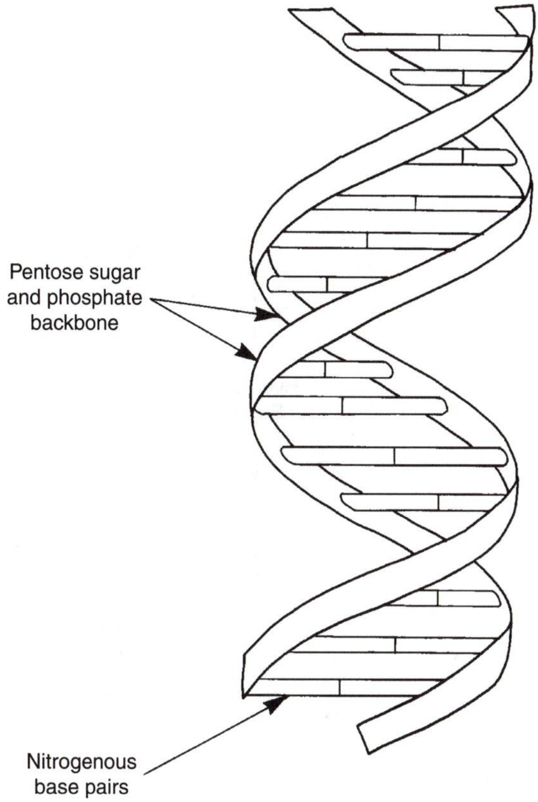

FIGURE 16.1 Representation of DNA showing the typical two-stranded form: the strands consist of pentose sugars and phosphate ions, which are connected by the nitrogenous bases.

Single-stranded forms of DNA do occur during DNA replication. A nucleotide consists of a pentose (five-carbon) sugar, a phosphate group, and a nitrogenous base. The nitrogenous bases are of two types: pyrimidines and purines. The pyrimidines consist of cytosine (C) and thymine (T), while the purines are adenine (A) and guanine (G) (Figure 16.2). The two strands are held together by hydrogen bonds between the bases (Figure 16.2). The nitrogenous bases are always paired. For example, a cytosine on one strand will always be paired with a guanine on the other strand. Likewise, an adenine on one strand will always be paired with a thymine on the other strand. These matched pairs of bases are called DNA base pairs. There are only two different DNA base pairs: AT and GC.

DNA occurs in either a single-stranded or double-stranded form at various times in the life cycle of a cell. During DNA replication the strands are separated (Figure 16.3a and b), and from each strand a new double-stranded DNA is synthesized. The second strand is synthesized by enzymes placing and binding into position a nucleotide containing the appropriate base to maintain the base pairs (Figure 16.3c). Thus, a nucleotide containing a guanine will always be placed opposite a nucleotide

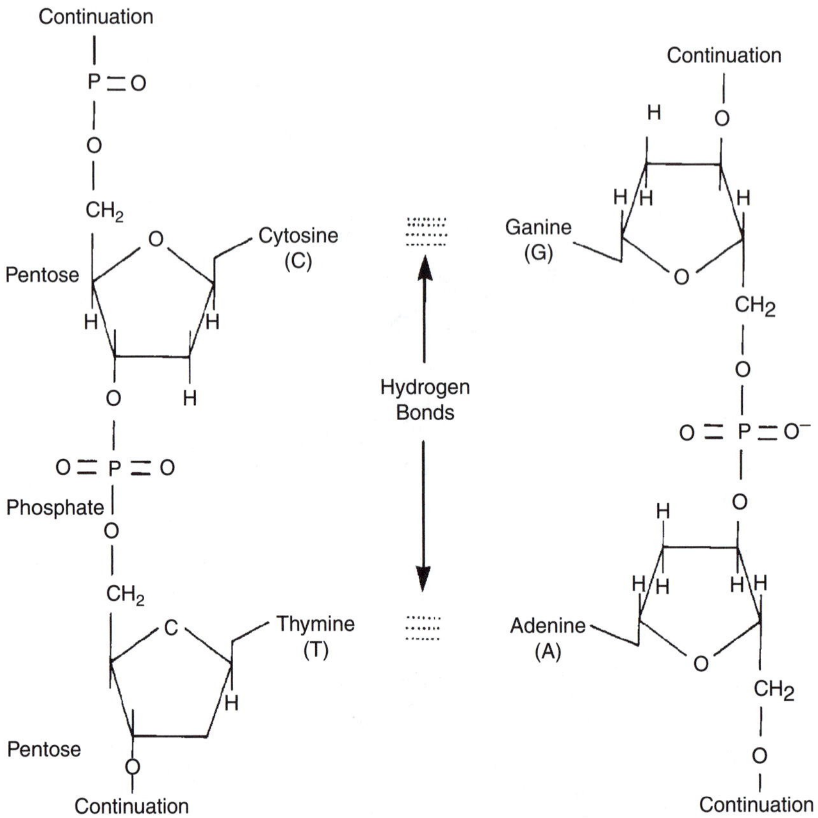

FIGURE 16.2 The chemical forms of the components of DNA and bonds between the pentose sugars and phosphate ions and between the nitrogenous base pairs.

containing cytosine in the existing DNA strand. Likewise, a thymine will always cause an adenosine to be placed in the other strand or vice versa.

The other function of DNA, besides simply replicating itself, is the synthesis of proteins. This process is called **transcription**, as the sequence of bases on the DNA is transcribed to a series of nucleotides in ribonucleic acid (RNA). The nucleotides of RNA are thymine (T), adenine (A), guanine (G), cytosine (C), and uracil (U). RNA is formed by a portion of a double-stranded form of DNA unravelling and the strands separating, as in DNA replication (Figure 16.3). Enzymes then move between the strands and use the sequence of bases present on one strand as the template for the binding of nucleotides to form RNA. The RNA is synthesized using the same base pairs used to synthesize DNA (Figure 16.4a), except that a nucleotide containing adenine on the DNA means a nucleotide containing uracil will be added to the RNA strand (Figure 16.4b).

The nucleotides in RNA are in turn converted to proteins in a process termed **translation**. Every DNA base codes for a specific RNA base, and every set of three RNA bases codes for one specific amino acid. However, not all the sets of three bases code for different amino acids; for instance, UUA, UUG, CUU, CUC, CUA,

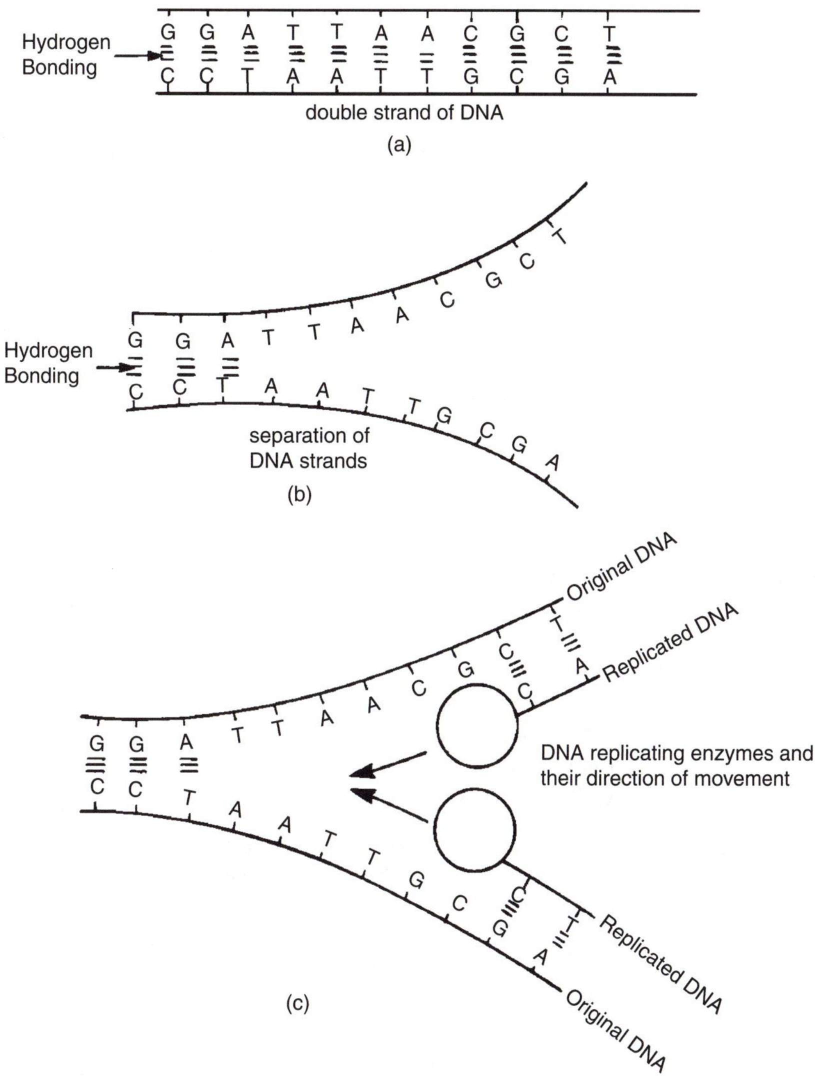

FIGURE 16.3 The process of DNA replication. The normal double-stranded form of DNA (a) has the two strands separated (b), and then enzymes commence using the original DNA strands as the template to synthesize the new replicated strand of DNA (c).

and CUG all code for the amino acid leucine. These groups of three bases, regardless of whether they are from DNA or RNA, are called codons, for obvious reasons. Some of the codons and the amino acids they code for are illustrated in Table 16.2.

Proteins are synthesized by the ribosome, a large complex biomolecule consisting of approximately 60 proteins and many enzymes. The ribosomes bind to the RNA at points where an AUG codon occurs. This codon is the initiation code for protein synthesis (Table 16.2). The ribosome then moves along the RNA and adds an amino acid for each codon it encounters onto the other amino acids already

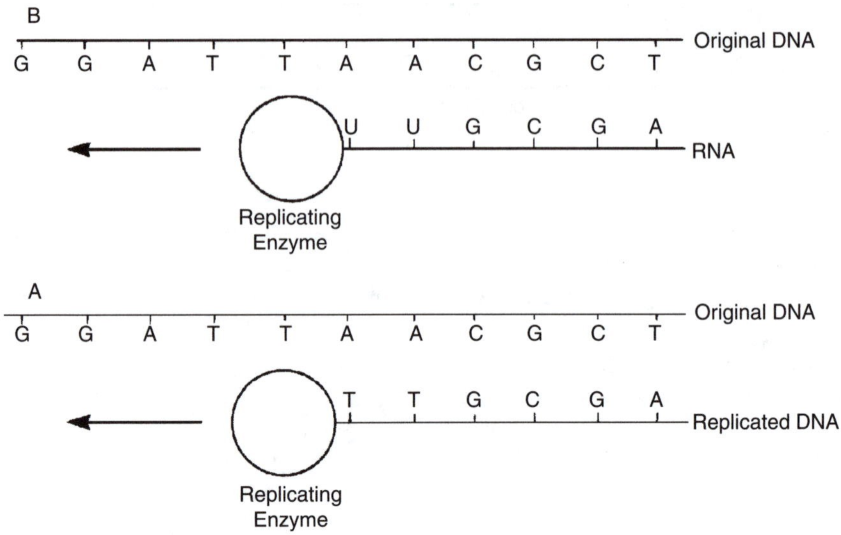

FIGURE 16.4 An example of the difference in the sequence of nitrogenous bases when DNA is replicated (a) and when RNA is synthesized (b). Both use the same DNA.

TABLE 16.2
Some Example Codons and the Amino Acids and Messages They Code For

Codon Composition	Amino Acid Coded For	Other Meanings
AUG	Methionine	Start
CUC	Leucine	
ACA	Threonine	
GAA	Glutamic acid	
CGU	Arginine	
UGU	Cysteine	
UAA, UAG, UGA	—	Stop
GGG	Glycine	

formed, thus forming a polypeptide. This process is analogous to the synthesis of RNA from DNA (Figure 16.4). The protein synthesis is terminated when the ribosome encounters the codon UAA.

Series of DNA bases linked together that code for one entire protein or enzyme are called genes. Generally, genes contain about 1000 base pairs. Groups of genes arranged in series form chromosomes. The number of chromosomes within all cells of an animal should be the same (except if a mutation has occurred), and they should be the same for all individuals of the same species. Humans normally have 23 chromosomes, which contain between 200,000 and 250,000 genes.

16.3 TERATOGENS AND TERATOGENESIS

16.3.1 Mechanisms of Action

Teratogens are chemicals that interfere with the normal reproduction process and cause either a reduction of successful births or offspring to be born with physical, mental, developmental, or behavioral defects. It is important to note the difference between teratogens and mutagens. Mutagens can cause changes in the DNA of cells regardless of cell type. Teratogens may cause defects by a variety of means, including DNA damage; however, they only affect somatic cells (all cells other than ovum and sperm). Deformities resulting from teratogen exposure cannot be passed on to further generations unless the offspring are similarly exposed.

Teratogens appear to have very specific mechanisms of action; however, they can be broadly classed as either **genetic** or **epigenetic**. Genetic teratogens exert their effects by gene mutation, chromosomal abnormality, or inhibiting mitosis by slowing DNA synthesis or preventing spindle formation (a vital step in mitosis). Epigenetic mechanisms of action include affecting the levels and forms of energy available, the supply of key metabolic substrates, and inhibition of enzymes, and modifying cellular membranes, thus altering membrane permeability.

The development of a fetus is not a smooth, even process. The gestation for humans is divided into three distinct periods: implantation (weeks 1 to 2), embryonic (weeks 3 to 7), and fetal (weeks 8 to 38) (Figure 16.5). During the implantation

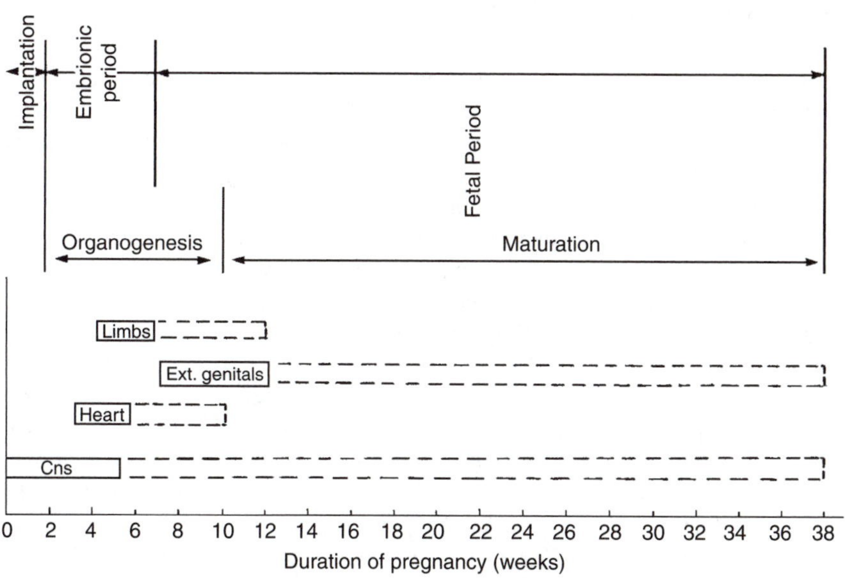

FIGURE 16.5 A time sequence of the development of some organs during human pregnancy. Solid bars indicate periods of greatest sensitivity to teratogens. Dashed bars indicate periods of reduced sensitivity.

period, fetal cells undergo rapid replication followed by differentiation into specialized cell types that will eventually develop into all the different types of tissue (e.g., heart muscle, liver, bone). Throughout the embryonic period and into the first 2 weeks of the fetal period, all the body's organs and limbs are formed and develop. This process is called **organogenesis**. Each organ or part of the body commences, developing at certain preset times during organogenesis. During the fetal period the organs and limbs become functionally mature. The time of completion of maturing is different for different organs/limbs. The particular organ or limb deformed by exposure to a teratogen depends on the period of gestation when the fetus was exposed (Figure 16.5).

When a particular organ or limb is forming and developing, the involved cells have greatly increased rates of DNA replication and cell division. If a genetic teratogen causes a mutation of the DNA of a cell in the early stages of organogenesis, then all cells derived from that will carry the same mutation. With the rapid cell division that occurs during organogenesis, a large number of cells could end up with the same mutation, the expression of which could cause deformity in the developing fetus.

One mechanism by which epigenetic teratogens cause deformities is to limit either energy or vital substrates necessary for the cell division — thus slowing down the development or even killing cells. The period in which organ and limb development can occur is very limited. Slowing cell metabolic rates means there may not be enough cell division to allow full development of the organ or limb before the period of organogenesis finishes. The resulting organ may be completely nonfunctional or partially functional, depending on how much the metabolism of the cells was affected. Cell death is a normal event during organogenesis that can be compensated for, and consequently, development of the organ or limb is unaffected. However, if too many cells are slowed down or killed, to be replaced by the normal mechanisms, insufficient cells may remain to allow the organ or limb to develop properly in the allotted period of organogenesis.

The earlier in the pregnancy the exposure to teratogens, the more severe the deformities for the same concentration. Exposure to teratogens generally leads to miscarriage if it occurs prior to organogenesis, which commences in week 3 of gestation. Exposure during organogenesis (weeks 3 to 9), however, leads to gross morphological deformities, and exposure during the fetal period (weeks 8 to 38) causes physiological and functional defects. However, miscarriage may occur during the latter two periods if exposure to teratogens is sufficient. A feature of many teratogens is that they can exert their effects very early in the pregnancy, in many cases before the woman knows she is pregnant. An example is ethanol, which has its greatest effect in the first 4 weeks of pregnancy.

By 1984, approximately 2000 chemicals had been tested for teratogenicity using test animals, of which 782 definitely caused teratogenicity and a further 291 were possibly teratogenic. However, only 30 of these chemicals are known to cause teratogenicity in humans. Examples of teratogenic chemicals include ethanol, methyl mercury, thalidomide, and chlorobiphenyls (Figure 16.6).

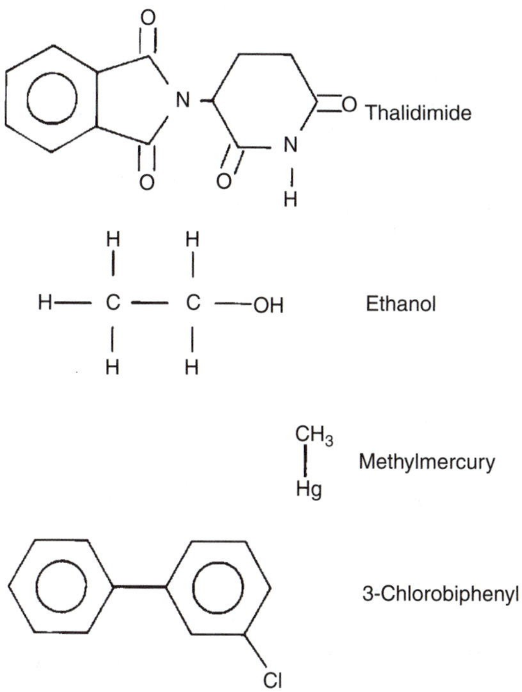

FIGURE 16.6 The chemical structures of some known teratogens.

16.3.2 Examples of Teratogens

16.3.2.1 Alcohol (Ethanol)

Continued or excessive exposure to alcohol can lead to the fetus developing fetal alcohol effect (FAE), progressing with increased exposure to fetal alcohol syndrome (FAS). Typical FAS symptoms include a range of facial deformities (small head, small eye openings that are widely separated, and a thin upper lip) and severe growth, developmental, and intellectual retardation. In many cases, it is difficult to identify patients with FAE or FAS without prior knowledge of the mother's drinking history, as similar symptoms can be caused by other factors.

16.3.2.2 Methyl Mercury

The discovery of the teratogenic properties of methyl mercury occurred during the investigation into the causes of Minimata disease: a case of the methyl mercury poisoning in Japan where many members of a community at Minimata were affected. Concurrent with the outbreak of Minimata disease was an outbreak of children born with quite severe mental retardation (approximately 6% of children born were affected). Typical symptoms included primitive oral and grasping reflexes, poor coordination, salivation, character disorders (unfriendly, shy, nervous, and restless), seizures and epilepsy, deformed limbs, and slow growth. Right from the onset it was felt, by some involved with the situation, that these children were also suffering from Minimata disease. However, it was several years later when autopsies could be performed on two dead infants that methyl mercury was proven to be the causative agent.

In 1974, it was discovered that women who gave birth to congenitally poisoned children had exhibited none of the early symptoms of methyl mercury poisoning. This was due to the ability of methyl mercury to easily cross the placenta and concentrate in the developing fetus. Concentrations in fetal brains were up to four times those in the mother, and fetal blood levels were 28% greater than the mothers'.

16.3.2.3 Rubella (German Measles)

Rubella was one of the first agents that was found to be teratogenic. This was discovered by Gregg in 1941 following a rubella epidemic in Austria. The deformities and outcome depend on when the mother was exposed. Generally, it leads to eye, heart, and ear defects and mental retardation. In an attempt to minimize the occurrence of such deformities, it is often recommended that sexually mature women be immunized against rubella.

16.3.2.4 Thalidomide

Possibly the most notorious teratogen is **thalidomide** (Figure 16.6), which was developed as a sedative/tranquilizer. By accident it was discovered that it was also a very powerful suppressor of morning sickness associated with pregnancy. The characteristic symptoms of thalidomide exposure are deformities of the limbs, predominantly the arms. These findings very promptly led to the total removal of the drug in developed countries; however, it remained on sale in many countries for a number of years. Estimates of the number of children affected range from 5000 to greater than 10,000.

From retrospective studies, it appears that exposure of the fetus between weeks 6 and 7 of the pregnancy led to the characteristic deformities. It appears that thalidomide exerts its teratogenic effect between weeks 6 and 7 by killing cells that were developing into limbs. The exact mechanism has not been resolved.

Many have claimed that if thalidomide was tested for teratogenic properties, that this episode would never have happened. This is not necessarily so since the teratogenic effects of thalidomide vary greatly with the test species. For instance, rats and mice exhibit very little effect even at concentrations 4000 to 8000 times those that affect humans. Only certain strains of rabbit and some species of primates have similar susceptibilities to thalidomide as humans. This problem of extrapolating the findings of tests from the test animals to humans is a problem in the use of data on mutagenesis and carcinogenesis tests.

16.4 MUTAGENS AND MUTAGENESIS

16.4.1 Types of Mutations

A mutagen is any chemical that causes damage to the DNA or inhibits or damages the DNA repair mechanisms. Whether the mutation this causes is inheritable depends on the type of cells that suffer DNA damage. Damage to DNA in cells involved in the production of ova or sperm (germinal cells) causes inheritable mutations, whereas damage to DNA from all other cells (somatic cells) is not inheritable. Mutagenesis is not solely due to exposure to synthetic organic chemicals. It is a natural process, a part of everyday life. In fact, mutation is the source of variation and change evoked to explain Darwin's theory of natural selection and evolution. Examples of natural mutagens include ultraviolet light, radiation from the decay of naturally occurring radioactive materials, alkaloids and flavonoids from plants, and mycotoxins from fungi. It has been estimated that every day several thousand mutations in the sequence of DNA bases occur in each mammalian cell.

Mutations can be beneficial, harmful, or have a neutral effect. However, with complex organisms that have evolved over long periods, most will not be beneficial. An interesting point is that through changes in our culture, we are now converting previously harmful mutations into neutral ones. For example, asthma and shortsightedness would have been harmful in hunter–gatherer societies, but are now no longer harmful in an evolutionary sense; i.e., these mutations do not reduce the ability of an individual to have children and thus perpetuate their genes.

Mutations of DNA can be classified into three main types: (1) where individual bases of DNA have been substituted, added, or removed (point mutations); (2) there is large-scale damage to chromosomes (clastogenesis); and (3) where an uneven distribution of DNA (aneuploidization) occurs, so that resulting cells have either too many or too few chromosomes.

16.4.1.1 Point Mutations

A substitution point mutation is when the mutation leads to the removal of a DNA base that is replaced (substituted) by another. There are three types of substitution point mutations: transition, translation, and frame shift. Where a DNA base pair is replaced by the other base pair (remember there are only two DNA base pairs), the purines and pyrimidine bases remain on the same strand of DNA (e.g., AT is replaced by GC); this is termed a **transition** (Figure 16.7b). Another form of point mutation of DNA bases is when a base pair is replaced by the other base pair, but the purine and pyrimidine bases swap DNA strands (e.g., AT is replaced by CG) (Figure 16.7c). This is termed **translation**, as the bases have been translated from one DNA strand to the other. The third form of point mutation is called **frame shift**. This is caused by the addition or deletion of one or more bases to the DNA. This may totally destroy the original message coded in the DNA, as the codons have all changed. The new sequence of nucleotides may code for different amino acids, be a nonsense code that corresponds to no amino acid, or code for the termination of protein synthesis, leading to the formation of incomplete proteins. The largest piece of DNA a point mutation can affect is a gene.

16.4.1.2 Mutations

Mutagens that cause chromosomal mutations can cause much more DNA damage than point mutations. The principal process causing such mutations is the linking

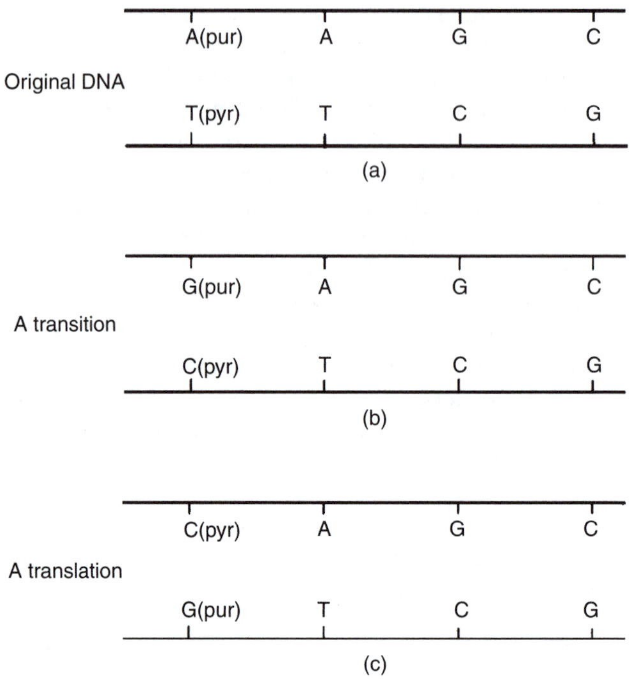

FIGURE 16.7 The types of point mutations of DNA. Transition b and transition c are illustrated by comparison with the original piece of DNA (a). Pur and Pyr indicate a purine base and pyrimidine base, respectively.

of DNA strands that belong to different chromosomes. Such a process is called **cross-linking**. During cell replication (mitosis), the two strands of DNA in each chromosome are separated. The separation of cross-linked chromosomes causes breakages in the DNA strands, which may lead to rearrangements, deletions, and additions of DNA. This is one means by which alkylating agents such as epoxides, aldehydes, alkane halides, alkyl sulfonates, nitorsoureas, and triazines (Figure 16.8) exert mutagenic effects.

16.4.2 Expression of Mutations and DNA Repair

The fact that a mutation has occurred does not mean it will be expressed as a biological effect. The mutation may be so minor as to not really have any effect. For instance, the changing of a single DNA base may not lead to any change in the message that is encoded in the DNA. Even if the mutation leads to a different amino acid being incorporated in an enzyme, to have an effect, the changed amino acid must be in the specific region that forms the receptor site (the site where the reactants bind to the enzyme and are converted to product).

Another reason that mutations may not be expressed is the presence of DNA repair mechanisms (predominantly based on enzymes). It is important that the repair of DNA mutations occurs before the next replication of the DNA or immediately

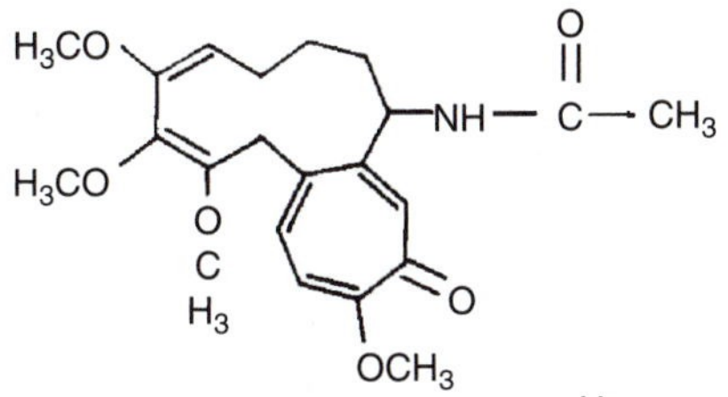

Colchicine

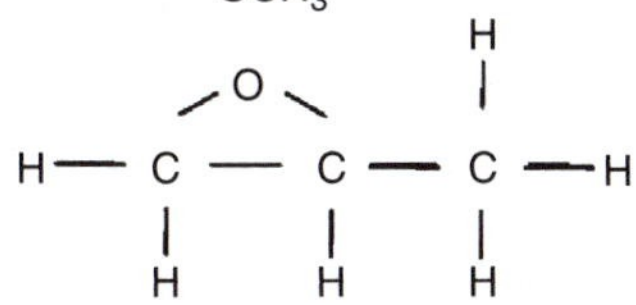

Ethylene oxide
(1,2 epoxyethane)

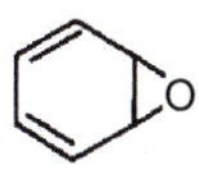

Benzene-1,2-oxide

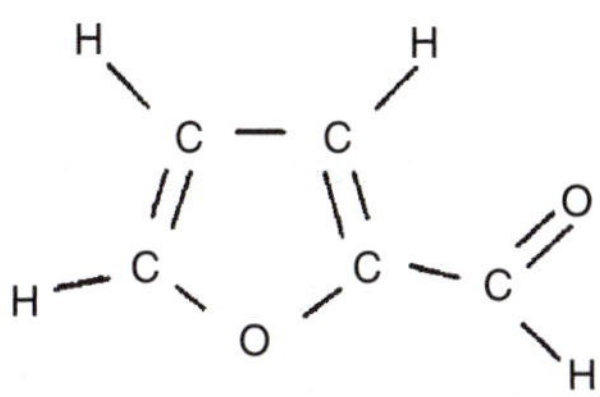

Furfural (aldehyde)

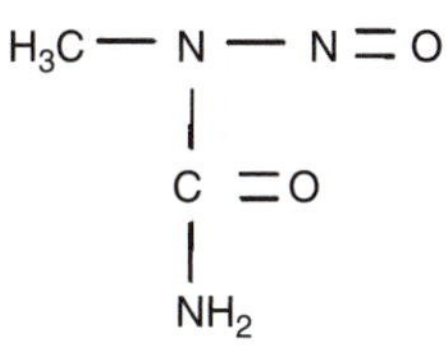

N-methyl-N-nitrosourea
(NMO)

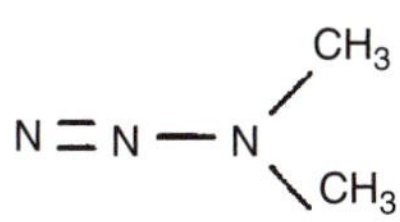

3-3-dimethyl-1-phenyl
triazene

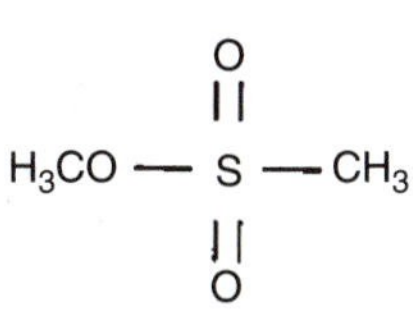

Methylmethane sulfonate

FIGURE 16.8 Chemical structures of mutagens known to cause chromosomal mutations by cross-linking chromosomes or by changing the number of chromosomes present in cells.

afterwards; otherwise, the mutation may become permanent. Repair enzymes either directly correct DNA damage (e.g., photoreactivation) or remove DNA surrounding the mutation and then use the sister strand of DNA to reconstruct the correct sequence of bases.

Photoreactivation is a repair mechanism that uses repair enzymes to correct DNA damage and remedy the most common mutation due to UV exposure: the formation of thymine dimers. A thymine dimer is two adjacent thymine bases on the same strand of DNA that are bound together by covalent bonds. Normally, there are no bonds between adjacent DNA bases (Figure 16.2). A range of enzymes recognize these dimers and break the covalent bonds, repairing the mutation.

16.5 CARCINOGENS AND CARCINOGENESIS

16.5.1 Types of Carcinogens

Carcinogens are compounds that lead to the formation of tumors, a tumor being a cluster of cells, all derived from one cell, that are undergoing continual growth in terms of both size and number of cells. Many chemicals have now been found to be carcinogenic or classed as suspected carcinogens to animals. These have been discovered in a number of ways, for example, during routine testing that is done before a chemical or new product can be released onto the market or during postrelease tests that are conducted due to evidence of carcinogenic action from observation and epidemiological studies.

Carcinogens can be subdivided into two main types based on the mechanism of action, which is genotoxic or epigenetic. Within these two categories there are several subdivisions. Genotoxic carcinogens irreversibly alter the DNA, leading to inheritable changes. Epigenetic carcinogens exert their cancer-forming effect by all means other than direct interaction with DNA. If a chemical has both genotoxic and epigenetic properties, it is assigned to the genotoxic category. If insufficient information on the mechanism of action is known, the chemical is simply called a carcinogen.

16.5.1.1 Genotoxic Carcinogens

Genotoxic carcinogens are mainly **electrophilic** reactants; i.e., they are electron deficient. In fact, most of the organic carcinogens are electrophilic genotoxic carcinogens. They include components of soot, coal tar, and tobacco smoke, such as polycyclic aromatic hydrocarbons (PAHs) and quinolines. Such compounds are highly reactive and form covalent bonds with nucleophilic compounds (compounds that tend to donate electrons in reactions) such as DNA and many other cellular compounds. When electrophilic genotoxic carcinogens react with DNA, they form adducts, where the chemical is either reversibly covalently bound or irreversibly bound. The formation of DNA adducts leads to damage to DNA in the form of mispairings of bases and breaks of the DNA damage. A reasonable amount is known about how DNA adducts are formed and the various types, yet little is known of how or even if these lead to cancer. One group of chemicals that readily reacts with

DNA to form adducts is the alkylating compounds in Figure 16.8. These chemicals form carbonium ions (CH_3^+), which can methylate the bases and phosphate groups of DNA.

Within the genotoxic group of carcinogens there is a further subdivision, which is based on whether they are innately carcinogenic and do not require biotransformation to induce cancers (activation independent) or they will not induce cancer unless they have been biotransformed (activation dependent). Most environmental carcinogens are activation dependent. Examples of activation-dependent carcinogens include benzene, PAHs, arylamine, and nitrosamine (Figure 16.9a). The activation-dependent process is analogous to the bioactivation of toxicants. The original non-carcinogenic compound is generally stable in the environment and is termed a **precarcinogen**. Metabolites that have higher carcinogenic activities are termed **proximate carcinogens**, while the final carcinogenic compound is the **ultimate carcinogen**. The ultimate carcinogen is generally an electrophile. Some examples of carcinogenic electrophiles are presented in Figure 16.10. In the case of benzene, the ultimate carcinogens have not yet been identified.

Activation-independent carcinogens generally have very short half-lives in the environment due to their high chemical reactivity. Active halogenated compounds, alkaline epoxides, and sulfate esters are examples of activation-independent carcinogens (Figure 16.9b). The extent of carcinogenic effect of activation-independent carcinogens depends on the proportion of the chemical interacting with DNA and that reacting with other non-DNA cellular material. The more that reacts with non-DNA cellular material, the lower the carcinogenic effect. The biotransformation of activation-independent carcinogens invariably leads to a reduction in genotoxic potency.

16.5.1.2 Epigenetic Carcinogens

As stated previously, epigenetic carcinogens exert the carcinogenic effect by any other means than reacting directly with the DNA. Mechanisms of action include:

1. Suppressing the immune system
2. Modifying hormone activity
3. Promoting the effects of other carcinogens (promoters)
4. Enhancing the effects of other carcinogens (cocarcinogens)

The concepts of cocarcinogens and promoters and how such compounds contribute to the development of cancer are important. Neither cocarcinogens nor promoters are carcinogenic when administered by themselves. A **cocarcinogen** is a compound that, when added either prior to or simultaneously with a carcinogen, leads to an increased carcinogenic effect (i.e., increased number of tumors or a decreased period before tumors appear). Known cocarcinogens include ethanol and components of tobacco smoke, such as catechol and phenolic compounds. A **promoter** is a compound that, when administered after a carcinogen, leads to an increased carcinogenic effect. If cells are exposed to promoters before carcinogens, there is no enhancement effect. Known promoters include organochlorine pesticides,

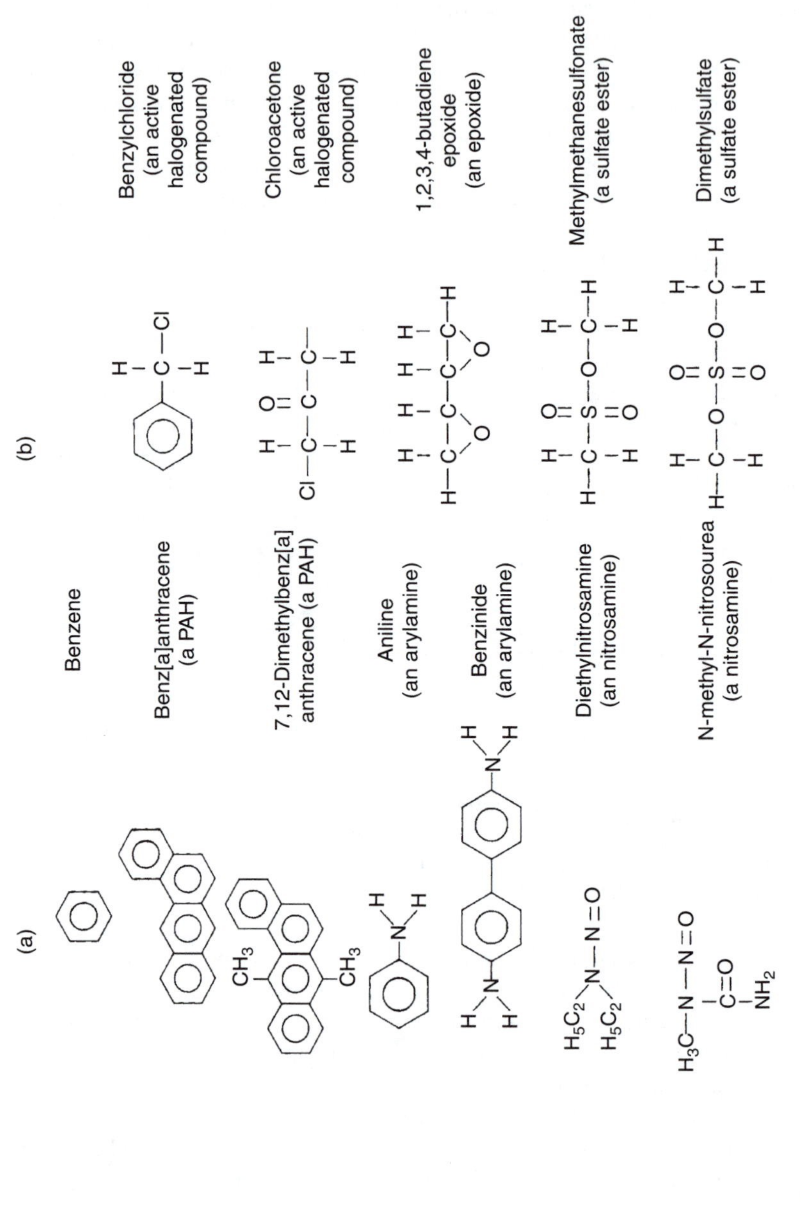

FIGURE 16.9 The chemical structures of examples of (a) activation-dependent genotoxic carcinogens and (b) activation-independent genotoxic carcinogens.

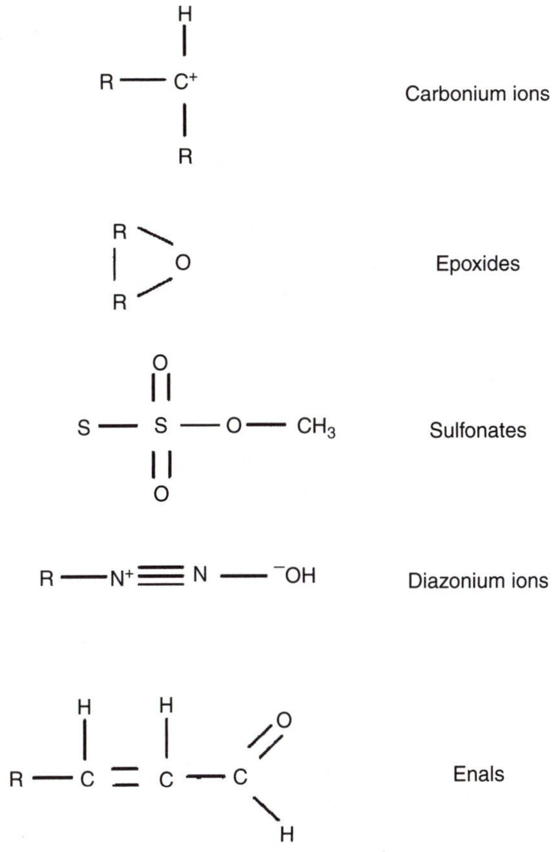

FIGURE 16.10 Examples of carcinogenic electrophiles.

saccharin, tobacco smoke, benzopyrene, and phorbol esters, particularly 12-0-tetradecanoylphorbol-13-acetate (TPA) and dietary fat. The role of promoters will be discussed in greater detail in Section 16.5.3.

16.5.2 Tumors

Tumors are a mass of cells that are all derived from one original cell and are growing in a manner not controlled by the body. As part of the cancer development, the normal cell regulatory processes become inoperative.

There are two principal types of tumors: benign and malignant. Benign tumors are slow growing, do not invade the surrounding tissue, and do not release cancer cells that are transported throughout the body and may form secondary cancers. This process of forming secondary cancers, described as **metastases**, is called **metastasis**. Benign tumors can be surgically removed and should not recur if all the tumor is removed. Unfortunately, benign tumors can, by as yet unknown mechanisms, be converted to malignant tumors.

Malignant tumors are the more dangerous type of tumor, as they grow rapidly, invade surrounding tissue, and form metastases. Malignant tumors are dangerous as even with the removal of an identified tumor, other secondary tumors may have already formed elsewhere, necessitating more operations and making total removal very difficult.

In general, tumors form wherever tissue has been exposed to carcinogens and promoters. Thus, common sites of cancers are the major sites of toxicant uptake (skin, the respiratory system, and the gastrointestinal tract) and the major organs of biotransformation (liver, lungs, and kidneys). Many carcinogens appear to almost exclusively cause cancers in one or two organs. This is at least partially due to the fact that generally enzymes can only react with a limited number of chemicals (precarcinogens), with different enzymes reacting with different chemicals. Other reasons include the uneven distribution of enzymes that activate precarcinogens, and that the number of enzymes and their rate of metabolic activity vary dramatically from organ to organ. For example, ethanol often causes cancer of the liver (called cirrhosis of the liver), as it contains high concentrations of enzymes that use ethanol as the substrate (e.g., alcohol dehydrogenase) with high metabolic rates of activity.

16.5.3 Development of Cancer

The order of the main steps involved with carcinogenesis, as they are currently understood, is given in Figure 16.11. The first step is **initiation**, where cells are exposure to initiators. Initiators are either activation-independent genotoxic carcinogens, ultimate genotoxic carcinogens, or epigenetic carcinogens that cause some damage to the cell. Epigenetic carcinogens generally require higher concentrations and more prolonged exposure in order to initiate cancer.

The cellular damage at this stage is still reversible. If DNA is damaged or mutated or the cell damaged, it does not automatically mean that cancer will occur or the mutation will be expressed. In fact, the great majority are not expressed. This is because the body has DNA repair mechanisms, which have already been discussed in this chapter.

In order for the damage to be made permanent and cancer to develop, the initiated cell must be exposed to promoters for prolonged periods. This step in the development of cancer is termed **promotion**. It appears from research that once a cell is initiated, exposure to promoters, regardless of the length of the intervening period, will still enhance the carcinogenic effect. For cancer to develop, the exposure to promoters must be long. A cessation of exposure to promoters dramatically decreases the chance of cancer developing. This is clearly illustrated by the fact that the probability of contracting lung cancer rapidly returns to that for nonsmokers when smoking is ceased.

At some point after prolonged exposure to promoters, the carcinogen-induced changes are made irreversible and the cell expresses new physiological and immunological properties and its replication is no longer controlled by the normal mechanisms. The final stage is **progression**; it is characterized by the growth of the tumor and the conversion from a benign to a malignant tumor.

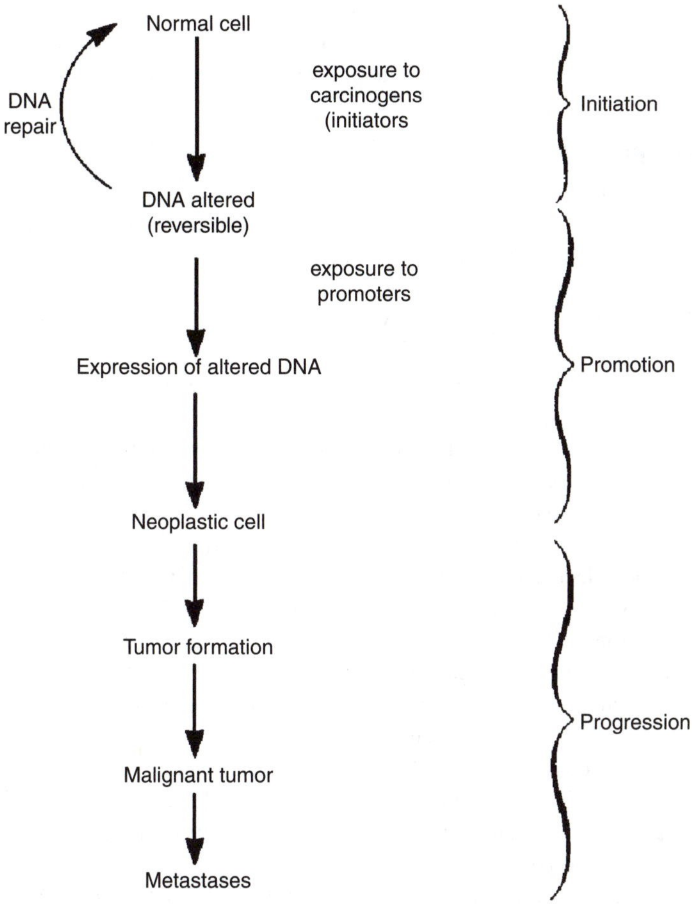

FIGURE 16.11 A flowchart of the steps involved and the stages in development of cancer.

16.5.4 Problems with Genotoxicity Tests

Testing is usually carried out on test animals, and there are problems in extrapolating from animals to humans. This is particularly the case with genotoxicity tests, where animals are used to derive safe levels of chemical exposure for humans. The most commonly used test organisms are mice and rats, which are relatively closely related. Yet the agreement of results from carcinogenic tests (i.e., carcinogenic or noncarcinogenic) for these two species is approximately 65%, and only in 50% of cases were the same organs affected. The genetic difference is much larger between humans and rodents than between mice and rats.

Even if one assumes that there are no problems with extrapolating from mice and rats to humans, there is still a significant problem. This is that the genotoxicity of chemicals is predominantly determined on individual substances, whereas humans, in fact all animals, are exposed to very complex mixtures of chemicals (Krewski and Thomas, 1992). Estimates of safe levels do not take into account the combined genotoxicity of chemicals. If synergistic interactions occur (the toxicity of the mixture is greater than the sum of the toxicities of the individual components), then the safe levels could quite significantly underestimate the effect.

16.6 KEY POINTS

1. Genotoxicity is the study of the adverse effects of substances on the genetic material of cells (DNA) and the subsequent expression of these changes as biological effects.
2. Genotoxic effects, particularly those expressed as cancer, are a major cause of death in many societies.
3. The genetic material that codes characteristics of life is DNA (deoxyribonucleic acid).
4. DNA consists of two long strands of nucleotides wound in a spiral form often described as a helix. The nucleotides consist of a pentose, a phosphate group, and a nitrogenous base.
5. DNA replicates by separating into single-nucleotide strands, on which enzymes place and bind the corresponding sequence of nitrogenous bases. Proteins are also synthesized utilizing DNA through transcription and translation.
6. Teratogens are chemicals that interfere with the normal reproduction process and cause a reduction of successful births or offspring to be born with physical, mental development, or behavioral defects.
7. The gestation of humans is divided into three distinct periods: implantation (weeks 1 to 2), embryonic (weeks 3 to 7), and fetal (weeks 8 to 38). The earlier in the pregnancy the exposure to teratogens occurs, the more severe the expressed deformities.
8. Examples of teratogens include alcohol (ethanol), methyl mercury, rubella (German measles), and thalidomide.
9. A mutagen is a chemical that causes damage to the DNA or inhibits or damages the DNA repair mechanisms. But mutagenesis is not solely due to synthetic organic chemicals; it is a natural process that occurs as a part of everyday life. Mutations can be beneficial, harmful, or have a neutral effect.
10. Mutations can occur as a result of the substitution, addition, or removal of bases of DNA, large-scale damage to chromosomes, and an uneven distribution of DNA in cells.
11. Occurrence of a mutation does not necessarily result in the expression of adverse biological effects. One reason for the lack of expression is the presence of repair mechanisms.
12. Carcinogens are compounds that lead to the formation of tumors, which are a cluster of cells, all derived from one cell, that are undergoing continual growth that is not controlled by normal body metabolic processes.

13. Genotoxic carcinogens are mainly electrophilic reactants, i.e., electron deficient, which react with DNA and other cellular components that are nucleophilic, i.e., electron donating. Examples of genotoxic carcinogens are soot, coal tar, and tobacco smoke, which contain polycyclic aromatic hydrocarbons (PAHs).
14. Most environmental carcinogens are activation dependent, which means they are converted into other compounds by biota to give the ultimate carcinogen.
15. Cancer develops through a number of steps: initiation, promotion, and progression.
16. Cocarcinogens and promoters enhance the action of a carcinogen while not being carcinogens themselves. A cocarcinogen is a compound that, when added either prior to or simultaneously with a carcinogen, leads to an increased carcinogenic effect. A promoter is a compound that, administered after a carcinogen, leads to an increased carcinogenic effect.

REFERENCES

Ames, B.N., Magaw, R., and Gold, L.S., Ranking possible carcinogenic hazards, *Science*, 236. 271–280, 1987.

Ames, B.N., Profet, M., and Gold, L.S., Nature's chemicals and synthetic chemicals: comparative toxicology, *Proceedings of the National Academy of Science of the United States of America*, 87, 7782–7786, 1990.

Marshall, E., Experts clash over cancer data, *Science*, 250, 900–902, 1990.

Krewski, D. and Thomas, R.D., Carcinogenic mixtures, *Risk Analysis*, 12, 105–113, 1992.

QUESTIONS

1. Pervival Pott in 1775 concluded that cancer of the scrotum was associated with soot. Most of his patients with this disease were chimney sweeps. Give a brief outline of the possible carcinogens and their mode of action.
2. Some observers have claimed that our society is facing an epidemic of cancer in humans as a result of the use and occurrence of residues in food, water, and the environment. List factors for and against this claim.

ANSWERS

1. **Possible carcinogens**: Common carcinogenic components in combustion products are polycyclic aromatic hydrocarbons (PAHs). These would be expected to occur in soot in chimneys.
   Mode of action: Soot, containing PAHs, would become entrapped in the chimney sweeps' clothing. The PAHs are lipophilic and would be expected to partition into the skin of the sweeps. These skin-associated PAHs could be activated by oxidative cellular processes forming the ultimate carcinogen. After a period, the tumor formation process would occur, leading to the development of scrotum cancer.
2. **For:**
   - There may be an increasing occurrence of synthetic chemicals in air, food, soil, and water over time.
   - A large proportion of the population dies from various types of cancer.
   - Many tests have indicated that synthetic chemicals are carcinogenic.

   Against:
   - It is not clear that exposure to synthetic organic chemicals and related compounds is increasing. The levels of pollution in air, water, food, and soil may be decreasing due to increased management.
   - The causative agents for the large proportion of the population affected by cancer are probably not environmental agents.
   - Tests are often carried out on very high concentrations of chemicals compared with environmental levels, and extrapolation from animals to humans is required, which introduces a range of uncertainty.

Part V

Management of Hazardous Chemicals

17 Monitoring of Chemicals in the Environment

17.1 INTRODUCTION

The growth in public concern regarding the occurrence and possible effects of chemicals in the environment is related to developments in methods for chemical analysis (Table 17.1). This effectively started in 1952 when James and Martin invented the gas–liquid chromatography (GLC) technique, which was later extended in sensitivity by the development of the flame ionization and electron capture detectors. Atomic absorption spectroscopy (AAS) was developed by Walsh in 1952, also opening the door to analysis of a wide range of environmental contaminants.

The detectors were designed to operate with the GLC allowing the detection of trace amounts (mg kg^{-1}) of organic chemicals, such as pesticides (e.g., DDT, dieldrin), polychlorinated biphenyl (PCBs), and petroleum hydrocarbons, in environmental samples. So, for the first time, trace contaminants could be reliably monitored in the environment. The invention of atomic absorption spectroscopy allowed analysis of the occurrence of trace levels of metals in environmental samples, further extending the analysis capability. These chemical analysis techniques have been further developed and extended to a wide range of techniques, now available for chemical analysis.

Environmental analysis should be seen in a framework that uses chemical monitoring to develop management activities to protect the environment from hazardous chemicals, as shown in Figure 17.1. It should be kept in mind that, usually, the

TABLE 17.1
Ananlytical Methods for Trace Environmental Contaminants and Environmental Management

Prior to 1952	No reliable methods and low level of concern on trace chemical contamination in the environment
1952	Gas–liquid chromatography (GLC) invented Atomic absorption spectroscopy (AAS) invented
1957/1958	Flame ionization and electron capture detectors for GLC invented
1962	Publication of *Silent Spring* by Rachel Carson
1967	Significant database on occurrence of DDT and other contaminants in the environment
1970	GLC was combined with mass spectrometry (MS) to give a new technique GC/MS which became commonly available and there was introduction of registration of chemicals and legislation to control hazardous chemicals

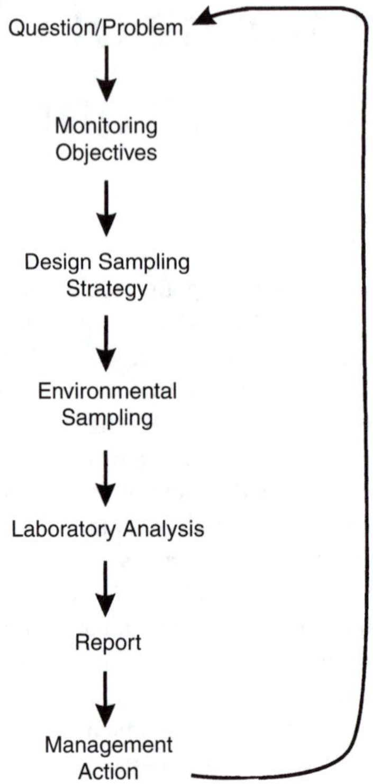

FIGURE 17.1 Outline strategy for monitoring chemicals in the environment.

principal objective is to monitor adverse biological effects in the environment resulting from the presence of contaminant chemicals. This could be achieved by using such techniques as monitoring human health changes, changes in species diversity in ecosystems, changes in the biochemical functions of individual organisms, and so on. However, although these techniques are very important and widely used, they can be difficult to apply in many situations, expensive, and often not sufficiently precise to allow their use in environmental management.

It would be an extremely attractive proposition to simply insert detection probes *in situ* into the sector of the environment under consideration and read off parameters of interest. This would eliminate the difficulties and expenses involved with sampling and analysis. At present, this can only be achieved with a few simple parameters, such as pH and conductivity, but a considerable amount of research is being carried out in this area and the use of *in situ* probes can be expected to increase in the future.

17.2 MONITORING AND SAMPLING THE ENVIRONMENT

The design of a monitoring program must be guided through a series of steps, as shown in Figure 17.2, by the specific problem or question that is being addressed (Keith, 1990).

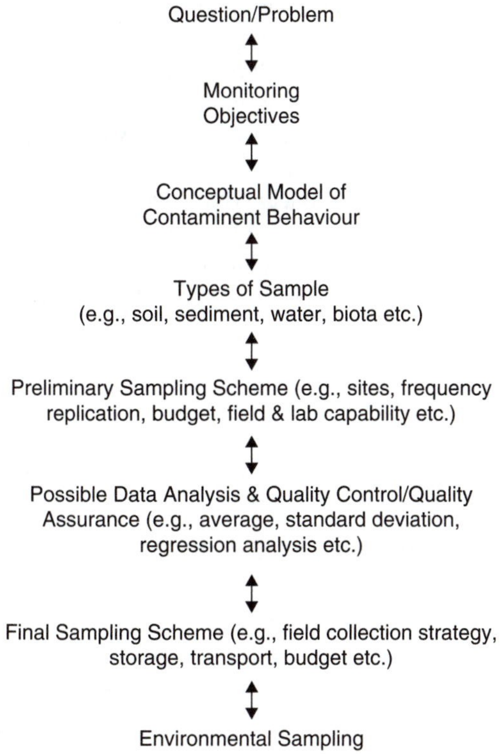

FIGURE 17.2 Steps involved in the design of a sampling strategy.

It could be such matters as:

- Compare levels of contaminants in water bodies, with guideline values.
- Compare levels of contaminants in one environmental system, e.g., a river or lake, with another.
- Benchmark particular environmental systems so that levels can be compared over time.
- Determine if a particular management action, e.g., control of a discharge, is effective.
- Exploratory investigation to obtain preliminary information on the occurrence of contaminants.

In a general sense, the basic objective is to accurately determine the concentration of a substance in a sample that is representative of the environment in order to monitor certain adverse effects of the substance in that environment. The concentrations that cause adverse effects have been previously investigated, using bioassays and related techniques, and evaluated, leading to the setting of guidelines, criteria, standards, and so on. It is assumed that adverse effects occur in proportion to the concentration of the chemical in accord with dose–response relationships. Incorrect sampling can generate samples for analysis that do not represent the environment under investigation. Even if the rest of the sampling and chemical analysis are strictly correct and accurate, the final results will be incorrect.

It is helpful, in these situations, to develop a conceptual model of the chemical in the environmental system under investigation. This provides an insight into the phases in the environment that contain the chemical and the dynamic processes that influence the occurrence of the chemical. Thus, the types and location of samples, e.g., water, soil, biota, etc., the timing and frequency of sampling, and the interpretation of results can be carried out in a more informed manner.

A sampling scheme can be designed using some fundamental strategies. These can be catagorized as:

- Random
- Systematic
- Judgmental
- Combined schemes

These are shown diagrammatically in Figure 17.3.

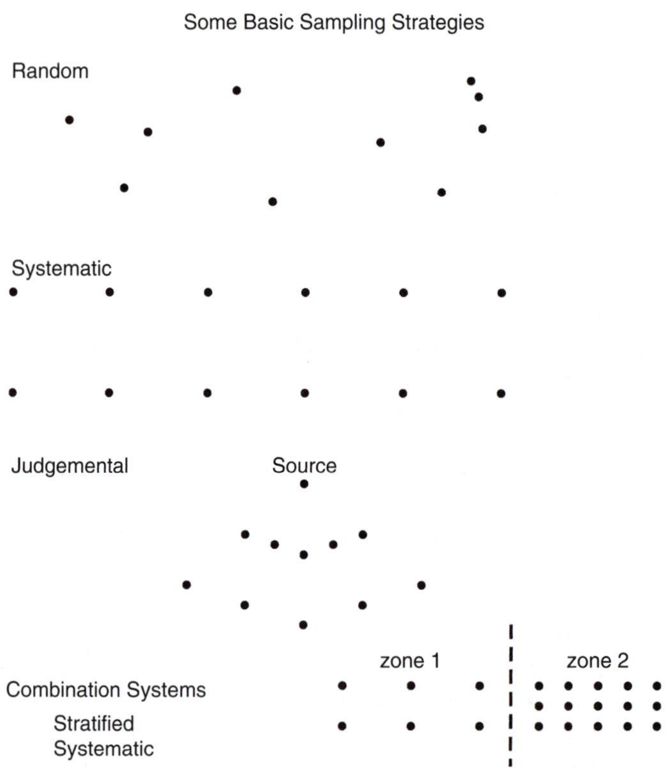

FIGURE 17.3 Some basic sampling strategies.

The **random** scheme has a random distribution of sampling points generated by a system designed for this purpose. The **systematic** scheme follows a fixed pattern of sampling, while the **judgmental** scheme has a specific pattern of sampling points related to a source of contaminant or discharge point. There can be combinations of these schemes for different sectors of an environmental system according to needs. For example, in Figure 17.3, a combination system, commonly described as a **stratified** scheme, is shown for two zones, with Zone 1 at a lower level of sampling intensity than Zone 2. Figure 17.4 shows how the different sampling schemes can give quite different sets of samples for the same environment. The sampling scheme that is appropriate is the one that provides samples in accord with the objectives.

Temporal factors need to be considered when the frequency of sampling is considered. The fluid environments of air and water usually need to be sampled relatively frequently, which could be daily, weekly, or seasonally, depending on the variability in the system. In some environments, random events may be important; for example, the flow of nutrients in rivers may be strongly influenced by storm events. More fixed-solid environments, such as soil and sediments, have relatively

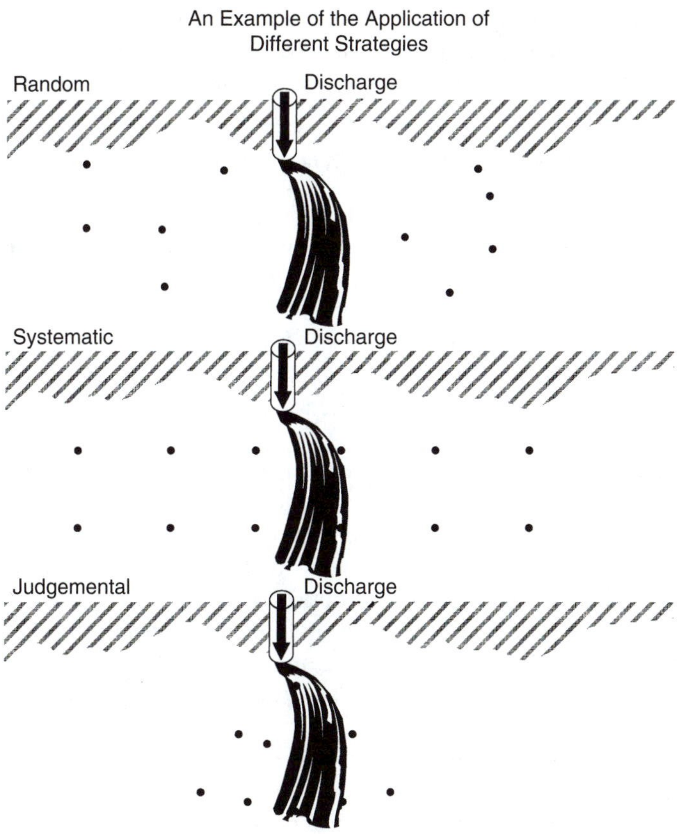

FIGURE 17.4 An example of the application of different strategies.

low temporal variability and so can be sampled on a much less frequent basis. On the other hand, these environments can have a high level of spatial variability, which can be accounted for in the spatial sampling scheme as outlined above. It can be difficult to decide the extent of the period during which sampling should be carried out. In principle, it should be continued until a representative sample is obtained or, more often, resources are depleted. Seasonal variations may need to be considered; thus, it could be argued that the period should be at least 3 years so that a minimal sample of three is obtained each season. With soil, which shows minimal variability, it could be suggested that a single sampling is sufficient. Each environment has to be considered individually and evaluated in terms of the objectives of the investigation.

The statistical analysis of the results produced by chemical analysis need to be considered at this stage also. Means, standard deviations, and regression analysis are some of the statistical techniques that could be used, and sufficient samples should be collected to allow the appropriate analysis. The quality assurance/quality control program needs to be considered as well. This depends on the specific nature and objectives of the program, but could involve blank samples to be chemically analyzed to reveal contamination due to the sampling equipment, sample containers, and so on. In addition, samples can be spiked in the field with a known concentration of the chemical of interest and samples can be collected from control sites of known low concentrations of chemical. This is covered in more detail in Section 17.4.

17.3 TECHNIQUES FOR CHEMICAL ANALYSIS

In the environmental monitoring procedure outlined in Figure 17.1, and in more detail in Figure 17.2, the environmental sampling step is followed by **laboratory analysis**. In this step, the samples are placed in appropriate containers, transported to the laboratory, and stored under conditions where deterioration does not occur; then analysis for the chemicals of interest is carried out by a range of different techniques (Lessley, 1997; Manahan, 2000). These techniques can be divided into three main categories: wet chemistry, spectrophotometry, and chromatography as shown in Figure 17.5.

Wet chemistry techniques are given this name since they are techniques used principally for the analysis of water and wastewater. They could be described as traditional techniques since they have been in use for a relatively long time and have well-developed and standardized procedures. Within this category are **titrimetric methods**, which are methods based on titrating, reacting one reagent with another, both in solution, and calculating the concentration of the chemical of interest from the volumes of the reactants. **Gravimetric analysis** is based on the isolation of a chemical reaction product, weighing it, and calculating the concentration of the chemical of interest from the weight.

Spectrophotometric techniques applied in the wet chemistry category are based on the principle of reaction of a chemical of interest in solution in water and wastewater, with a colorimetric reagent. The reagent gives a color in proportion to the amount of chemical of interest. The intensity of the color (the absorbance) is measured in the ultraviolet-visible range of the spectrum using a spectrophotometer. This is compared with a set of standards with color developed with known

Wet Chemistry	Spectrophotometry	Chromatography
• Titrimetric methods	• Atomic absorption	• Gas - liquid
• Electrode methods	• Atomic emission	• Gas - mass spectrometry
• Spectrophotometry	• Inductively coupled plasma - atomic emission	• Ion exchange
• Gravimetric Methods		• High performance liquid
Analytes:	Analytes:	Analytes:
• chloride	• sodium	• chlorohydrocarbon pesticides
• ammonia	• magnesium	• PCB's
• nitrate	• potassium	• hydrocarbons
• phosphate	• mercury	• dioxins
• sulfide	• arsenic	
	• lead	
	• many metals	

FIGURE 17.5 The various techniques for chemical analysis that are commonly used for environmental samples and some common analytes.

concentrations to give a linear plot of absorbance vs. concentration. This standard curve is known as a Beer–Lambert plot, as shown in Figure 17.6.

The Beer–Lambert's law states that

$$A = \varepsilon . b . c.$$

where A is absorbance (–log I_t/I_o, as in Figure 17.6); ε, molar absorptivity; b, path length (cm); and c, the concentration of chemical in solution. Since ε and b are constants, A is directly proportional to c and a linear plot is obtained.

Atomic absorption spectroscopy (AAS) is widely used for the chemical analysis of metals, including many of environmental importance, such as mercury and lead in soils, water, and other samples. The principles of Beer–Lambert's law apply, and the basis of the technique is outlined in Figure 17.7.

The environmental sample can be liquid (e.g., water) or solid (e.g., soil), but they all are dissolved in strong acid solutions, e.g., hydrochloric and nitric acids, to release the metals into solution free of other materials. The solution is then drawn through a nebulizer into a mixing chamber. Here the sample is thoroughly mixed with the combustion gas before being burnt in a high-temperature flame (e.g., air–natural gas, air acetylene at 1700 to 2000°C). The flame absorbs radiation from the light source, which emits the unique spectrum of the specific metal since the critical lamp components are deliberately constructed from that metal. The transmitted radiation is measured using a monochromator adjusted to a unique wavelength emitted from the lamp and a detector. The concentration of metal is proportional to the light absorbed.

The trace organic chemical contaminants in environmental samples are usually extracted with a volatile solvent or by a supercritical fluid extraction procedure, and the extracting solvent is then removed, leaving the environmental chemicals in

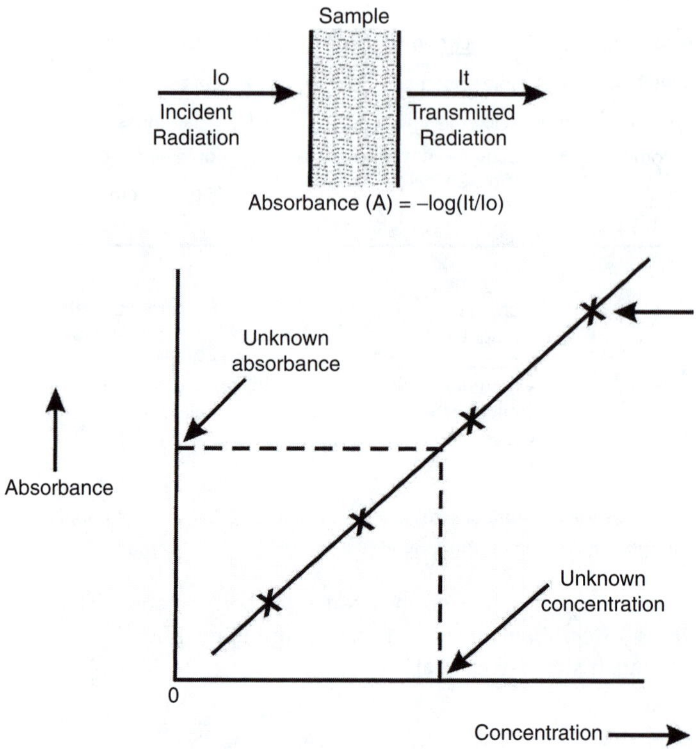

FIGURE 17.6 Plot of absorbance of standard solution absorbance against concentration to give a Beer–Lambert plot.

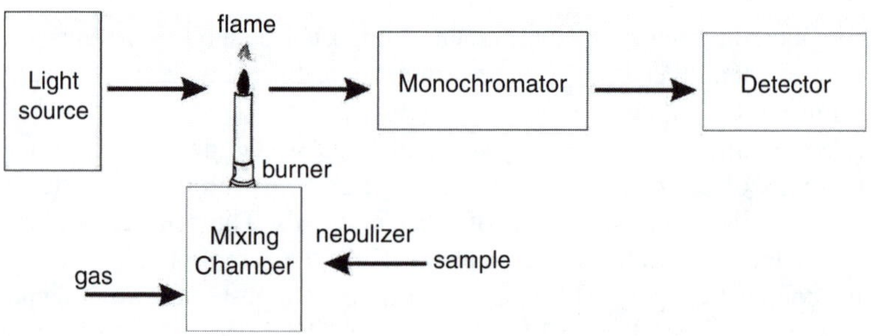

FIGURE 17.7 A diagrammatic illustration of the principal components of an atomic absorption spectrophotometer.

concentrated form. This procedure results in the removal of materials that cannot be subject to chromatography, such as soil and biological tissues, and concentrates the trace chemical contaminants so that there are levels that can be subject to analysis. These samples can then be subject to chromatography.

Chromatography is a totally different technique than spectrophotometry. It is a separation technique that can partition the components of mixtures of organic substances into the individual organic compounds. The individual compounds can then be subject to a detection technique that is sensitive to organic compounds, allowing the separated individual compounds to be quantified. Thus, **chromatographs** have two important components: the separation system and the detection system.

In chromatographs, the separation system is usually a high-boiling-point, low-solubility liquid, such as a silicone grease, which is coated onto a porous material, such as diatomaceous earth, and packed into a hollow metal or glass tube. This high-boiling-point liquid is described as the stationary phase since it does not move from its location in the porous support. A liquid, such as methanol, or a gas, such as nitrogen, under pressure, is passed through the packed column, entering at one end and exiting at the other. If organic compounds are placed at the entry, they will be flushed through the column by the mobile phase but retained to different degrees by the stationary phase. The strength of the stationary phase retention is related to the partition coefficient between the stationary phase and the mobile phase. Compounds that are strongly partitioned into the stationary phase are strongly retained and take a relatively long time to pass through the column. On the other hand, compounds that are strongly partitioned into the mobile phase are weakly retained and pass quickly through the column. Thus, if three organic compounds were placed at the entry to a chromatography column, then the compounds would pass through the column at different speeds and would be separated as shown in Figure 17.8. If a scan for organic compounds in the column were made, a peak for each of the compounds would be obtained, as shown in Figure 17.8. The peak would result from the distribution of the individual molecules in the column and would usually be a normal distribution.

A chromatograph is set up utilizing the separatory system outlined above, with an injection system to place organic compound mixtures on the column without upsetting the system and a detector to detect organic compounds in the mobile phase exiting the column, as shown in Figure 17.9. Optimum conditions for analysis can be obtained by adjusting the column temperature (using the oven), the gas flow, the length of column, the type of liquid phase, and so on. The mobile phase (gas or liquid) is kept continuously flushing through the column until all of the organic compounds have passed through the detector. As each compound passes out of the column and is detected, the data-recording system registers a peak, as shown in Figure 17.9 and Figure 17.10. Two types of information can be obtained from the chromatogram as outlined below:

1. A compound can be tentatively identified from the time it takes to pass through the column, its retention time (t_r), since this time is a characteristic of the compound under the specific conditions used. For example, in Figure 17.10 the compounds producing Peak 1, Peak 3, and Peak 6 are due to methanol, *n*-butanol, and *n*-heptanol since the retention times correspond with the standards. However, it should be kept in mind that identifications made in this way are tentative due to the possibility of other

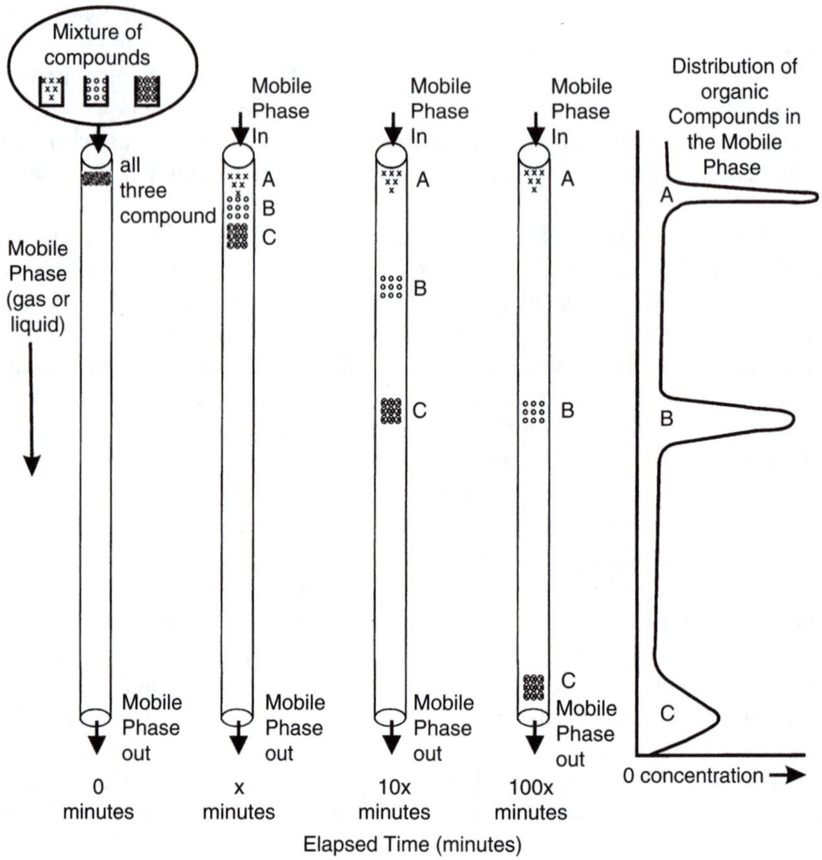

FIGURE 17.8 A chromatography column at progressively longer elapsed periods (x, 10x, and 100x) showing the separation of three compounds.

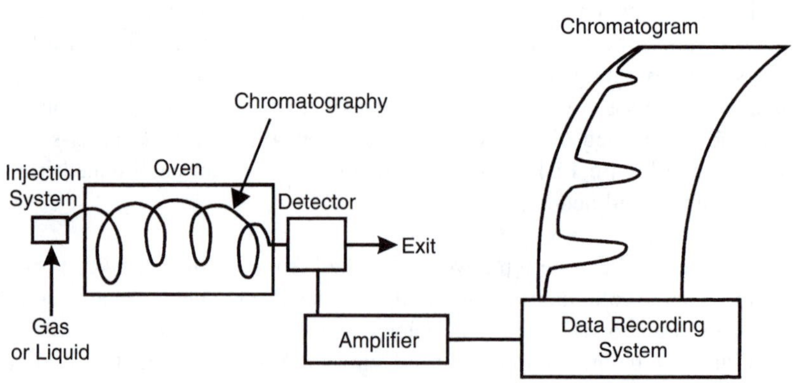

FIGURE 17.9 A diagrammatic illustration of a chromatograph.

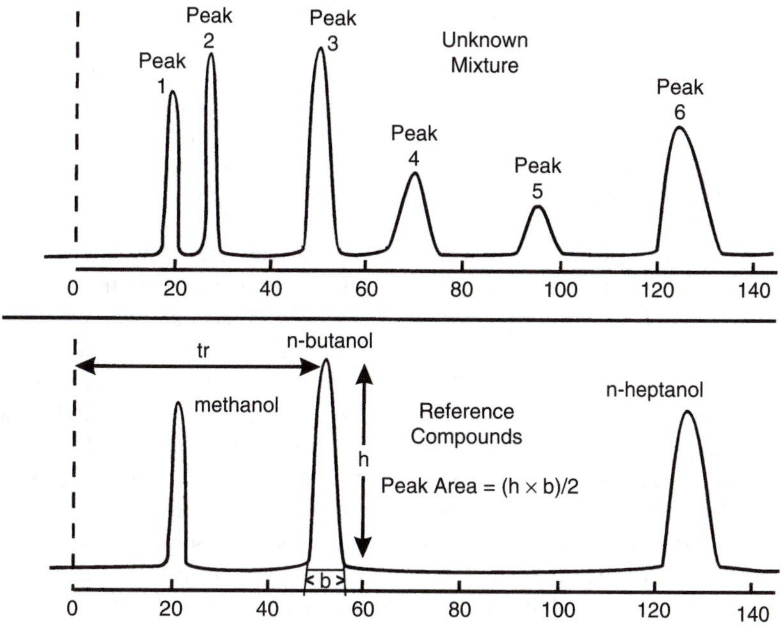

FIGURE 17.10 Typical chromatographs produced by a gas–liquid chromatograph.

compounds having the same retention time. For a clear and firm identification of compounds, usually further confirmatory evidence is required.

2. The amount of compound can be quantified from the peak areas produced on the chromatogram. To do this, the areas produced by each compound are calibrated using authentic reference compounds. A mass spectrometer can be used as the detector which can produce mass spectra of separated compounds used to establish their identity.

17.4 QUALITY ASSURANCE/QUALITY CONTROL (QA/QC)

17.4.1 Background

After a set of environmental samples are collected, chemical analysis for contaminants is carried out using various techniques in the laboratory, as shown in Figure 17.1. The sampling and analysis are conducted to meet certain objectives that have been formulated to address specific questions and problems. As a result of the analysis, a report is prepared and management action is taken. Management actions include such matters as reduction of discharge of contaminants for additional waste control, restrictions in usage of agricultural chemicals, and changes in chemical manufacturing methods. These actions need to be based on accurate information; otherwise, significant economic costs and damage to the environment can occur. To be sure the chemical analysis information is as correct as possible, a system of quality assurance/quality control is needed.

Quality assurance (QA) is a system of activities whereby a consumer or user is informed that analyses meet satisfactory standards by comparison with standards, setting acceptable variability and detection limits, and that appropriate auditing procedures are in place. Quality control (QC) is a management system of laboratory procedures and day-to-day activities to control and assess the results obtained.

17.4.2 QA/QC Procedures

Some typical QA/QC procedures used in laboratories are shown in Table 17.2. Blanks are samples that are collected in such a way as to reveal the background levels of contaminants in reagents used in the analysis, background contamination in the laboratory, background contamination on equipment used for sampling, and so on. Also, analytical procedures are fully calibrated before use by analysis of laboratory control samples so that the standard deviation, precision, and detection limit are known. **Laboratory control samples** are samples held by the laboratory of known composition, often from previous analyses. The results should be sufficient to allow statistical analysis. In addition to these internal samples, there are **standard reference materials** obtained from an external authoritative source that authenticates the composition of the material. These can be analyzed according to a statistically appropriate procedure and the results processed and presented. A well-known check on results is to submit duplicate (two) or replicate (more than two) samples for analysis and then compare the results obtained. The chemical of interest can also be added in known concentrations to a matrix of the environmental sample, known as a **matrix spike**, and submitted for analysis. The results from analysis of a known reference material can provide a useful check on the output of a laboratory over time. The analyses are conducted on a regular basis over time and can be plotted as a chart. When a significant number of analyses have been carried out, the chart can have an average line added with standard deviation. A warning is noted when results exceed ±2 standard deviations, with a control limit at ±3 standard deviations.

Laboratories can contact other laboratories conducting the same analyses and organize an interlaboratory calibration program. Laboratories can be registered or accredited by appropriate organizations through on-site evaluation by independent assessors, taking into account many of the matters shown in Table 17.3.

TABLE 17.2
Some Typical Quality Control Procedures

Blanks (background, laboratory, etc.)
Calibration of equipment and analytical procedures: standard deviation, detection limit, precision
Laboratory control samples: could be previously analyzed samples
Standard reference materials
Duplicates
Matrix spikes
Control charts (upper warning limit, upper control limit, etc.)
Internally organized interlaboratory calibration program

TABLE 17.3
Independent Laboratory Registration

Qualified personnel
Adequate facilities and equipment
Inspection of procedures
Validation of methods
Adequate documentation of results
Internal QA/QC
Independent interlaboratory calibration program

17.5 REPORTING OF RESULTS

In making measurements of any type, it is essential to consider the number of significant figures that are appropriate in reporting the results. In making any measurement, only the numbers that have a meaning should be recorded. For example, making measurements with a ruler, a set of measurements could be 1, 1.1, 1.11, or 1.111 cm. In fact, 1.1 cm is correct because with 1.11 and 1.111 cm, the last two figures have no meaning since the ruler cannot measure to this level of accuracy. Recorded numerical values should be rounded off to the significant figures according to the following:

1. Retain no digits beyond the first uncertain one.
2. If the number in the place beyond the last significant one is less than 5, round off to the existing last significant figure; if it is 6 or greater, round off to the next highest number. Thus, 1.11 cm becomes 1.1 cm and 1.16 cm becomes 1.2 cm.
3. Round off after calculations have been completed.
4. Reporting to more figures than appropriate is incorrect.

The units of reporting can vary according to the objectives of the sampling and monitoring program. Solids, such as soil, can be in weight/weight, e.g., mg g^{-1}, ng g^{-1}, etc., but can also be in terms of dry or wet weight of soil. Often dry weight is preferred since this value is more consistent and does not vary with moisture content. If more theoretical chemical considerations are being taken into account, then the chemical can be reported as moles or related measures, e.g., nm g^{-1}. With liquids, such as water, weight/volume, e.g., mg l^{-1}, is usually used, although nm l^{-1} and related molar units can also be used.

Means and standard deviations are commonly used to present data on analyses. Small standard deviations reflect a high precision in the analysis, which means that when the analysis is repeated, the results are close together and have a high level of repeatability. However, this may not necessarily indicate that the results are highly accurate, i.e., very close to the true value. Usually analytical results that are not accurate are adjusted by the calibration procedure to give as high an accuracy as possible.

In reporting analytical results, it is important to report the **detection limit** and the **limit of quantification (LOQ)**. This is necessary because many chemical contaminants occur in the environment in very low concentrations. The detection limit is usually considered to be three times the concentration equivalent to the background standard deviation at zero concentration. At some low concentrations a chemical can be reported as simply being present as a **trace** since it may not be possible to quantify the actual concentration. The limit of quantification is usually considered to be at a concentration equivalent to 10 times the standard deviation at zero. Thus, a chemical at concentrations between the detection limit and the limit of quantification may be reported as trace. If a chemical cannot be detected at all by the analytical technique, it is not possible to report that the concentration is zero since it may be present in concentrations below the detection limit. In these situations, concentrations are usually reported as less than the detection limit.

Low concentrations in environmental analyses can be important. The guideline concentrations set for different purposes, e.g., maximum residue levels (MRLs) in food, maximum levels for drinking water, and so on, are based on risk assessment procedures outlined in Chapter 18. This process utilizes dose–response data and exposure data derived from experiments, models, actual measurements, etc., and does not include any considerations of the detection limits and LOQs of analytical methods. As a result, a guideline value can be set that may be lower than the detection limit and LOQ for the standard analytical method for a chemical contaminant. This could lead to a situation where the outcome of an investigation is that no conclusion could be reached regarding this chemical. This could result in the analytical method being upgraded, use of nonstandard analytical methods, or simply no consideration of any concentrations below the detection limit or LOQ.

17.6 KEY POINTS

1. Development of interest regarding trace contaminant chemicals in the environment has run parallel to the development of methods for chemical analysis of environmental samples. This effectively started in 1952 with the invention of gas–liquid chromatography (GLC) by James and Martin and atomic absorption spectroscopy (AAS) by Walsh. These techniques, and developments based on them, have allowed the determination of the occurrence of low levels of contaminants in the environment.
2. Environmental analysis should be seen in a framework that uses chemical analysis to develop management actions to protect the environment from hazardous chemicals.
3. The design of a chemical monitoring program for the environment should proceed through a set of steps guided by the specific problem or question that is being addressed.
4. In general sense, the basic objective of a monitoring program is to accurately determine the concentration of a substance in a sample that is representative of the environment in order to monitor certain effects and properties of that substance in that environment.

5. Environmental sampling schemes can be designed using four fundamental strategies:
   - Random — sampling locations decided on a random basis
   - Systematic — sampling locations decided according to a systematic scheme, e.g., a grid pattern
   - Judgmental — sampling locations decided on the basis of a judgment as to the most appropriate location
   - Combined — situations where a combination of strategies are used
6. The techniques for the chemical analysis of environmental samples can be classified into three categories:
   - Wet chemistry — methods used for the analysis of water and wastewater, many of which are older and have a relatively long history of usage
   - Spectroscopy — methods based on the measurement of the intensity of radiation
   - Chromatography — methods based on the separation of mixtures by the chromatography technique into individual compounds that can be quantified and identified
7. The wet chemistry techniques include titrimetric methods, electrode methods, spectroscopy, and gravimetric methods.
8. Spectroscopic techniques for analysis utilize the Beer–Lambert's law, which states that

   $$A = \varepsilon.b.c.$$

   where A is absorbance; ε, molar absorptivity; b, cell length (cm); and c, concentration of the chemical in solution. This law means that concentration of a chemical in solution is directly proportional to absorbance.
9. A major spectroscopic technique used in environmental analysis is atomic absorption spectroscopy (AAS), used principally for the measurement of heavy metals in environmental samples.
10. Chromatographic methods are utilized in several different forms, principally gas–liquid and liquid–liquid, in combination with detectors such as flame ionization, electron capture, and mass spectrometer. These methods are used mainly with organic compounds such as chlorohydrocarbon pesticides, PCBs, hydrocarbons, and dioxins.
11. An organic compound in an environmental sample can be tentatively identified and quantified from its characteristic retention time (t_R) and its peak area, but its identity should be confirmed by other evidence.
12. The quality assurance/quality control (QA/QC) program is an integral part of an environmental monitoring program and is needed to ensure that the results are correct.
13. Typical procedures used in QA/QC programs include blanks; measurement and calculation of standard deviation, etc., on laboratory control samples and standard reference materials; and use of duplicate and replicate samples, spiked samples, laboratory charts, and an interlaboratory calibration program, as well as independent laboratory registration and accreditation.

REFERENCES

Keith, L.H., Environmental sampling: a summary, *Environmental Science and Technology*, 24, 610–617, 1990.

Lessley, S.D., Laboratory methods of analysis, in *Sampling Analysis*, Ostler, N. and Holley, P., Eds., Prentice Hall, New York, 1997, chap. 10, pp. 119–159.

Manahan, S.E., *Environmental Chemistry*, 7th ed., Lewis Publishers, Boca Raton, FL, 2000.

QUESTIONS

1. **The scenario**: A developer has lodged an application with a government agency to build a childcare center on a former service station site on which leaded fuel was stored and sold. The site is bound by two roads. Road A is a road with heavy traffic, while Road B has 10% of the traffic flow of Road A. You have been asked to assess the health risk from contaminants in soil from the site. A site plan has been provided in Figure 17.11. The plan indicates the location of the underground fuel storage facility (which has been removed) and the two boundary roads.
   The task: Using the format in Figure 17.2, set out the steps involved in the development of a procedure to address the question posed in the scenario above. On a sketch map show the possible sampling site locations.

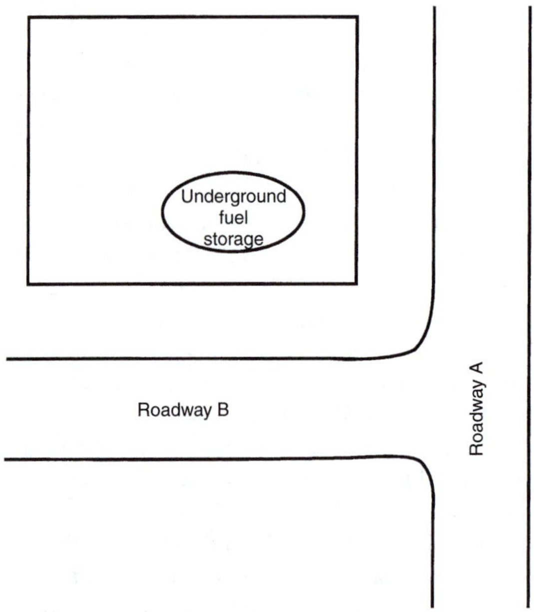

FIGURE 17.11 Site plan for the proposed childcare facility.

2. A preliminary investigation into the possible exposure of an industrial worker to zinc has been carried out by the chemical analysis of zinc in the worker's hair. Four separate hair samples were collected, and from these samples four representative subsamples, each 0.20 g, were prepared. Each sample was digested in 100 ml of 5% acid and aspirated into an AAS set to a Zn line to give the following absorbances: 0.75, 0.70, 0.80, and 0.76. The AAS was calibrated using four zinc standard solutions at 0, 2, 4, and 6 mg l^{-1} to give absorbances of 0, 0.33, 0.56, and 0.84.
   - Calculate the zinc content of the worker's hair in units of g g^{-1}.
   - What is the purpose of digestion of the hair with 5% acid?
3. An exploratory survey of the lead content of soil in an urban area has been carried out. Five surface (down to 6 cm) soil samples were collected using a grid sampling strategy and submitted to the laboratory for analysis. Each soil sample was oven dried at 110°C to remove moisture, the sample was ground, and a subsample (5.0 g) of each sample was prepared. The subsamples were each digested in concentrated nitric acid (50 ml), and a 5-ml sample was removed and diluted to 50 ml with water. These solutions gave absorbances of 0.05, 0.25, 0.35, 0.60, and 0.85 at a lead line (217 nm) in an AAS.
   Standard solutions of lead (10, 15, 20, and 25 mg l^{-1}) gave absorbances of 0.12, 0.18, 0.24, and 0.28, respectively, under the same conditions in the AAS.
   - Calculate the lead concentrations in the soil samples.
   - Comment on the variability in the soil concentrations.
   - Explain the purpose of oven drying, acid digestion, and dilution of the acid digest (5 ml) with water to a volume of 50 ml.
4. A sample was collected from a liquid waste discharge as part of a preliminary investigation. A volume (1 l) was measured out and extracted with 10 ml of solvent, which was then concentrated down to a volume of 500 µl by evaporating the solvent off the solution, leaving the extracted components of the liquid waste behind. A sample of this concentrate (10 µl) was injected into a gas chromatograph to give the chromatogram in Figure 17.10. The solvent remaining in this solution does not give a peak on the gas chromatogram. The gas chromatograph was calibrated with a mixture of reference compounds — methanol, *n*-butanol, and *n*-heptanol — and the detector response to each of these compounds is the same. The chromatogram produced is shown in Figure 17.10.
   Tentatively identify the compounds in the mixture and quantify the composition of the components in percentage terms and in terms of concentration in the liquid waste discharge. The t_R of normal alcohols not included in the reference compounds can be obtained using the knowledge that a plot of the log t_R against carbon number for a homologous series is linear.

ANSWERS

1. To assess the health risk from soil contaminants at the proposed childcare facility:

   Monitoring Objectives: To collect representative samples for analysis and compare the results with guidelines for human health, including children.

   Conceptual Model of Contaminant Behavior: The known sources of contaminants are the exhaust from cars on the road and fuel being handled on the site. The exhaust from cars, including past periods, contains lead and polycyclic aromatic hydrocarbons (PAHs). The fuel contains hydrocarbons, including PAH, and lead.

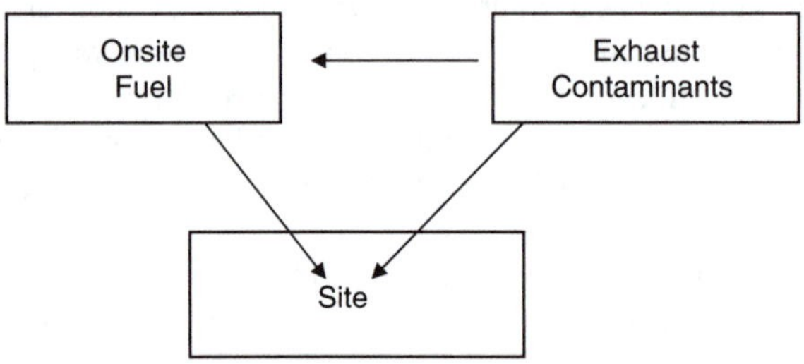

   Types of Sample: The most appropriate sample would be soil.

   Preliminary Sampling Scheme:

   Frequency: One occasion.

   Sampling Strategy: Stratified systematic, with a higher intensity of sampling focused on the areas of likely contamination.
   Surface samples except for the sites close to the underground storage, where samples are taken down to 1 m depth.

   Possible Data Analysis: Cross section plots of the concentrations across different parts of the site can be made and compared with standards.

   QA/QC Scheme: This could include:

   - Background blanks of laboratory analysis and sampling equipment
   - Submission of one sample five times to indicate standard deviation of analysis technique
   - Submission of samples to another independent laboratory for comparative purposes

   The possible sampling site locations are shown in Figure 17.12.

   Final Sampling Scheme: The scheme as outlined in the Preliminary Sampling Scheme above, but incorporating the samples required in the QA/QC Scheme.

   Environmental Sampling and Laboratory Analysis: These are carried out according to appropriate guidelines.

   Evaluation of Health Risk: This is carried out by comparing the results obtained with guidelines for human health, including children.

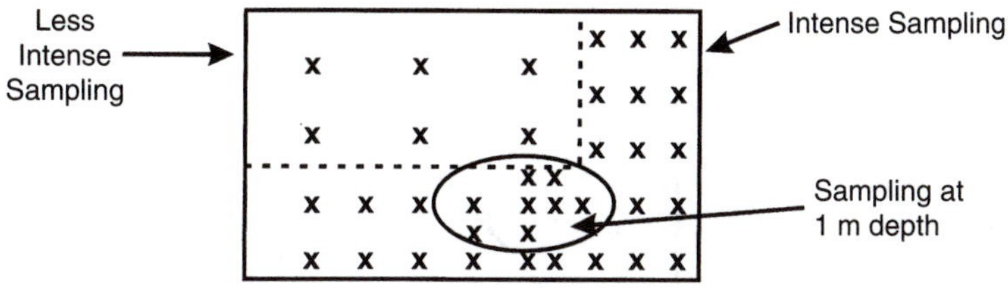

FIGURE 17.12 Possible sampling site locations.

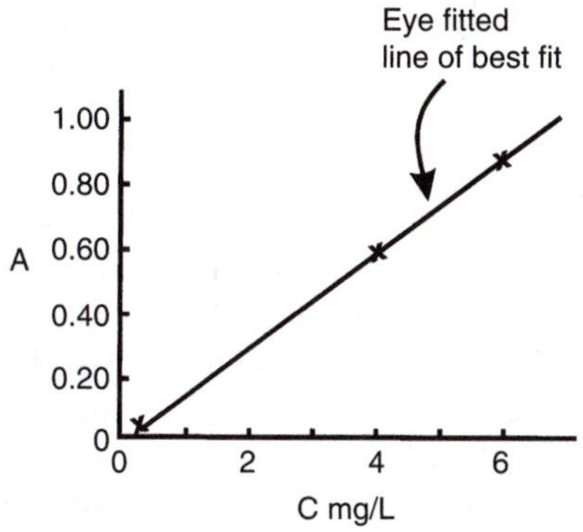

FIGURE 17.13 Standard curve for zinc.

2. <u>**Step 1**</u>: Prepare Standard Curve
   The first step is to use the calibration data for the AAS to plot a calibration curve for zinc, as shown in Figure 17.13.
   <u>**Step 2**</u>: Calculate Zn Concentrations in Digest Solutions
   Using the absorbances of the four acid digest solutions (0.75, 0.70, 0.80, and 0.76), zinc concentrations can be read from the plot as 5.3, 4.9, 5.7, and 5.4 mg l^{-1}.
   <u>**Step 3**</u>: Calculation of Zn Concentrations in Hair
   The concentrations of Zn in hair can be calculated as follows (using 5.3 mg l^{-1} as an example):

Zn concentration in digest solution	= 5.3 mg l^{-1}
Amount of Zn in 100 ml of solution	= 5.3 × 100/1000 mg
Same as the amount in the 0.20 hair sample	= 0.53 mg
Thus, 0.20 hair contains 0.53 mg of Zn	
Zn concentration in hair	= (0.53/1000) g in 0.20 g of hair
	= 0.00265 g g^{-1} of hair
Rounding off to two figures	= 0.0027 g g^{-1} of hair
Similarly, the other hair concentrations are	0.0025, 0.0029, and 0.0026 g g^{-1} of hair
Mean	= 0.0027
Standard deviation	= 0.0002
Zn concentration in hair	**= 0.0027 ± 0.0002 g g-1**

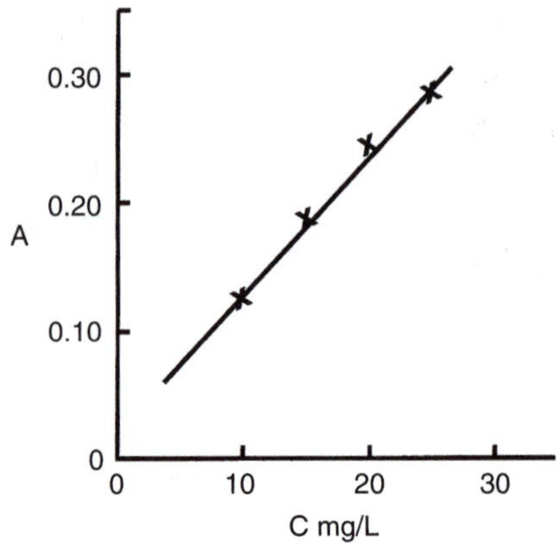

FIGURE 17.14 Standard curve for lead.

The purpose of the acid digestion is to chemically break down the hair into simple substances from which the Zn, which is chemically bound in hair, can be released into solution. The solution of Zn can then be aspirated into the AAS and the Zn absorbance measured.

3. **Step 1**: Prepare Standard Curve
   The data on the concentrations of standard solutions of lead and their corresponding absorbances are plotted as a standard curve, as shown in Figure 17.14.
   Step 2: The concentrations of the acid digest solution can be read from the curve as 4.4, 22, 31, 53, and 75 mg l^{-1}.
   Step 3: The concentration in soil can be calculated as follows, using 22 mg l^{-1} as an example:

5.0 g of soil digested in 50 ml of acid solution, then 5 ml of the acid digest is diluted to 50 ml with water

Thus, the amount of soil in the 50 ml of diluted solution	$= \left(\frac{5.0\text{ g} \times 5}{50}\right)\text{g}$
	$= 0.50$ g
Pb concentration in dilute solution	$= 22$ mg l^{-1}
∴ Amount of Pb in 50 ml of dilute solution	$= \left(\frac{22 \times 50}{1000}\right)$ mg
	$= 1.1$ mg
Thus, the Pb concentration in soil	$= 1.1$ mg in 0.50 g of soil
	$= 2200$ mg kg^{-1}
Pb concentrations in soil are 440, 2200, 3080, 5280, and 7480 mg kg^{-1}	
∴ Average Pb concentration is 3696, but rounded off to two significant figures	$= 3700$ mg kg^{-1}
The standard deviation is 2740 and rounded off is 2700 mg kg^{-1}	
Thus, the Pb concentration is 3700 ± 2700 mg kg^{-1} soil (dry weight)	

Variability: The variability is high, suggesting that there may be many factors affecting the concentration, such as soil type, site location, and so on.

Purpose of Various Procedures:

- Oven drying removes water as an unpredictable variable and provides a more consistent basis to calculate Pb concentrations.
- Acid digestion results in chemical breakdown of all soil components so that chemically bound lead is released into solution and can be measured by AAS.
- The acid digest is too concentrated to be accurately measured by the AAS, and dilution brings the Pb concentration into the range of the instrument.

4. **Step 1**: Identification of Unknown Compounds in the Mixture

This can be done by making a plot of the log t_R against carbon number for the normal alcohols. From the chromatogram for reference compounds in Figure 17.10, the following data can be obtained:

	C Number	t_R (minutes)	log t_R
Methanol	1	20	1.30
n-Butanol	4	50	1.70
n-Heptanol	7	126	2.10

Using this data, a plot of log t_R against carbon number has been made and is shown in Figure 17.15. From this, the log t_R of the missing alcohols can be read off to give the results in the table below.

	C Number	t_R (minutes)	log t_R
Ethanol	2	28	1.45
n-Propanol	3	37	1.57
n-Pentanol	5	71	1.85
n-Hexanol	6	96	1.98

Now by comparison of the full set of t_R values in the two tables above with those of the compounds in the unknown mixture, identified by peak number, the following compounds can be identified as giving the chromatogram shown in Figure 17.10:

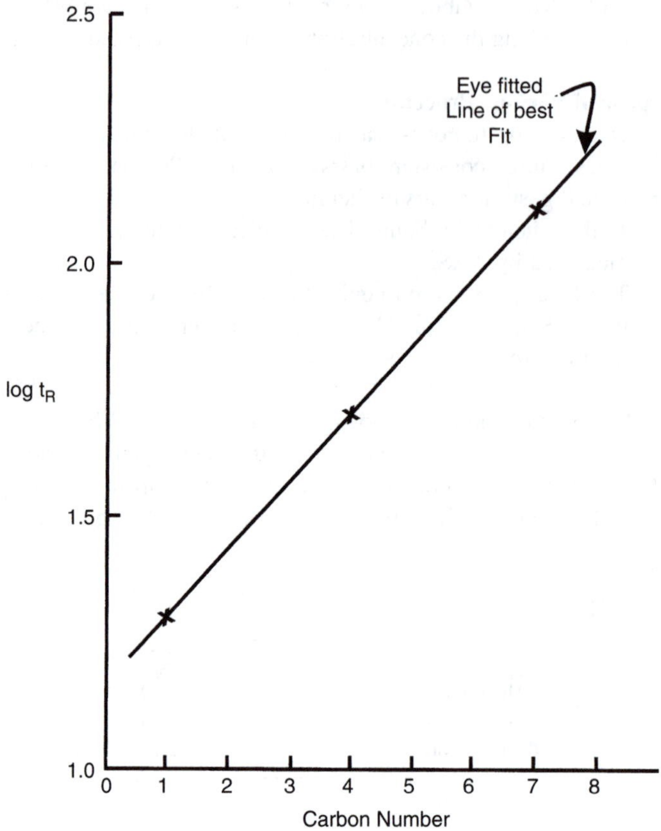

FIGURE 17.15 Plot of log retention time against carbon number for normal alcohols.

Peak Number	t_R (minutes)	log t_R	Identification
1	20	1.30	Methanol
2	28	1.45	Ethanol
3	50	1.70	*n*-Propanol
4	71	1.85	*n*-Pentanol
5	96	1.98	*n*-Hexanol
6	126	2.10	*n*-Heptanol

To fully confirm these identifications, further evidence would be required.

Step 2: Quantification of the Composition of the Mixture

The quantity of compound is proportional to the peak area in the chromatogram and all compounds have the same response to the detector. This means the percentage composition can be calculated directly from the peak areas, which can be estimated by triangulation. The peaks are assumed to be triangles and the peak area is half the base x height (Figure 17.10). These have been calculated, and the percent composition derived from this is shown below.

The peak areas of the reference compounds come to a total of 620 mm^2 (methanol, 84; *n*-butanol, 243; and *n*-heptanol, 294 mm^2), which is the response from a 1-μl injection. Thus, the quantity in microliters yielding the peak areas due to the unknown mixture can be calculated as shown in the table below.

The total amount of the alcohols in the concentrate injected into the GC is 1.4 μl in a 10-μl injection, with the remainder being solvent not detected by the GC.

$$\text{The total amount in a 500-μl concentrate} = \left(\frac{1.4}{10} \times 500\right) \mu l = 70\ \mu l\ .$$

This is the same amount that is present in the 10-ml solvent used to extract the discharge and is the same amount as in the 1-l waste discharge sample (assuming 100% extraction efficiency). Thus, the concentration in the discharge is 70 μl l^{-1} of total normal alcohols. The concentrations of each individual alcohol can be calculated in a similar manner.

Peak Number	Identification	Peak Area (mm^2)	Percent Composition	Quantity in Concentrate Injected into GC (μl)	Concentration in Discharge (μl l^{-1})
1	Methanol	84	9.5	0.1	5
2	Ethanol	104	12.0	0.2	10
3	*n*-Butanol	240	27.0	0.4	20
4	*n*-Pentanol	147	17.0	0.2	10
5	*n*-Hexanol	66	7.5	0.1	5
6	*n*-Heptanol	238	27.0	0.4	20
Total		**882**	**100.0**	**1.4**	**70**

18 Human Health and Ecological Risk Assessment Due to Exposure to Chemicals

18.1 INTRODUCTION

The chemical industry has grown enormously in recent decades. It provides petroleum fuels, antibiotics and other drugs, plastics, pesticides, food preservatives, agricultural fertilizers, and so, without which our society cannot survive. About 100,000 chemicals are estimated to be in daily use, and of these, approximately 7000 are produced commercially in comparatively large quantity. Most of these substances have little or no adverse environmental effects, but some may be harmful to human health or the natural environment (Connell & Miller, 1984). Often these effects only become apparent after wide and prolonged usage, and then control measures are introduced. Clearly, there should be an effective testing and evaluation program to determine those chemicals that present a potential environmental hazard before use. Many countries have developed hazard evaluation programs for new chemicals. Some have indicated that this control program will be their major environmental management program for the foreseeable future.

Chemicals are used in an immense variety of ways for the benefit of human society. The use of many chemicals has a direct risk involved; for example, medicinal compounds such as antibiotics are consumed directly. These substances have a risk of resultant adverse effects on some members of the population. Usually only very small proportions of people are involved, but with all of these chemicals a certain amount of risk is involved in their usage. Also, the use of pesticides in agriculture results in the exposure of nontarget organisms to these active biological agents. In this way, human populations and organisms in the natural environment can both be exposed to biologically active chemicals. Unintentional, and often unplanned, risks can also be involved in the disposal of waste chemical products. Waste chemicals are often disposed to land, water, and air and may contain agents that can have adverse effects on human health and the natural environment.

Government agencies throughout the world have had difficulties in placing the risk involved in all of these situations into a clear perspective. For example, efforts may be made to manage the risks due to a particular discharge containing chemicals when in fact other discharges may pose a greater risk. As a result, there has been a need to develop techniques that give a quantitative evaluation of the risks involved in various chemical discharges (Corvello & Morkhofer, 1993). These evaluations can then be used to gain an insight into the needs for management of the risks involved and allow the allocation of resources in the most effective way to reduce or eliminate that risk. This gives:

1. Greater confidence that efforts in environmental protection from chemicals will be successful in protecting human health and natural ecosystems.
2. Resources will be used in the most effective manner in controlling chemicals.
3. Industry and other activities will be managed in the most cost efficient manner without wastage of resources on chemical management problems that may not be of high priority.

18.2 THE RISK ASSESSMENT: PROCESS AND PRINCIPLES

In recent years, increasing use is being made of risk assessment procedures in government agencies, industry, and elsewhere for management of chemicals. Risk assessment is part of the overall approach to managing chemical hazards, as shown in Figure 18.1. **Risk assessment** consists of identifying the hazard, then quantifying the risk posed by that hazard by risk characterization. This information is then used

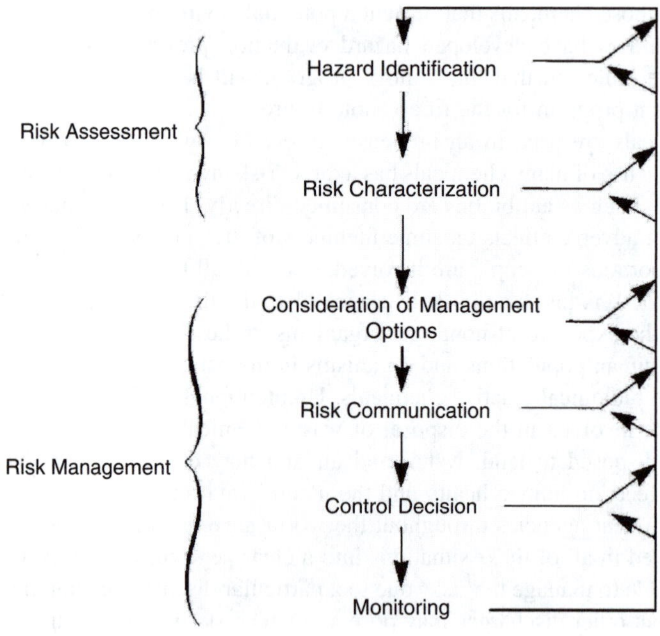

FIGURE 18.1 Overall concept of the risk assessment and management process.

in the risk management process, which consists of consideration of options, communication of risk to those involved, control decisions, and then monitoring to ensure that a correct decision has been made. All of these steps can feed back into previous steps in the process or into the initial hazard identification step, and this procedure may occur several times before a satisfactory result is obtained. A learning procedure may be involved when new information is developed and fed into the system in later steps, which could modify the outcome of preceding steps.

This chapter will be focused on the risk assessment process. Risk assessment is concerned with two principal areas:

1. Human health
2. Natural ecosystems

Somewhat similar general approaches are used in both cases, but there can be considerable difference in the applications in detail.

Toxicology is the science of poisons, and it has been successful in addressing the measurement of the toxic effects of chemicals and many related aspects. In recent times, there has been concern that chemical agents in very low concentrations over relatively long periods of time present a hazard to human health and the natural environment. However, toxicology techniques have been mainly applied in measuring short-term effects at relatively high concentrations of chemicals. Long-term effects usually result from a very low exposure over lengthy periods of time and are not amenable to normal toxicology approaches.

The basic concept of risk itself has many difficulties in interpretation and understanding. **Risk** can be defined as the probability of realization of a **hazard** resulting from exposure to a chemical or other agent. However, the public perception of risk involves many social and cultural factors. Some hazards are particularly dreaded and rank higher than comparable hazards as a result. For instance, hazards from nuclear power generation are rated very high by the public but are ranked relatively low by risk experts who have studied the data involved. On the other hand, the use of alcoholic beverages is regarded by experts to be a reasonably high risk activity, but in the public mind it is rated considerably lower. To minimize these problems, the risk assessment process is usually a quantitative technique, so risk is usually defined in terms of probability of a particular effect resulting from exposure to a chemical. It has the dimensions of frequency of occurrence or incidence coupled to an exposure, for example, with human health a risk of 1 in 1 million (1 in 10^6) of an individual contracting cancer resulting from exposure to a chemical in air at a certain concentration breathed 24 hours/day for a lifetime, usually 70 years. It is commonly used to set acceptable levels of contaminants in the environment in regard to both human health and the natural environment. With ecological risk assessment the risk can be defined in terms of an adverse effect on an ecosystem, e.g., occurrence of fish kills.

18.3 HUMAN HEALTH RISK ASSESSMENT

18.3.1 Background

In the human health risk assessment procedure (Figure 18.2), a hazard identification is required in which an adverse effect or possible adverse effect needs to be identified (Scala, 1991). This may occur in a variety of ways. For example, a chemical may be evaluated for use in a new or unknown situation and would be identified as the hazard due to its observed properties in the laboratory. In existing situations, the hazards and possible adverse effects may not be known. With human populations epidemiological techniques may give indications of adverse effects on mortality, reproduction, neurotoxicity, cancer incidents, and so on. Other ways of identification of adverse effects and hazards could be clinical surveys and the conclusions resulting from research conducted in the laboratory. The adverse effects then need to identified as being due to a chemical or some other specific agent. This can be done by utilizing laboratory and field data on relationships. So the hazard, or several possible hazards, may be identified in this manner. In workplace and site contamination, the hazard may be identified by an evaluation of the activities involved and contaminants released or disposed of as a result.

To characterize a risk quantitatively, we need to know the exposure to the chemical, **exposure assessment**. Of particular concern is the exposure of the target, which could be the human body to the chemical or the amount of the chemical at the target organ. Also, we need to know the potency of the adverse effect. Thus,

$$\text{Risk} = \text{Exposure} \times \text{Toxicity or Adverse Effect}$$

This principle leads to the general process for evaluating risk to human health and the natural environment, as shown in Figure 18.2. The first step is the hazard identification followed by the establishment of dose–response relationships. There

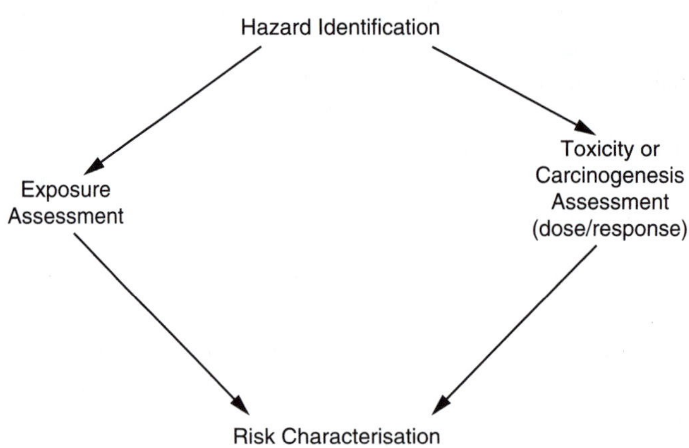

FIGURE 18.2 The human health risk assessment process for a hazardous chemical.

are two main sources of data for carrying this out: (1) epidemiological data from actual studies on human populations and (2) laboratory, or toxicology, data, based on laboratory experiments on test animals. Both sources of data have limitations in being applied to risk assessment, and techniques are used to compensate for these problems, as explained in Section 18.3.3. Finally, the risk can be characterized in quantitative terms using the data on exposure and dose–response relationships obtained as outlined above. This allows acceptable levels in food, air, water, and soil to be established and the management of risk carried out in a methodical manner. As new information becomes available, this process may be repeated, giving rise to revised levels of protection in the environmental media.

18.3.2 Exposure Assessment

The generalized set of steps shown in Figure 18.3 allows an evaluation of the intake by the human population. Of course, this intake will not be the same for all sectors of the population. Some, for example, children, may be exposed to larger quantities of the contaminant than other sectors of the population, for example, adults. Contaminant releases can occur from a wide range of activities, including mining, motor vehicles, contamination of soils, rural activities, and so on. So initially, sources of contamination should be identified. Often the quantities involved can be estimated to allow further calculations to be made on possible levels in the environment. Exposed populations must be identified initially. This could be, as mentioned before, children or adults, or it could result from the activities of particular groups of people

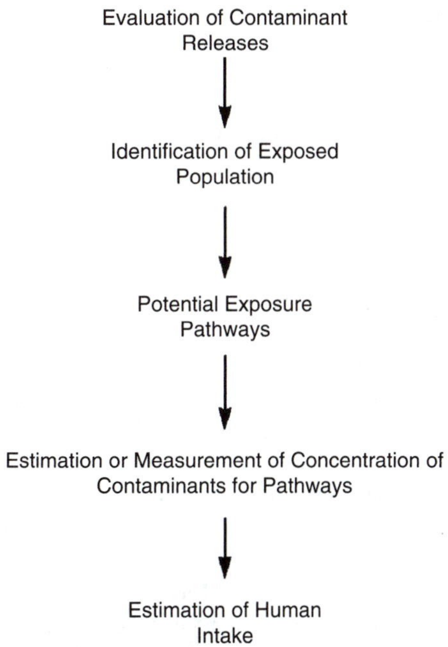

FIGURE 18.3 Evaluation of human exposure to a toxicant.

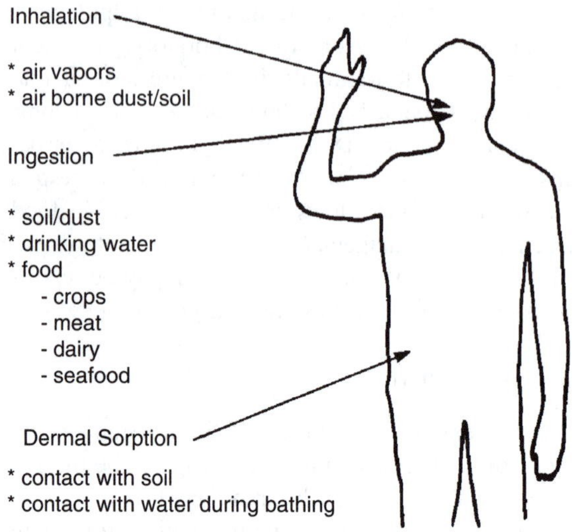

FIGURE 18.4 Pathways of intake of chemicals by humans.

on sites or in work situations. In other words, the exposure could be related to whether the main activities of an individual or group are related to industry, domestic situations, sport, and so on. When the group has been identified, the usage pattern resulting in exposure needs to be quantified. Exposure pathways can be developed for particular exposed groups of the population. Generally, these pathways will fall into the following classes:

- Inhalation of vapors and dust
- Skin contact
- Ingestion from food and water
- Direct ingestion of contaminated materials

Figure 18.4 shows potential exposure routes in a diagrammatic manner.

Once the exposure pathways have been identified, it is then necessary to estimate quantitatively the amount of exposure resulting from each particular pathway. The simplest form that this can take is by the direct measurement of the amount and concentrations of contaminant, the amounts of contaminant taken up by the population, as well as the periods of exposure. This gives a direct measurement of the concentrations and amounts of chemicals involved in a particular pathway. Of course, these data may not be available, and so alternative methods may be used. The use of a variety of models is now common for exposure assessment. With this technique calculations are made of the concentrations and amounts of chemical involved in different pathways to the population.

Another useful technique is the use of **biomarkers**. This technique utilizes the concentrations of the chemical contaminant that occurs in biota already using the area as a habitat or specimens of biota that are deliberately placed in the habitat to measure biological exposure. Examples of biota that can be used in this way are rats, mice, earthworms, and a variety of other organisms. Analysis of these biota give an indication of the concentration of the chemical likely to occur in humans and in other sectors of the environment that may be involved in pathways to biota. The general equation for intake by humans is

$$\text{Daily intake (DI)} = \Sigma\,(\text{intake by the individual pathways})$$

This means that DI = $\Sigma\,(C_i.A_i.BA_i)$, where C_i is the concentration in the medium involved (mg/kg), A_i the amount of medium involved (kg), and BA_i the bioavailability of the contaminant in the medium to the human being (unitless proportion). For daily intake expressed in terms of body weight (bw), then

$$DI = [\Sigma\,(C_i.A_i.BA_i)/bw]\ \text{mg/kg bw}$$

This may be modified to suit pathways with particular characteristics. For example, with the dermal pathway from the soil, an exposure factor is used that is estimated from the period of contact of the soil with the skin. Another example is with the particulates taken up by inhalation from the atmosphere, where an evaluation of the particulates retained by the human body is needed.

Example: Adults Exposed to Dieldrin in Soil

The significant pathways for uptake of dieldrin from soil by human beings are:

- Dermal
- Inhalation
- Ingestion of contaminated soil

It can be assumed that drinking water, food, and bathing water will generally not be contaminated by the contaminants in soil in a particular location and will be not be significant sources of contamination at a particular site. So looking at the daily intake, then

$$DI = DI_d + DI_{ih} + DI_{ig}$$

where DI_d is dermal absorption, DI_{ih} inhalation, and DI_{ig} ingestion of contaminated soil.

By expanding these daily intake characteristics, the following equation can be obtained:

$$DI = [(C_{sd}.A_{sd}.BA_{sd}.EF) + (C_{sih}.A_{sih}.BA_{sih}.P) + (C_{sig}.A_{sig}.BA_{sig})]/bw$$

where C_{sd}, C_{sih}, and C_{sig} are the concentrations in dermal, inhalation, and ingested soil, respectively; A_{sd}, A_{sih}, and A_{sig} the amount of soil involved; and BA_{sd}, BA_{sih}, and BA_{sig} the bioavailability in the different pathways.

In this equation an **exposure factor** (EF) is needed for dermal contact since this occurs on a noncontinuous basis. The exposure factor can be calculated from periods of exposure obtained by observation. In the case of dieldrin, this factor has been estimated at a value of 392, which has no units. With inhalation, some of the particles are retained while others are passed out of the lungs without having any significant effect. This proportion (P) must be taken into account as indicated above to evaluate exposure from this pathway. Exposure factors are not needed with inhalation and ingestion since these occur on, or can be considered to occur on, a continuous basis.

The **bioavailability** of dieldrin in soil through the dermal route is 0.05, or 5%. The amount of soil in contact with the skin is about 8 mg, so on average A_{sd} is 8×10^{-6} kg. Turning to inhalation, the amount of particulate absorbed can be considered to be generally about 0.3 mg/day (A_{sih}), and of this, 0.35 can be considered to be the proportion retained by the body (P). The bioavailability of dieldrin from airborne particulates can be considered to be 1, or 100%. Looking at ingestion, A_{sig} can be considered to be 0.1 mg/day and the bioavailability (BA_{sig}) 0.1, or 10%.

By using these data, and the expression above, the total soil daily intake (DI) by adult human beings can be calculated as

$$DI = [C_{sd}(8 \times 10^{-6})\ 0.05 \times 392 + C_{sih}(0.3 \times 10^{-6} \times 0.35 \times 0.1) + C_{sig}(0.1 \times 10^{-6})\ 0.1]/70\ \text{mg/kg.bw}$$

In this case, adult average weight is considered to be 70 kg. Calculating this through, then

$$DI = [(C_{sd}\ 157 \times 10^{-6}) + (C_{si}\ 0.02 \times 10^{-6}) + (C_{si}\ 0.01 \times 10^{-6})]/70\ \text{mg/kg.bw}$$

This indicates that the ingestion and inhalation pathways are relatively unimportant. This means that

$$DI = C_{sd}\ 2.24 \times 10^{-6}\ \text{mg/kg.bw}$$

Thus, by substituting values for C_{sd}, which can be assumed to be the same as the concentration of contaminant in soil, the uptake from soil can be calculated.

The data produced by this exposure assessment gives information on the total intake of contaminant by a general adult population group. It can be reapplied to specific groups of the population with different exposure characteristics. In this way, assessment of exposure characteristics for different groups in a population can be obtained.

18.3.3 Dose–Response Relationships

18.3.3.1 Data Available on Dose–Response Relationships

Exposure to toxic chemicals gives rise to a variety of possible responses in a population depending not only on the nature of a chemical, but also on the dose and period of exposure that occurs. A generalized sequence of the effects of a chemical on a population is shown in Table 18.1. At high and very high doses, the exposure period is short and death is a common response. At intermediate doses, a major proportion of a population survives, but the survivors exhibit severe effects in many

TABLE 18.1
Range of Effects of a Toxicant on a Population

	Dose	Exposure Period	Response
Increasing dose ↓	Very low	Very long (many years)	No detectable effects
	Low	Long (months/years)	Death of sensitive individuals Sublethal effects in survivors
	Intermediate	Intermediate (days)	Equal numbers of deaths and survivors Severe effects in some survivors
	High	Short (hours/days)	Few resistant individuals survive
↑ Increasing exposure period	Very high	Very short (hours)	Death to all members of the population

cases. The lethal levels (dose and concentration) for 50% of the population (LD_{50} and LC_{50}) are used for toxic characteristics of a chemical to evaluate effects in the very high, high, and intermediate ranges of concentration. At low doses, exposure can be for months ranging into years, and the LD_{50} becomes less useful since lethality occurs only with sensitive individuals. But in addition, a range of sublethal effects occurs in survivors that are not evaluated by the LD_{50} test. Other adverse biological responses occur apart from lethality. At very low doses and exposures for many years with many biota, there may be no effects that can be detected by current techniques.

In environmental toxicology and risk assessment the concentrations generally involved are in the low, probably more often very low, dose range, and long exposure periods of months to many years. So in these cases, the conventional toxicological techniques for evaluating toxicity are more difficult to apply.

The adverse responses of biota, particularly human beings, are needed to utilize the exposure data, which can be calculated as indicated in the previous section. By combining the exposure with an indication of the adverse effects of that exposure, a characterization of the risk posed by a particular chemical can be made. The dose–response data available to carry out this risk characterization fall into two broad groups:

- Epidemiological evaluations of the effects on human populations
- Experiments conducted under controlled conditions on organisms in the laboratory

Both sets of data must be used according to availability and suitability. One one hand, the data on human populations would be considered to be the most useful, since these data relate directly to the organism of interest, human beings in the case of human health evaluations. On the other hand, the laboratory experiments on animals, such as rats, would be expected to be less useful since the two organisms, human beings and rats, are considerably different in many of their characteristics. But the use of epidemiological data is difficult for the following reasons:

- The human population is exposed to numerous agents that may cause adverse biological effects, not only the chemical of specific interest.
- Temporal patterns of exposure of humans are usually not known and can also be variable over long periods.
- The population includes many subgroups, such as children, adults, occupational groups, and so on, which may have different responses and exposures to the contaminant.
- The data available on epidemiological studies on specific chemicals are very limited.

Animal experiments also have a range of limitations, as outlined below:

- Experimental animals and humans may respond in different ways to the same chemical agent.
- Dose rates and exposure concentrations with experimental animals are much higher than with most actual or potential human environments.
- It is not practical to conduct experiments over long periods, for example, 15 to 20 years, with very low dosages.
- The exposure route in laboratory experiments may differ from that in the human exposure situation.
- A significant incidence of adverse effects in a human population could be an effect on 1 in 10^4 or 1 in 10^5 or lower. The exposure of human populations to chemicals can be on the order of millions of individuals, with chemicals in air, food, water, and other media. Thus, experiments would be needed on very large numbers of animals to observe the low incidences that are of significance in the human population.

For the reasons outlined above, the LD_{50} and LC_{50} data are not commonly used in evaluating environmental effects, although these characteristics may be used in the absence of other data. The most common toxicity data used are the **lowest observable adverse effect level (LOAEL)** and the **no observable adverse effect level (NOAEL)** as shown in Figure 18.5. These characteristics were described in Chapter 4. These values are more useful in environmental applications since they approach the low-level long-term effects usually of interest more closely than the LD_{50} and LC_{50} data. Despite the many difficulties in using the available toxicity data, both the epidemiology and animal experiment data are used according to availability and suitability in the evaluation of human health effects, as outlined below.

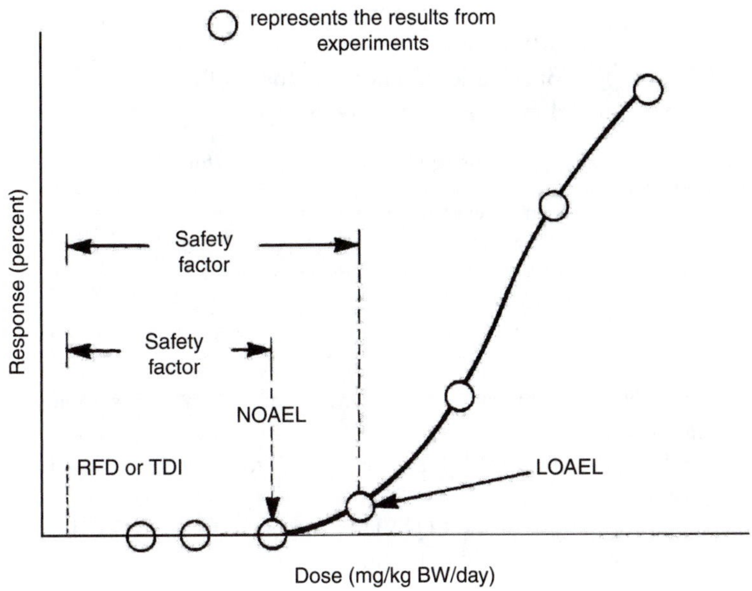

FIGURE 18.5 Diagrammatic illustration of the application of the safety factor (SF) to experimental data.

18.3.3.2 Safety Factors

Most data are available from experiments conducted on animals in the laboratory, since this is a consistent and practical source of information. If data are lacking, they can be obtained by the conduct of experiments on animals with the chemical of interest. However, such experiments cannot be carried out on human populations, and epidemiology data may not be available or able to be obtained when required. In addition, epidemiology data usually originate by the chance exposure of a population, and the design and the significance of the epidemiological results may not be optimal. Even when epidemiology or other data on human beings are available, many of the limitations mentioned previously apply.

This has forced environmental toxicologists to seek ways to apply laboratory experimental data despite the limitations. Of course, when epidemiological data on human are available and appropriate, they should be used. The most common approach in the utilization of the experimental data and, to a certain extent the epidemiology is to take these data and apply a **safety factor**, or **uncertainty** or **application factor**, to them, which will increase the toxicity as it is applied to evaluate safety. This has been done for many years and has achieved considerable success in protecting public health.

Briefly, the purpose of the safety factor is to:

- Account for the different sensitivities of individuals within the human populations

TABLE 18.2
Magnitude of Safety Factors with Human Health Assessments

Extrapolation	Factor
Average human to sensitive human	10
Animal to human	10
Short-term to long-term exposure	≤10
LOEL, not NOEL, used	≤10
Limited database	≤10

- Allow for the possible increased sensitivity of human beings as compared to animals
- Extend short-term toxicity and exposure to long-term toxicity and exposure
- Account for the use of the LOAEL as compared to the NOAEL
- Account for the availability of a limited amount of data

The overall safety factor is obtained by multiplying the individual factors that apply to the data set according to the criteria set out in Table 18.2.

With human health assessment, the actual values of the safety factor range from 10 to 10,000. Values of less than 10 are occasionally used where humans are believed to be less sensitive than the test animals. It is considered that values greater than 10,000 indicate that the information is very imprecise and may not produce reliable results. As a general rule, safety factors greater than 1000 indicate that more information is needed to obtain a lower safety factor. The lower the safety factor, the higher the confidence in the result, and if the safety factor were unity, our information would be precise and apply to the population being considered exactly.

18.3.4 Acceptable or Tolerable Daily Intake (or Reference Dose) for Human Populations

In many situations an **acceptable** or **tolerable daily intake** (ADI and TDI, respectively) is derived as a basis for risk characterization. The TDI is also referred to as the reference dose (RfD). The TDI is the benchmark dose derived from NOAEL by the use of safety factors, as shown in Table 18.2. The actual values for the safety factors are derived to reflect the type of data available. The application of the safety factor to the NOAEL and LOAEL is indicated diagrammatically in Figure 18.5. The LOAEL is a higher value than the NOAEL, and so a larger safety factor is applied with LOAEL data. The use of safety factors, as described here, assumes that there is a threshold concentration below which no adverse effects will occur and the application of the safety factor will provide a measure that is below that threshold. So the TDI (or RfD) is expressed as

$$\text{TDI} = [\text{NOAEL/SF}] \text{ mg/kg.bw/day}$$

In practical terms, the TDI is a value that has a degree of uncertainty to its numerical expression. It indicates the daily exposure a human population can experience without an appreciable risk of adverse effects.

18.3.5 Risk Characterization

The final step in the risk assessment process is to characterize the risk involved in numerical terms. This requires the use of the exposure assessment to evaluate the actual amounts of chemical to which a human population is exposed, and relating this exposure to the toxicity and other information obtained from dose–response relationships. If the daily exposure is less than the TDI, then there should not be any significant adverse effects on the health of the human population. These effects relate to toxic effects rather than carcinogenicity, which is a separate evaluation. Sometimes the outcome of this process is quantified as the hazard quotient (HQ), where

$$\mathrm{HQ} = \frac{\mathrm{Dose}}{\mathrm{TDI}}$$

HQs above unity have the possibility of adverse effects in human populations. The risk of cancer induction in the population is not taken into account in this procedure, which is concerned with toxic effects only. A similar, but characteristically different, procedure is used for carcinogenic risk assessment.

18.4 ECOLOGICAL RISK ASSESSMENT AND ECOTOXICOLOGY

18.4.1 Background

Techniques to evaluate aspects of the biological effects of chemicals have been described previously in Chapters 3, 4, and 15. These are all concerned with the various aspects of the behavior and effects of a chemical in the environment. However, previously many of these aspects were considered as separate factors and the biological effects related principally to individual species. There is a need to draw these aspects together into an overall approach that addresses the effects on whole ecosystems (Hoffman et al., 1995). Our capability to evaluate the effects on whole ecosystems is not particularly strong, but approaches based on **ecotoxicology** have provided a systematic approach (Carlow, 1993).

18.4.2 The Ecotoxicology Concept

Scientific investigation of the effects of pollutants in the natural environment has focused on the effects on individual biological species. Such investigations can only give suggestions as to the likely effects on the complex ecosystems existing in the natural environment. Thus, there has been an increase in interest in the effects of pollutants on whole ecosystems, since these are of primary management concern. This has led to the development of **ecotoxicology**.

Ecotoxicology is concerned with the fate and toxicity, and related effects, of chemicals in natural ecosystems. The ecotoxicology of a chemical can be seen to be based on a sequence of interactions and effects controlled by the physical, physicochemical, and biological properties of a chemical, as indicated in Figure 18.6. A chemical discharged to the environment can then be subject to distribution in the atmosphere, water, or soils and sediment, depending on its physical-chemical properties, as described in Chapter 15. At the same time, it can be chemically modified and transformed by abiotic processes or more often by microorganisms in the environment. The organisms present are then exposed to the toxicant in its original form and in its degraded or transformed state and at concentrations resulting from its dispersal. Uptake of the chemical and its degradation products occurs, and organisms can exhibit a variety of different reactions, from negligible to sublethal effects such as reduced growth, reproduction decline, behavioral effects, or ultimately death. The complex natural ecosystem of which the organisms are an integral part can react in a variety of ways to the effects on the component organisms. Food chain relationships, energy flows, and so on, may be altered. Thus, the ecotoxicology of a chemical can be considered as a sequence of steps, starting with a source and following through to the ecosystem response, with different properties of a chemical involved at each step, as illustrated in Figure 18.6.

Initially, the physical state of the contaminant, whether it is a solid, liquid, or gas, has a major influence on its physical dispersal from the source. Following that, physical-chemical properties influencing movement into different environmental phases become important. In later stages, interactions with organisms, populations, communities, and ecosystems are produced by the biochemical and physiological properties of a chemical. As the subsequent effects of a chemical enter more complex levels of biological organization, such as the whole ecosystem, the direct effects of the chemical on individuals becomes less significant. The biological effects generated at simpler levels of biological organization flow through the system to impact on whole ecosystems, rather than the direct effects of the chemical itself at this level of organization.

18.4.3 Types of Toxicant Discharged to the Environment

The types of toxicant discharged to the environment are quite numerous. Toxic gases and nongaseous compounds originate from motor vehicles, electricity generation, industry, and numerous other sources. Toxic materials are often discharged to waterways in sewage, storm water runoff, and industrial discharges. Soils can be contaminated from most of the sources indicated above. A summary of the sources and types of chemical groups involved and the environment affected is given in Table 18.3.

A wide variety of different types of chemicals are involved that have an ecotoxicological effect in the environment. However, with the organic toxicants, the groups commonly involved are hydrocarbons, aromatic hydrocarbons, chlorohydrocarbons, oraganophosphorus compounds, polychlorinated biphenyls (PCBs), polychlorodibenzo dioxins (PCDDs), polychlorodibenzo furans (PCDFs), and surfactants. With the inorganic substances involved, the toxic metals are probably the most common toxicants.

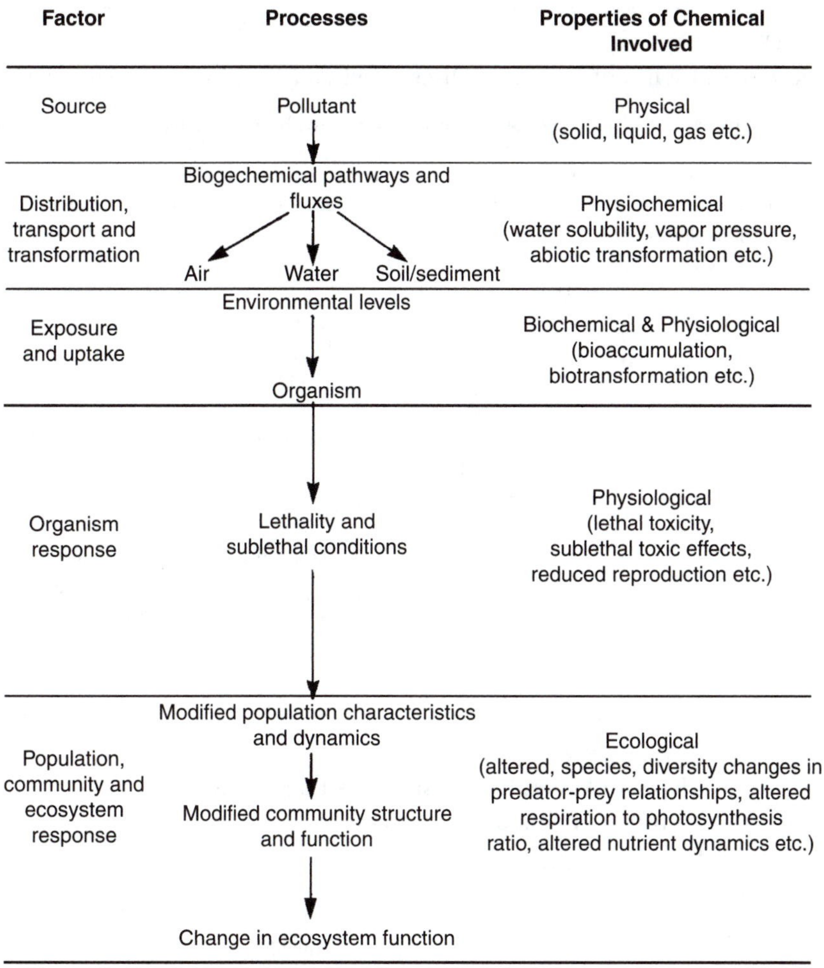

FIGURE 18.6 Diagrammatic illustration of the impact of pollutants on the components and properties of ecosystems, together with the properties of the chemical involved.

18.4.4 Laboratory Testing for Environmental Effects

The approach to the prediction of possible adverse effects is often to use laboratory tests to evaluate key parts of the system, which can then be used to give an indication of the overall effects on the whole ecosystem. An outline of the types of the laboratory tests that are used to evaluate chemicals is given in Table 18.4. In broad terms, the tests relate to the various stages involved in the ecotoxicology of a chemical. A comparison with the properties of the chemical involved, indicated in Table 18.4, shows the general relationship of the types of tests used to the ecotoxicology approach to the behavior of chemicals in the environment. The initial physical-chemical properties relate to dispersal from sources and the distribution in

TABLE 18.3
Sources and Types of Toxicants Discharged and Environments Affected

Source	Some Chemical Groups Involved	Environments Affected
Motor vehicle exhausts, electricity generation, and industrial discharges to the atmosphere	Lead and other toxic metals, carbon monoxide, carbon dioxide, aromatic hydrocarbons, sulfur dioxide, hydrocarbons, PCDDs, PCDFs, PCBs	Human and natural terrestrial systems
Sewage	Aromatic hydrocarbons, chlorohydrocarbons, toxic metals, surfactants	Aquatic systems
Storm water runoff	Aromatic hydrocarbons, hydrocarbons, lead and other toxic metals	Aquatic systems
Industrial discharges to waterways	Acids, toxic metals, salts, hydrocarbons, PCDDs, PCDFs	Aquatic systems
Urban and industrial discharges to soil	Toxic metals, salts, hydrocarbons, PCDDs, PCDFs, PCBs	Human and natural terrestrial systems
Rural industries	Chlorohydrocarbons, organophosphorus compounds	Human, natural terrestrial, and aquatic systems

TABLE 18.4
Types of Laboratory-Based Tests Used to Evaluate Adverse Effects on Ecosystems

Category	Examples
Physical-chemical properties	K_{OW} value Aqueous solubility Adsorption/desorption
Degradation and accumulation	Biodegradability Fish bioaccumulation
Effects on biotic systems	Alga — growth inhibition Fish — acute toxicity Terrestrial plants — growth test

environmental phases. The degradation and accumulation of a chemical give an indication of the actual exposure of biota to the chemical in the different phases in the environment. Tests on individual organisms such as alga, fish, and terrestrial plants yield an indication of effects on natural ecosystems. These laboratory tests are conducted under highly specified conditions, which define the species and test conditions to be used and the manner in which the results are to be recorded and processed.

18.4.5 Effects on Ecosystems

Tests on whole ecosystems are generally not fully effective due to several factors. For example, in a stream, conditions can vary from pool environments to rocky riffles over a short distance. Second, there can be considerable variations in environmental conditions, such as temperature throughout the seasons, which are difficult to reproduce in test situations. Also, variations in the biotic composition of ecosystems due to natural factors are not well understood, making it difficult to interpret variations due to the effects of contaminants. Nevertheless, attention has been given to direct investigation of the effects of toxicants on test sections of large natural ecosystems (mesocosms) and small artificial simplified ecosystems (microcosms). There are some general principles that can be applied to the effects of chemicals on ecosystems.

Organisms in an ecosystem exhibit differing susceptibilities to toxicants. For example, herbicides are selectively toxic to plants and insecticides have a major impact on insect populations. Even within these broad groups of organisms, there are wide range of different susceptibilities with different species. Thus, when any toxicant enters an ecosystem, it will selectively remove those susceptible species from the range of organisms present. This generally translates into a reduction in the species diversity of organisms present and a change in the community structure.

Figure 18.7 indicates the general pattern of response of the components of an ecosystem to a single application of a toxicant causing herbivore mortality, as indicated by simulation modeling. Initially, both species diversity and biomass fall and the community structure is altered, but all tend toward the original situation as toxicant concentration decreases over time. The initial effects on carnivores and top carnivores are relatively low, but there is a delayed effect that extends over a comparatively long period. With chronic pollution over extended periods, situations somewhat similar to the initial effects of a single application are noted. For example, a decrease in species diversity and alteration of community structure may become apparent.

In addition to toxic effects on different components of an ecosystem, other important secondary effects usually result. In natural ecosystems, some animals and plants have adapted to the existing environmental conditions. There is a wide range of interacting and interdependent organisms, including microorganisms, plants, invertebrates, vertebrates, and so on. But the interrelationships between these groups are usually not well understood, which can lead to difficulties in understanding the effects of toxicants. However, in many cases the effects due to a toxicant are large compared with natural variations and can lead to more obvious ecological effects.

Selective removal or alteration of a population of organisms in an ecosystem results in a modified food web, as illustrated in Figure 18.8. This results in a change in energy and matter flow patterns. In an ecosystem, such characteristics as total respiration, total primary production, respiration-to-primary production ratio, nutrient cycling rates, predator–prey relationships, and so on, are changed.

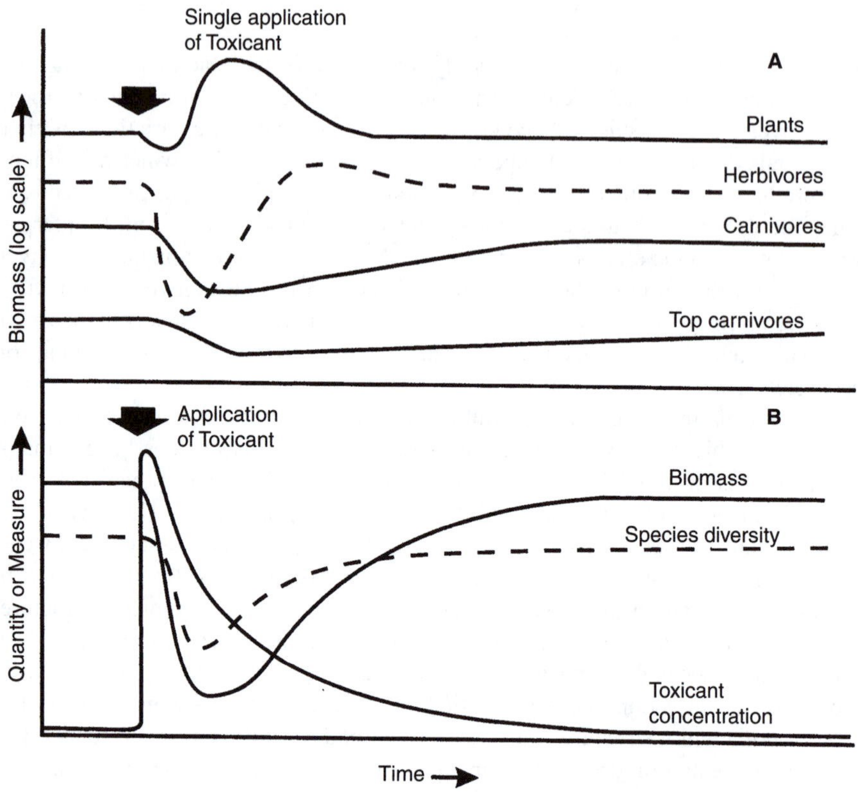

FIGURE 18.7 Some generalized characteristics of a toxicant causing herbivore mortality in an ecosystem, indicating (a) changes over time to some general groups of biota and (b) changes over time to the toxicant concentration and related overall biological measures.

18.4.6 The Ecological Risk Assessment Process

The ecological risk assessment framework was formulated by the U.S. Environmental Protection Agency (EPA) in 1992, and this process is followed, in general terms, by many government agencies throughout the world. It generally follows and fits within the overall concept of risk assessment and risk management shown in Figure 18.1, but has specific differences.

The characterization of risk in natural ecosystems can be directed to the integrity of the whole ecosystem or toward the effects on specific biotic components. Thus, with whole ecosystems we are concerned with the maintenance of the species present, whereas with specific biotic components we are concerned with rare and endangered species, important game species, and so on. Data on dose–response relationships may not be available, and extrapolation from other ecosystems will often be needed, but here data are limited also.

The framework shown in Figure 18.9 has a similar structure to that for human health risk assessment shown in Figure 18.2. The first step in human health risk assessment is the identification of the hazard, and similarly with ecological risk

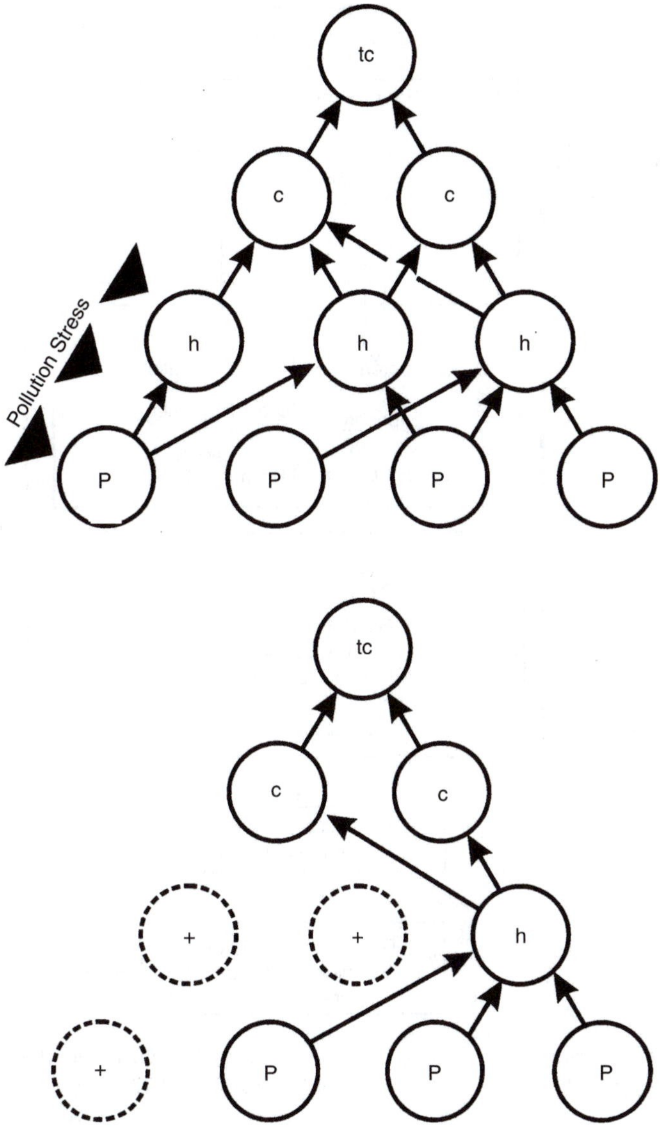

FIGURE 18.8 A hypothetical food web illustrating modifications as a result of elimination of some members due to pollution stress by toxicants. TC = top carnivore; C = carnivores; h = herbivores; p = primary producers.

assessment, there is the identification of the problem. With human health risk assessment, the objective is to evaluate the risk to human health, but there is no similar simple answers with ecological risk assessment. It could be to protect endangered species, protect species diversity of fish, prevent fish kills, and so on. It needs the available information to be collected and collated to provide a sound basis on which risk assessors, risk managers, state agencies, voluntary groups, etc., can formulate

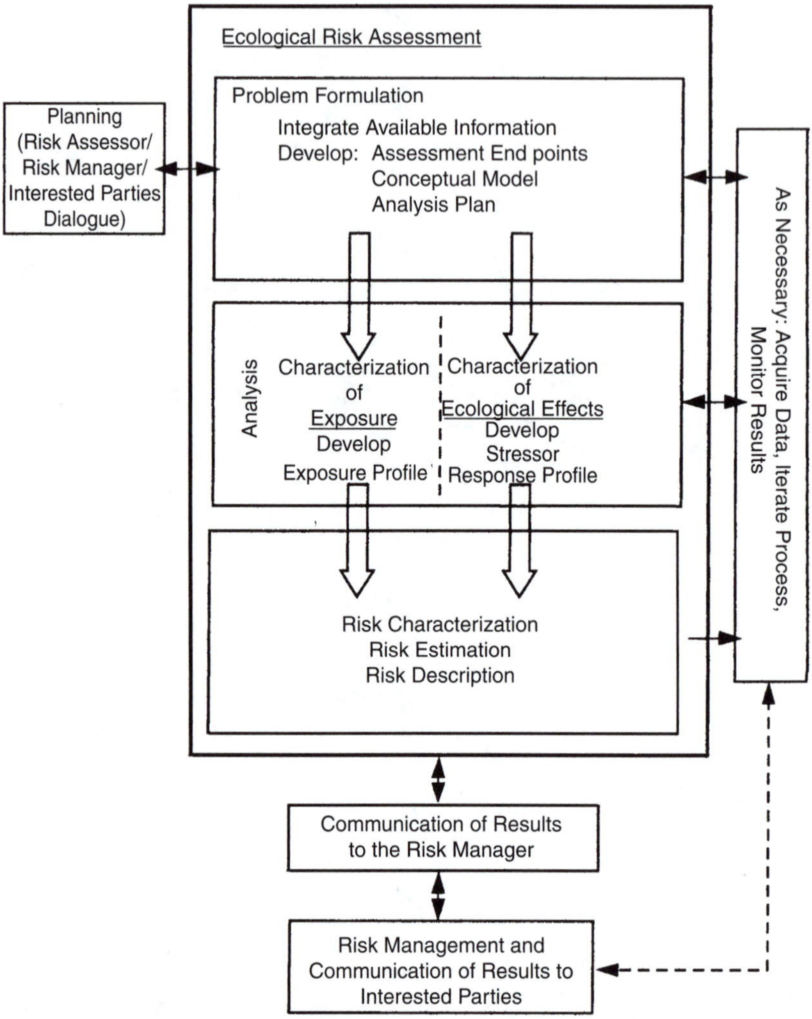

FIGURE 18.9 A framework for ecological risk assessment. (Modified from U.S. EPA, *Guidelines for Ecological Risk Assessment*, Document EPA/630/R-95/002F, U.S. EPA, Washington, DC, 1998.)

the problem in quantitative terms, e.g., no more than 5% loss of invertebrate species, no more than one fish kill in 5 years. Conceptual models can be of considerable value in providing a visual presentation of the ecosystem and its stressors that can integrate all the information, allowing clear communication between the parties involved.

The characterization of the exposure of the ecosystem to chemical stressors can be carried out using a number of techniques, including direct measurement in the ambient environment and modeling of the chemical distribution from input data,

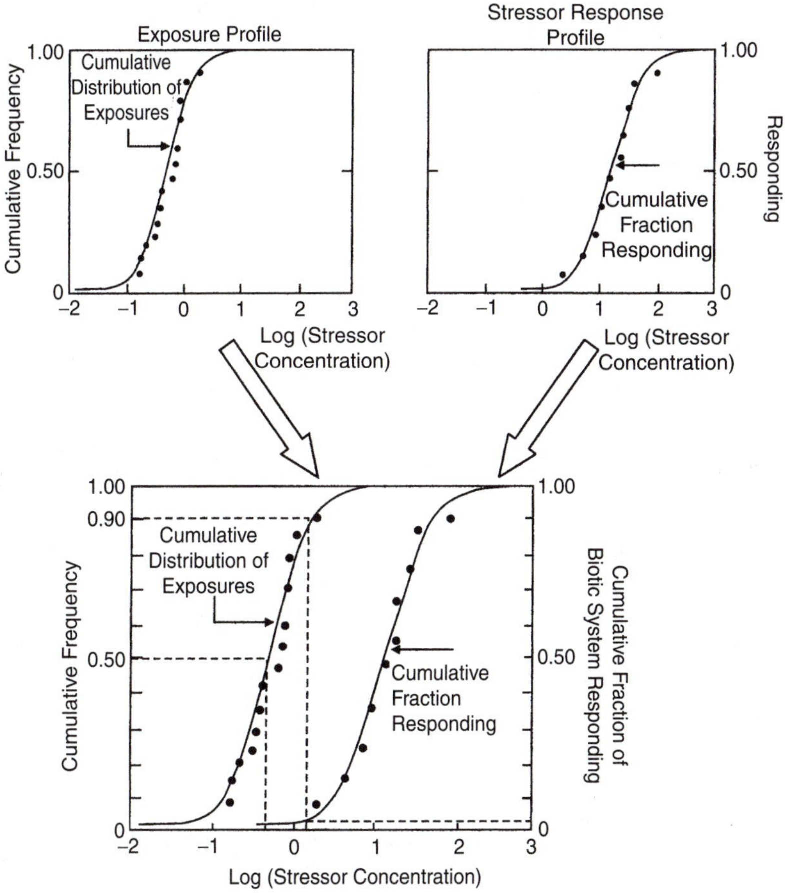

FIGURE 18.10 Relationship of exposure and stressor response profiles.

measurement of biotic body residues, and so on. The exposure data are specific to a particular system.

A **cumulative frequency** curve can be plotted to give a profile of the exposure, as shown in Figure 18.10 (top left-hand corner). These profiles usually approximate linear normal distributions and have considerable utility. For example, they can be used to estimate the possible frequency of occurrence of different concentrations in the future, presuming the existing conditions persist.

In addition, the linear cumulative frequency scale on the vertical axis can be converted into a probabilistic scale, which allows more of the data points, particularly those at very high and low levels, to be plotted to fit a straight line. Thus, extrapolations and interpolations can be made to estimate the frequency of extremely high and extremely low concentrations.

The characterizations of ecological effects can occur at the level of the individual organisms, but are often measured at the level of a population, community, or ecosystem. Individuals of a single species are said to constitute a **population** occupying a defined space, although the boundaries can sometimes be difficult to define. Populations of different species that occur together form a **community**. The habitat consists of abiotic components such as soil, air, and water. It is the habitat with its community of living organisms that constitutes an ecosystem. One characteristic of natural ecosystems is that they exhibit a varying degree of change in response to normally encountered conditions (temperature, rainfall, etc.) and nutrient availability (varying runoff). A problem that sometimes occurs is whether an observed change is due to a particular pollutant or to fluctuations inherent in the ecosystem itself. The ecological effects of concern are defined in the **problem formulation** stage as **assessment endpoints**, and these provide a focus for development of the stressor response profile. These data are principally derived from the scientific literature. This means that in applying this information to specific systems, differences will occur and need to be considered. These differences include:

- Between taxa in the natural system and in the scientific information available
- Between the assessment endpoints for the risk assessment and the test endpoints available in the scientific information, e.g., mortality as compared to growth reduction
- Between the conditions and other factors in the laboratory and natural system
- Between short exposure in the laboratory data and long-term exposure in the natural ecosystem

The characterization of ecological effects can be developed in several different ways. These methods are somewhat similar to the methods used with human health risk assessment. The NOAEL, LOAEL (Figure 18.5), LC_{50}, EC_{50} (effective concentration at 50%), and LD_{50} (see Section 4.7) can be divided by an appropriate uncertainty factor, considering the differences listed above, to yield a concentration related to an assessment endpoint, sometimes described as the **critical toxicity value (CTV)**.

The stressor response profile can also be expressed as a cumulative frequency curve, as shown in Figure 18.10 (top right-hand corner), if sufficient data are available. This type of curve is usually an approximately normal distribution and is commonly used to express toxicity data (see Figure 4.6, Figure 4.7, Figure 4.10, and Figure 18.5). This profile can be used to estimate the fraction of a biotic system that will respond to a specific concentration. The biotic system could be an individual species, a population, a community, or an ecosystem, provided the data are available. The values can be those commonly available, such as the NOAEL, LOAEL, LC_{50}, EC_{50}, or LD_{50}. The risk characterization can be carried out using the hazard quotient (HQ), as is done with human health risk assessment (see Section 18.3.5). Thus,

$$HQ = \frac{\text{Observed Level}}{\text{Critical Toxicity Level}}$$

The units in the two factors used in this equation should be the same. If it was possible to develop cumulative frequency curves for the exposure and stressor response profiles, a **probabilistic risk assessment** can be made. These two profiles can be integrated into one figure, as shown in Figure 18.10 (bottom section). Using this, the risk associated with many possible different levels of the toxicant in the system can be calculated. For example, the concentration of 1.6 units (log 1.6 = 0.2) occurs at a cumulation frequency of 0.90 (90%) and about 0.03 (3%) of the population would be expected to respond, and at a lower concentration of 0.4 units (log 0.4 = –0.4), which occurs at a cumulative frequency of about 50%, <0.02 (2%) of the population would be expected to respond. This method allows different exposure scenarios to be evaluated according to appropriate assessment endpoints.

The techniques described above do not take into account secondary effects, such as bioaccumulation, changes in predator–prey relationships, and so on. These introduce additional uncertainties, which add to the differences between the data from the scientific literature used to develop the stressor response profile and the natural system under investigation outlined.

18.5 KEY POINTS

1. **Risk assessment** is a procedure for identifying hazards and quantifying the risks presented by contaminants to human health (human health risk assessment) and natural ecosystems (ecological risk assessment).
2. The risk assessment process consists of identifying the hazard involved or identifying the problem, quantifying the exposure of human populations and natural ecosystems to the contaminant, evaluating dose–response relationships available for the chemical, and finally characterizing the risk involved by integrating the exposure and the dose–response information.
3. Environmental contaminants present particular problems in the evaluation of dose–response relationships since exposures are at low concentrations for long periods and the data available are generally for relatively high concentrations and relatively short periods.
4. Data are often not available on exposure of human populations and natural ecosystems for long periods, and so extrapolations might be needed to obtain suitable data for risk assessment.
5. Exposure can be evaluated by measurement of the amounts of contamination in the various pathways and calculations of quantities available made using a range of models for chemical distribution.
6. With human health evaluations, dose–response data can be obtained principally from experimental data on animals. To utilize these data, differences between the test animals and the human population being evaluated need to be taken into account. A safety factor is applied in toxicity evaluations to increase the apparent toxicity and thereby account for these factors.
7. Safety factors are used in human health evaluations to account for the different sensitivities of individuals in a population, the differences

between the test animals and the human population (mentioned in point 6 above), temporal factors, and so on, and can range from 10 to 10,000. Safety factors greater than 1000 indicate considerable uncertainty in the data available and the need for more definitive information.

8. The tolerable daily intake (TDI), also termed acceptable daily intake (ADI) and reference dose (RfD), can be calculated from the NOAEL by division by the safety factor. Thus,

$$\text{TDI} = \frac{\text{NOAEL}}{\text{SF}} \text{ mg/ kg.bw/ day}$$

9. Risk can be characterized in terms of the TDI as a hazard quotient (HQ), where HQ = DOSE/TDI. Levels of $HQ < 1$ indicate no significant adverse effect in a human population, whereas levels of $HQ > 1$ indicate possible adverse effects.
10. Ecotoxicology is concerned with the effects of chemicals on ecosystems and involves the evaluation of a sequence of interactions of the chemical with the natural environment. Thus, ecotoxicology is concerned with the following:

Sources
↓
Distribution, transport, and transformation
↓
Exposure and uptake
↓
Organism response
↓
Population, community, and ecosystem response

11. Chemicals can enter the environment in motor vehicle exhaust, electricity generation discharges, industrial discharges to the atmosphere, sewage, storm water runoff, industrial discharges to waterways, industrial contamination of soil, and rural activities.
12. Hydrocarbons, aromatic hydrocarbons, chlorohydrocarbons, organophosphorus compounds, PCBs, PCDDs, PCDFs, and surfactants are relatively common organic compounds discharged to the environment.
13. Laboratory-based tests for physicochemical properties, degradation and transformation, and bioaccumulation effects on organisms provide an indication of the effects on key aspects of the natural and human system, but are not effective in indicating whole ecosystem effects and sublethal effects.

14. General effects of toxic chemicals on ecosystems include reduced species diversity, reduced biomass, a change in the types of biota present, and changes in the energy and nutrient flows in ecosystems.
15. Risk assessment processes can quantify risk in terms of the hazard quotient, and with ecological risk assessment, probabilistic information can be obtained.

REFERENCES

Carlow, P., Ed., *Handbook of Ecotoxicology*, Vols. 1 and 2, Blackwell Scientific Publications, Oxford, 1993.

Connell, D.W. and Miller, G.J., *Chemistry and Ecotoxicology of Pollution*, John Wiley & Sons, New York, 1984.

Corvello, V.T. and Morkhofer, M.W., *Risk Assessment Methods: Approaches for Assessing Health and Environmental Risks*, Plenum Press, New York, 1993.

Guidelines for Ecological Risk Assessment, Document EPA/630/R-95/002F, U.S. Environmental Protection Agency, Washington, DC, 1998.

Hoffman, D.J., Rattner, B.A., Burton, G.A., and Cairns, J., *Handbook of Ecotoxicology*, Lewis Publishers, Boca Raton, FL, 1995.

Scala, R.A., Risk assessment, in *Casarett and Doull's Toxicology: The Basic Science of Poisons*, 4th ed., Amdur, M.O., Poull, J., and Klaassen, C.D., Eds., McGraw-Hill, New York, 1991, pp. 985–997.

QUESTIONS

1. On searching the scientific literature on dieldrin, you find two separate sets of data on rats, which indicate NOAELs of 0.25 mg/kg.bw/day in one and 0.05 mg/kg.bw/day in the other. Each set of data was obtained from one experiment conducted over 2 months. Estimate the safety factor that should be used in applying these data to human health evaluations, and then calculate the tolerable daily intake (TDI).
2. In the text an expression was derived for the daily intake (DI) of dieldrin by a human population exposed to this substance in soils. Using this and the TDI derived above, develop a guideline for contamination of soil by dieldrin below which no significant adverse effects would be expected.
3. The chlorohydrocarbon pesticide DDT was widely used in agriculture in the 1950s and 1960s with subsequent adverse effects. This persistent compound with a high log K_{OW} value (approximately 6.0) was found to be a potent fish poison and has caused eggshell thinning and a resultant decrease in the population of some birds. Briefly outline the ecotoxicology of DDT using the information above and elsewhere in this book as appropriate.

4. Laboratory tests are used to evaluate the ecotoxicology of a chemical. Briefly indicate how the tests in Table 18.4 can be used to evaluate the ecotoxicology of a chemical in a particular usage situation.
5. If we assume the data in Figure 18.10 are representative of a biotic system that is experiencing a chemical stressor having the cumulative frequency exposure profile shown, then deductions can be made regarding the risks to the biota. If we decide that it was necessary to protect 90% of the biota, then quantify the level of risk in terms of the exposure. If a decision was made to give the same level of protection to 95% of the biota, what change in the stressor concentration would be required?

ANSWERS

1.

Factor Involved (from Table 18.2)	Value
Rat-to-human extrapolation	10
Two months-to-many years extrapolation	10
One experiment is a limited database	10

Both experiments would involve a safety factor of 1000; the lower experimental value should be used to calculate the TDI; thus,

TDI = 0.05/1000 mg/kg.bw/day

∴ TDI = 0.00005 mg/kg.bw/day

2. The expression for the TDI is

$$\text{TDI mg/kg.bw/day} = C_{sd}\ 2.24 \times 10^{-6}\ \text{mg/kg.bw/day}$$

where C_{sd} is expressed in mg/kg of soil and is the concentration in the soil. Thus,

$$C_{sd} = \left(\frac{\text{TDI}}{2.24 \times 10^{-6}}\right) \text{mg/ kg soil}$$

$$= \frac{5 \times 10^{-5}}{2.24 \times 10^{-6}}$$

$$C_{sd} = 22.3\,\text{mg/ kg soil}$$

Thus, the concentration of dieldrin below which no significant adverse toxic health effects would be expected is 22.3 mg/kg of soil. Of course, further information on dose–response relationships or the exposure calculation could give revised values.

3. **Source**: Used as an agricultural pesticide in broadcast applications in rural areas.
   Distribution: Because of the high log K_{OW} value of approximately 6.0, this substance would be expected to accumulate in soils and then to be swept into waterways on soil particles. The amount of the pesticide actually dissolved in water and swept into waterways in this form would be expected to be small. This substance is resistant to degradation, which would contribute to the accumulation of the pesticide in soils and sediments in aquatic areas.
   Exposure/Uptake: Since this substance will be present in sediments in waterways, a small quantity would dissolve giving low concentration in the overlying water, and this would be accumulated by the aquatic organisms. The level of bioconcentration in aquatic organisms would be expected to give a bioconcentration factor (K_B) on the order of about 1 million, which could produce relatively high biotic concentrations.
   Organism Response: This substance is toxic to fish and other aquatic organisms, and lethal effects could be expected in some situations. In addition, the well-known sublethal effect, resulting from bioaccumulation in birds, would be the occurrence of eggshell thinning, resulting in a lack of breeding success in some species.
   Population, Community, Ecosystem Response: In general terms, it would be expected that some populations, particularly of aquatic organisms, would be reduced, and possibly some species totally removed from the system.
4. **Source**: The amounts of chemical involved can be derived from usage patterns.
   Distribution: The distribution of the chemical can be evaluated using physical-chemical properties such as log K_{OW} and the amounts involved derived from the source information.
   Exposure/Uptake: The uptake can be evaluated using physicochemical properties and the information from bioaccumulation tests. The actual amount of uptake that occurs can be derived using the distribution information obtained above. Degradation can be evaluated to some extent by the usage of laboratory information on degradation and transformation tests.
   Organism Response: Organism response can be evaluated from the alga, fish, and terrestrial plant laboratory testing information, which would be dose–response data, and the information obtained above on distribution exposure and uptake.
   Population, Community, Ecosystem Response: There are no tests available to provide this information.
5. If a level of protection of 90% of the biota is the objective, this means that 10% (0.10) would be expected to respond. Using Figure 18.10 (bottom section shown in Figure 18.11), if we extrapolate this to the stressor response curve, we can then estimate the maximum exposure concentration of 3.2 units per volume (log 3.2 = 0.5) or lower, which occurs at a

cumulative frequency of about 95%. So the damaging concentrations of 3.2 units per volume or higher only occur with a frequency of about 5% (Figure 18.11). Similar extrapolations from 0.05 cumulative fraction responding (95% protected or not responding), we derive the lower concentration for protection of 2.0 units per volume (log 2.0 = 0.30), which occurs at a cumulative frequency of about 90%, and the damaging concentrations occur at 10% frequency. Thus, it would require a concentration reduction from 3.2 to 2.0 units, or a factor of about 60% reduction, to change the exposure profile to give 95% protection.

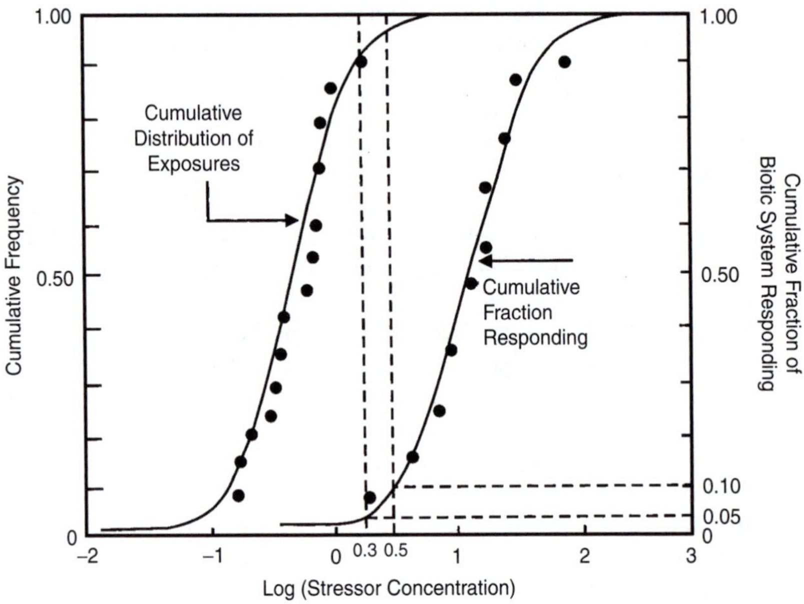

FIGURE 18.11 This figure is based on Figure 18.10 and shows the extrapolations involved in answering question 5.

Index

A

B

C

D

E

H

P

Q

R

S

T

U

V

W

X

Y

Z